Advance Praise for *Techno-Fix*

Science and technology have advanced impressively over the past century, but we have learned that often the benefits of new technology are accompanied by unexpected and deleterious consequences that we then attempt to solve using more technology with unpredictable effects. Thus, for example, we have altered the chemistry of the atmosphere with our use of fossil fuels and now hope that we can maintain our lifestyles, economies and consumption with the promise of geo-engineering like carbon capture and storage, stimulating ocean algal growth or spraying aerosols of sulfur dioxide in the sky. As this book shows, it is suicidal to put our hopes in such promises.

— DAVID SUZUKI, Canadian environmental activist, Professor Emeritus, University of British Columbia, host of CBC's The Nature of Things, author of 43 books, recipient of 25 honorary doctorates as well as numerous awards, including the Order of Canada.

We need to be mature enough to understand that technology alone won't be our salvation. This book helps explain why we need to think more deeply than that.

— BILL MCKIBBEN, journalist, environmental activist, Schumann Distinguished Scholar at Middlebury College, author of many influential books and articles, including The End of Nature and Eaarth: Making a Life on a Tough New Planet.

In *Techno-Fix*, Michael and Joyce Huesemann show us how unsustainable and destructive technologies, shaped and driven by the profit motive, have emerged as a major cause of harm to the health of people and the Earth. We need to go beyond a blind techno-religion. We ourselves need to choose the tools that shape our lives. A vibrant and vital democracy needs people's participation in technology choice. *Techno-Fix* shows how.

— VANDANA SHIVA, New Delhi-based environmental and anti-globalization activist, philosopher, author of 20 books including Soil not Oil and Staying Alive, and recipient of numerous international awards.

This book is outstanding, the most thorough, clear, systematic refutation that I've seen of the absurd idea that new technology will be our savior against advancing ecological breakdown. The authors show that technology is far more the problem than the solution, and no big new techno-utopian tinkering will change that. Far more than a technological revolution, the authors argue, we need a conceptual revolution; one that recognizes that our fantasies of dominion over nature, and the remake of the Earth's natural systems, and all the assumptions that brought us here will not get us out. This is a must-read for anyone seeking realistic pathways forward, rather than more of the same.

— JERRY MANDER, founder of the International Forum on Globalization, program director at the Foundation for Deep Ecology, and author of influential books on the social, cultural and environmental effects of mega-technologies.

This is the new "age of unreason." Environmental politics is now dominated by mysticism, myth and magical thinking. Even as the impacts of technology destroy the ecosphere, the faithful preach that technology alone can salvage civilization. Enter Michael Huesemann and Joyce Huesemann. With *Techno-Fix: Why Technology Won't Save Us or the Environment*, the Huesemanns have produced one of the most well-researched and possibly the best myth-busting book the environmental movement has ever seen. In a better world, it would be required reading for all elected officials and every student in every program at every university everywhere.

— WILLIAM REES, Professor, University of British Columbia, originator of the concept of the ecological footprint and author of *Our Ecological Footprint*, fellow of the Post Carbon Institute and the Royal Society of Canada.

Techno-Fix is a powerful and well-researched challenge to the widespread belief that modern technology alone will give us a clean environment, health, peace and happiness. The authors convincingly show that a paradigm shift and a corresponding change in the direction of science and technology are needed to create a more humane, just and sustainable world.

— JOHN ROBBINS, environmentalist, animal activist and author of numerous best-selling books, including *Diet for a New America* and *The Food Revolution*. Robbins has received the Rachel Carson Award and the Albert Schweitzer Humanitarian Award.

Salvation by technological advance and unlimited growth is the blind dogma of our age. The Huesemanns provide a devastatingly cogent and well-referenced critique of this modern gnosticism, as well as some good alternative ideas. Highly recommended.

— HERMAN DALY, former Senior Economist at the World Bank, Professor of Ecological Economics at the University of Maryland, and author of many influential books, including *Steady-State Economics* and *Ecological Economics*.

The nuclear disaster of Fukushima tragically confirms how right the authors are.

— ERNST ULRICH VON WEIZSAECKER, German scientist, academician, parliamentarian and author of influential books, including *Earth Politics* and *Factor Four*, founder and former president of the Wuppertal Institute and member of the Club of Rome.

It has frequently been proclaimed, especially by certain economists, that our problems—whether economic, environmental, social or political—can be resolved by technological wizardry. Julian Simon, an extreme proponent of this view, urged us to ignore the warnings of environmentalists and to stride toward a shining new future created by technologies. For the most part, techno-optimists have been simply misinformed and stand in urgent need of some extensive homework on the issue. All the more welcome, then, is this first-rate book. If only it could have emerged sooner, it might have saved us much trouble and much money.

— NORMAN MYERS, British professor and fellow at Oxford University; member of the US National Academy of Sciences, the World Academy of Art and Science and the Royal Society of Arts; advisor on environmental issues to the United Nations, the World Bank, as well as to scientific academies and various governments worldwide. In 1997 he was appointed by HRH Queen Elizabeth II to be a Companion of the Order of St. Michael and St. George for "services to the global environment."

Techno-Fix explains why science and technology will not save the economy and the environment. Drs. Michael and Joyce Huesemann have written an outstanding book that is most timely.

— DAVID PIMENTEL, Professor emeritus, Cornell University, author with Marcia Pimentel of *Food, Energy, and Society* and numerous other works on ecological integrity, pesticides, biofuels, energy flows in food production, human population growth and the environment.

Believers in unending growth argue that technology can overcome any environmental penalties of growth. This is the most detailed scholarly rebuttal of that view that I have seen. The Huesemanns systematically take on the pro-growth arguments, and their rebuttals are persuasive. Indeed, I would suggest that, in meeting the technophiles on their own turf, the Huesemanns may be too gentle. The formula

$$\text{Impact} = \text{Population} \times \text{Affluence (or Consumption)} \times \text{Technology}$$

was originated by environmentalists, but it has been adopted by technophiles because it equates technology with population as equal determinants of the environmental impact of growth and change. There are two problems with that: (1) It assumes a linear relationship among the variables, while the real world is non-linear, marked by thresholds of damage and changing impacts as the scale changes; (2) The range of impacts of the three variables is dramatically different. Population change affects most biological, social and economic interactions, whereas a given technological change by itself influences a much narrower band, and the net impact of technological change as a whole is incalculable.

Even using the original formula, the assumptions of which could be made considerably stronger, the book successfully refutes the belief that technology is a cure for growth.

The authors write from broad practical and academic qualifications. The book is also valuable in offering not just the authors' views but the viewpoints of many well-known writers on population change and its consequences. That in itself should make it valuable as a textbook for courses on population, environment or resources — indeed for courses in current history, which too often ignore those critical determinants of our future. Teachers in all those areas will find it useful for their own education. The authors hope that it will be read by "anyone intending to create a better future." Perhaps that is too optimistic, but certainly it will be valuable for anybody who hopes to be grounded in the topics it covers.

— LINDSEY GRANT, former US Deputy Assistant Secretary of State for Environment and Population, author of the classic *Juggernaut: Growth on a Finite Planet*, other important books and many articles on population issues.

Techno-Fix deals with a wide range of issues at the core of the sustainability crisis, showing that these problems are not going to be solved by technical advances that leave the fundamental structures and values of rampant consumer society in place. It presents a detailed and powerful case, based on extensive references to supporting studies and evidence, and expressed in a clear and easily readable style.

The alarming problems now accelerating on many fronts are not basically technical problems. They are mostly created by the mistaken goals we have chosen to pursue, especially the pursuit of limitless affluence and economic growth. Unfortunately mainstream thinking seems to be about as far from recognizing this as it ever was, and its main support derives from faith in technology. It seems to be generally believed that our wizard scientists will come up with ways whereby we can all go on merrily getting richer with no end in sight, and the poor of the world would rise to live as we do. This book provides a very effective refutation of this willful, deeply entrenched and rarely questioned delusion.

Among the book's virtues is that it goes beyond technical issues to consider the social, economic and philosophical dimensions of our predicament, pointing convincingly to many areas where radical rethinking of goals and means is urgently required.

— TED TRAINER, Australian environmental and sustainability activist at the University of New South Wales, author of *The Conserver Society* and *Transition to a Sustainable and Just World* as well as numerous other books and articles.

Technology has become our near-universal object of faith: new machines will solve all our problems! Few of us seem to understand that machines have in fact exacerbated most of our current environmental and social problems, and it is *people* who must provide the answers. Our appetites and economic arrangements —not our gadgets—must change if we are to survive. *Techno-Fix* argues this controversial thesis clearly, convincingly and entertainingly, and should be read and discussed in every home, school, and legislature.

— RICHARD HEINBERG, American journalist, Senior Fellow at the Post Carbon Institute, and author of ten influential books, including *The Party's Over, Peak Everything* and *The End of Growth*.

TECHNO-FIX

Why Technology Won't Save Us
or the Environment

Michael Huesemann and Joyce Huesemann

NEW SOCIETY PUBLISHERS

To Hermione, Lu and Rex

Cover design by Diane McIntosh.
Image © iStock (Ljupco)

Printed in Canada. First printing Sept 2011.

Paperback ISBN: 978-0-86571-704-6
eISBN: 978-1-55092-494-7

Inquiries regarding requests to reprint all or part of *Techno-Fix*
should be addressed to New Society Publishers at the address below.

To order directly from the publishers, please call toll-free
(North America) 1-800-567-6772, or order online at www.newsociety.com

Any other inquiries can be directed by mail to:

New Society Publishers
P.O. Box 189, Gabriola Island, BC V0R 1X0, Canada
(250) 247-9737

LIBRARY AND ARCHIVES CANADA CATALOGUING IN PUBLICATION

Huesemann, Michael
Techno-fix : why technology won't save us or the environment /
Michael Huesemann and Joyce Huesemann.

Includes bibliographical references and index.
ISBN 978-0-86571-704-6

1. Technology—Social aspects. 2. Technology—Moral and ethical aspects.
3. Technology—Environmental aspects. I. Huesemann, Joyce II. Title.

T14.5.H835 2011   303.48'3   C2011-904342-4

New Society Publishers' mission is to publish books that contribute in fundamental ways to building an ecologically sustainable and just society, and to do so with the least possible impact on the environment, in a manner that models this vision. We are committed to doing this not just through education, but through action. The interior pages of our printed, bound books are printed on Forest Stewardship Council®-certified acid-free paper that is **100% post-consumer recycled** (100% old growth forest-free), processed chlorine free, and printed with vegetable-based, low-VOC inks, with covers produced using FSC®-certified stock. New Society also works to reduce its carbon footprint, and purchases carbon offsets based on an annual audit to ensure a carbon neutral footprint. For further information, or to browse our full list of books and purchase securely, visit our website at: www.newsociety.com

NEW SOCIETY PUBLISHERS
www.newsociety.com

FSC
www.fsc.org
MIX
Paper from
responsible sources
FSC® C011825

Contents

Acknowledgments xv
Foreword, by Paul Ehrlich and Anne Ehrlich xvii
Introduction xxiii

PART I: TECHNOLOGY AND ITS LIMITATIONS

1. The Inherent Unavoidability and Unpredictability of
   Unintended Consequences 3
   Interconnectedness 3
   Human Improvement upon Nature 5
   Unavoidable Negative Effects of Technology 7
   Irreversible Consequences 8
   The Limitations of Reductionism 11

2. When Things Bite Back: Some Unintended
   Consequences of Modern Technology 17
   Unintended Environmental Consequences 17
   Unintended Consequences of Industrialized Agriculture . . . 23
   Unintended Side Effects of Genetic Engineering 25
   Unintended Consequences of the Automobile 28
   Intended and Unintended Consequences of
       High-Technology Warfare 31
   Unintended Consequences of High-Tech Medicine 33
   Unintended Consequences of Technological Revolutions . . . 43
   The Decline in Fitness of Future Generations 46

3. Technology, Exploitation and Fairness 49
   Technology and Exploitation 50
   Technological Exploitation of Nature 53
   Human Domination of Nature 56
   Machines and the Control and Exploitation of Workers . . . 60
   Television: A Powerful Tool for Social Control
       and Manipulation 64
   Military Technologies 68

4. In Search of Solutions I: Counter-Technologies
and Social Fixes   71
Counter-Technologies   73
Social Fixes   75
Environmental Counter-Technologies   77
Military Counter-Technologies   83
Medical Counter-Technologies   86
Unintended Consequences of Counter-Technologies
and Social Fixes   88

5. In Search of Solutions II: Efficiency Improvements   91
Technological Progress and Rising Material Affluence   94
Efficiency Improvements and Limited Resources   98
Inherent Limits to Efficiency Improvements 109
Unintended Consequences of Efficiency Solutions 112

6. Sustainability or Collapse? 117
Sustainable Development and Eco-Efficiency 119
Three Conditions for Long-term Sustainability 122
Challenge #1: Serious Environmental Impacts of
Large-Scale Renewable Energy Generation 125
Challenge #2: Replacement of Non-Renewable Materials
with Renewable Substitutes 133
Challenge #3: Complete Recycling of
Non-Renewable Materials and Wastes 135
Sustainability or Collapse? 137

PART II: THE UNCRITICAL ACCEPTANCE
OF TECHNOLOGY

7. Technological Optimism and Belief in Progress 145
Belief in Progress: A Brief History 148
Comparison of Belief in Progress to Religious Faith 152
Ignorance: The Basis of Most Techno-Optimism 154
Medical Techno-Optimism 159
Techno-Optimism and the Mass Media 167
The Decline of Techno-Optimism 169

8. The Positive Biases of Technology Assessments
   and Cost-Benefit Analyses. 173
   An Overview of Cost-Benefit Analysis 174
   Problem #1: Boundary Selection and Externalization of Costs 176
   Problem #2: Prediction of Potential Impacts and Selection
      of Appropriate Indicators 180
   Problem #3: Institutional Biases and the Perception of
      Costs and Benefits 183
   Problem #4: Monetization of Non-Market Values. 185
   Problem #5: The Ethics of Cost-Benefit Analyses 187
   The Uncritical Adoption of the Automobile 189
   The Hidden Costs of Biofuels 192
   Limited Testing of the Effectiveness of Medical Therapies . . 194
   Gross Domestic Product (GDP): A Biased Indicator
      of Economic Progress 200

9. Happiness. 207
   Technological Innovation, Consumerism and Materialism . . 208
   Material Affluence and Happiness. 214
   Explaining the Paradox 216
   Sources of Happiness 224
   The Destruction of Traditional Sources of Happiness 226

10. The Uncritical Acceptance of New Technologies 235
    The Myth of Value-Neutrality 235
    The Myth of Autonomous Technology. 241
    The Technological Imperative 243
    Technological Dependency and Loss of Freedom 245
    The Undemocratic Control of Technology. 248

11. Profit Motive: The Main Driver of
    Technological Development. 253
    Technological Development as a Social Process 253
    Understanding the Meaning of Profit 255
    Profit Maximization and the Development
       of New Technologies. 258
    Profit Maximization: Agriculture and Food 261

Profit Maximization: Medical Care 263
Profit Maximization: Military Technologies
    and Foreign Policy. 266

PART III: THE NEXT SCIENTIFIC AND
TECHNOLOGICAL REVOLUTION

12. The Need for a Different Worldview 271
The Power of Worldviews and Paradigms 271
Conflicting Worldviews and Paradigm Shifts 273
A Different View of Reality. 277
A Different View of the Economy 279
A Different View of Science and Technology. 285
A Different View of Medicine 286
The Need for Increased Awareness. 289

13. The Design of Environmentally Sustainable and Socially
Appropriate Technologies 295
Design Criteria for Environmental Sustainability. 295
Design Criteria for Social Appropriateness 300
The Prevention of Unintended Consequences. 304
The Democratic Control of Technology 306
Local Organic Agriculture: A Model of Environmentally
    Sustainable and Socially Appropriate Technology. 309

14. Critical Science and Social Responsibility 313
The Myth of Value-Neutrality 313
A New, Critical Science. 320
The Question of Responsibility 325
The Problem of Professionalism 328
The Need for Comprehensive Professional Ethics. 330
Toward a Critical Science and Engineering 334

For Further Thought. 339
Bibliography. 355
End Notes. 383
Index 415
About the Authors. 435

Acknowledgments

We would especially like to thank Anne and Paul Ehrlich for their kind encouragement and for writing an excellent foreword for our book. We would also like to thank David Suzuki, Bill McKibben, Vandana Shiva, Richard Heinberg, Herman Daly, Jerry Mander, Norman Myers, Ernst Ulrich von Weizsaecker, William Rees, John Robbins, David Pimentel, Lindsey Grant and Ted Trainer for their excellent endorsements. We would like to thank our dear friend and colleague, John Benemann, for his encouragement and suggestions. There are so many who have contributed to this work, albeit indirectly, whom we would like to thank, but the list would be very long indeed. We also wish that we could thank friends and mentors who, long before this book was written, passed from this world but are always remembered: Erna Huesemann, Alberta Cantwell Morris, Anton Rogstad and Edwin F. Carpenter.

We would like to thank Heather Nicholas, Sue Custance, Ingrid Witvoet, Sara Reeves, and E.J. Hurst at New Society Publishers and Audrey Dorsch at Dorsch Editorial for their fine work in bringing this book to the public. We would especially like to thank Diane McIntosh for her excellent cover design.

Foreword

Technology:
Not A Panacea, Maybe A Poison Pill

by Paul R. Ehrlich and Anne H. Ehrlich

TECHNOLOGY BY ITSELF can't save us. That is the basic message of this important book. The two greatest challenges facing humanity are interrelated. One is the giant and growing scale of the human enterprise, which is destroying our life-support systems. The other is inequity, especially the maldistribution of wealth, which helps prevent us from taking effective action to end that destruction. These are, of course, extraordinarily complicated issues.

The claim that "technology will solve the problem"—whatever that problem may be—is part and parcel of Western culture. It has been especially prominent in the past half-century, as both the scientific community and the popular media have given prominence to technology-related problems from silent springs, widespread hunger and oil spills to climate disruption, fisheries collapses and the Fukushima disaster. The record of "cures" for these problems promoted by technological optimists gives little room for cheer. Over those five decades, the putative advantages of claimed "fixes" have usually failed to appear or proved to be offset by unforeseen nasty side effects. The number of hungry people in the world has roughly doubled, but we are not feeding anyone on leaf protein, whales "farmed" in atolls or algae grown on sewage sludge, as was once proposed. Neither are nuclear-powered agro-industrial complexes solving human energy and hunger problems. Instead, toxic chemicals now are virtually ubiquitous, oil and gas wells are drilled in increasingly precarious situations, crops for energy compete with crops for food, and "guaranteed" emergency core cooling systems in nuclear power plants fail.

What appeared to be an exception—the "green revolution," the transfer of the technology of industrial high-yield agriculture from rich to poor countries—hasn't banished hunger, but has undoubtedly prevented countless deaths from starvation. However, the final environmental verdict on the green revolution is far from in. Indeed, as time goes on, it looks more problematic as pest and climate problems increasingly cut into harvests, and food and energy prices rise. A related accomplishment of technology has been its contributions to longer life expectancies, especially in developing nations, and to increased control over reproduction. Contraceptive technology, though, is a weirdly mixed bag: effective but underdeveloped for social reasons; inexpensive but underutilized, also for social reasons. Technology also has helped to ameliorate some forms of environmental damage in industrialized countries by developing pollution controls, advanced sewage treatment, hybrid vehicles and the like—but all in response to problems technology itself has helped to generate.

Yet as some of the symptoms have declined, the disease itself is becoming catastrophically worse. Climate disruption, losses of biodiversity and ecosystem services, and toxification of the planet are all quietly but inexorably and rapidly increasing. So are the social and economic inequities that impede solutions to those escalating environmental problems. Even the hard-won gains in higher life expectancies and lower fertility rates are showing some local reversals. Furthermore, while human life-support systems are deteriorating under the impact of technologies that are chiefly serving the consumption "needs" of the affluent, polls repeatedly show that the citizens of industrialized nations are *not* becoming happier despite the plethora of goods available to them, from fattening fast food, mood-altering pharmaceuticals and highly advertised electronic gadgets to "new" (read "slightly modified to promote sales") versions of everything from athletic shoes to automobiles. Meanwhile, billions of people not only cannot share in that overconsuming binge, they lack even basic necessities.

As we indicated, most environmental scientists conclude, and many studies have shown, that humanity has already reached a state of overshoot—we have exceeded Earth's long-term carrying capacity for people. On the other hand, the general public, businesspeople, governments and most economists appear to believe that population and per capita consumption can grow indefinitely. They think that the rich can get richer and the poor can catch up. They are convinced that human ingenuity leading to technological innovation can solve the problems associated with growth, and that eventually all economic inequities can be eliminated by growth itself. To us and our colleagues, this is an assumption entirely unwarranted by the evidence – and debunking it may be the single most important task of the scholarly community. *Techno-Fix* is a powerful and well-researched contribution to this effort.

Indeed, a substantial portion of environmental scientists are convinced that more than enough is understood about the human predicament, both its environmental and its social dimensions, to know what is needed to start civilization toward sustainability. It is clear that solutions to the predicament lie primarily in the domain of human behavior. Together with social scientists and scholars in the humanities, environmental scientists are organizing a Millennium Assessment of Human Behavior (MAHB),* the goal of which is to generate a global discussion of the most basic topics pertinent to sustainability—how people act individually and corporately in relation to population, consumption and technology. A critical aspect of the latter, central to the issues raised in this book, is whether technological ingenuity can be redirected to help solve rather than exacerbate environmental and equity issues.

How values associated with science and technology can lead us down questionable paths is rarely considered, but that they do is obvious. One incident that brought this home to us long ago occurred at a conference on energy technology in England. Present were some very bright young scientists working on fusion power—trying to find a way to generate steam in a power plant using a mechanism similar to that which naturally occurs in the interior of the sun. We pointed out to them that solar radiation could be used directly to supply energy for many uses. But the fusion scientists were not interested. After all, hot water for dishwashing and showers could be supplied by putting a black-painted 50-gallon drum on the roof, piping cold water into the bottom and drawing hot water from the top. Hardly a technical challenge compared with what those scientists were confronting: keeping magnets at a temperature near absolute zero close to a plasma (an ionized gas) that needed to be kept at temperatures as high as in the sun's interior so that hydrogen nuclei could fuse and release the desired energy.

Simple solar power, on the other hand, can be provided by widely distributed collectors and solar cells on roofs, in backyards or in local fields. Fusion generation of electricity, like conventional generation in electric power plants, must be centralized. Distributed solar power is far less susceptible to routine or disastrous failure than is any system—fusion, nuclear fission, fossil-fuel-based or hydro-powered, that depends on moving energy from a central plant through extensive transmission grids to large numbers of end users. But solar, unlike fusion, is less likely to make scientists enthusiastic, CEOs richer or politicians more able to control peoples' lives. Furthermore, after decades of research, billions of dollars invested, and many promises, fusion generators have yet to produce a single kilowatt-hour of commercial energy, and they show no sign of producing any for at least several decades, if ever.

Can humanity find ways to minimize the negative imperatives that grow out of such values of the scientific community and technologies based on those values—imperatives that are driving civilization toward collapse? As you'll see in reading *Techno-Fix*, our technological system is as much a part of the problem as a part of the solution. If properly selected and deployed (and that's a big "if"), technologies could help avoid a collapse of civilization. But a careful look at the empirical evidence shows that their record to date in this respect is generally dismal. When we published *The Population Bomb* in 1968, there were 3.5 billion people on Earth. Our critics said we were alarmists, that technology could feed, house, clothe, educate and provide great lives for even 5 billion people. Well, technology didn't, and centralized technological systems are moving the human enterprise ever more toward a crash. Today there are 7 billion people; a billion of them are hungry, and a couple of billion more are living in misery. And the "growthmaniacs," never learning, still insist that humanity, with its clever techno-fixes, *could* care for many billions more people.

It would make more sense to halt population growth and demonstrate that technology can give decent, sustainable lives to all the people already here and, equally important, to their descendants. If proponents of technology can show that technology really can accomplish this, then that will be the time to consider the costs and benefits of trying to develop new technologies and social-economic-political systems that can adequately support additional happy, healthy, free people. As *Techno-Fix* so persuasively points out, it will require a nearly complete revision of attitudes toward technologies and toward the silly but pervasive notion that science and technology are "value free." Re-examining human assumptions about technological systems and their relationships to power structures and equity is obviously one of the most important tasks of our time. When you've read *Techno-Fix*, we hope you'll participate in this crucial endeavor!

— Paul R. Ehrlich and Anne H. Ehrlich

PAUL R. EHRLICH is Bing Professor of Population Studies and professor of biology at Stanford University, a fellow of the Beijer Institute of Ecological Economics, and president of the Center for Conservation Biology at Stanford University. He is also a fellow of the American Association for the Advancement of Science, the United States National Academy of Sciences, the American Academy of Arts and Sciences and the American Philosophical Society.

ANNE H. EHRLICH is a Senior Research Scientist in the Department of Biology at Stanford University, Associate Director of the Stanford Center for Conservation Biology, a member of the American Academy of Arts and Sciences and coauthor of ten books and many articles on population biology and environmental policy.

The Ehrlichs are the authors of many influential books, including *The Population Bomb, The Population Explosion, One With Nineveh: Politics, Consumption, and the Human Future*, and *The Dominant Animal: Human Evolution and the Environment*.

The Ehrlichs are currently organizing the Millennium Assessment of Human Behavior (MAHB).

*The Millennium Assessment of Human Behavior (MAHB, pronounced "mob") is a comprehensive initiative being developed by the Ehrlichs and other environmental and social scientists. An introduction is available at http://igbp-portugal.org/mahb-mission-statement.html.

Introduction

Techno-optimism is pervasive in our society but hardly justified. In one form or another, we are repeatedly assured that "More efficient technology will solve the problem," "Continued economic growth is environmentally sustainable," "High-tech medicine and miracle drugs will abolish disease," "More military spending will ensure global peace and security," "Biofuels and nuclear power are the solution to global warming," "Overpopulation is not a problem—we will employ genetically engineered crops to feed an unlimited number of people," "Greater material affluence will increase happiness," "We have no choice anyway: technology is an autonomous force. Whatever can be done technologically, should be and will be done," "You can't put the genie back in the bottle."

Techno-Fix confronts these beliefs and many others. It questions a primary paradigm of our age: that advanced technology alone will extricate us from an ever-increasing load of social, environmental and economic ills. *Techno-Fix* shows why negative unintended consequences of science and technology are inherently unavoidable and unpredictable, why counter-technologies, techno-fixes and efficiency improvements do not offer lasting solutions and why modern technology, in the presence of continued economic growth, does not promote sustainability but instead hastens collapse.

Despite the serious shortcomings and consequences of past technologies, the public often uncritically accepts new technology, believing that additional and more advanced technology will eventually provide satisfactory solutions. *Techno-Fix* analyzes this paradox and asserts that technological optimism and the unrelenting belief in progress are based on ignorance, that most technological cost-benefit analyses are biased in favor of new technologies and that increasing consumerism and materialism, which have been facilitated by science and technology, have failed to increase happiness. The common belief that technological change is inevitable is questioned; the myth of the value-neutrality of technology is exposed; and the ethics of the technological imperative "what can be done, should be done" is challenged. Instead, the profit motive of corporations is identified as the main determinant of the direction of technological change. *Techno-Fix* asserts that science and technology, as currently practiced, cannot solve the many serious problems we face and that a paradigm shift is needed to reorient science and technology in a more socially responsible and environmentally sustainable direction.

Techno-Fix is one of the few, if not the only, comprehensive discussions of modern technology written not by philosophers, historians or journalists but by two inside observers of the technological scene. Michael holds a doctorate in chemical engineering and has an extensive background in environmental science, economics and business, as well as more than 25 years' experience in environmental research. Joyce holds a doctorate in applied mathematics and a master's degree in anthropology. Being educated and experienced in science and engineering, the authors are uniquely positioned to deliver an insightful and powerful critique of modern technology.*

The readers of *Techno-Fix* will learn a number of inconvenient truths about science and technology, topics that are rarely, if ever, covered in the media or discussed among professionals. Readers will be challenged to re-examine their current worldviews, their paradigms and assumptions about the so-called promises of modern technology. But they will also enjoy their newly gained knowledge and will feel empowered and inspired by the fact that most problems confronting humanity have inherently simple, low-tech solutions that do not rely on excessive technology.

Who should read *Techno-Fix*? Anyone interested in protecting nature; anyone concerned about the effects of technology on society and the environment; anyone teaching or studying science, engineering, medicine or related disciplines; anyone intending to create a better future.

The following is a brief overview of the chapters.

Part I: Technology and its Limitations addresses a number of important questions regarding modern technology: What kind of unin-

---

* The opinions expressed in this book are solely those of the authors and do not reflect the views of the authors' current or previous employers, their clients or the US government.

tended environmental and social consequences are associated with advanced technologies? Could they have been predicted and avoided? Are counter-technologies, social fixes and efficiency improvements really effective in solving the problems brought about by modern science and technology? Does increasing technology promote sustainability or accelerate collapse?

Chapter 1: The Inherent Unavoidability and Unpredictability of Unintended Consequences postulates that there are always positive and negative effects of any technology. It is impossible for humans to substantially modify natural systems without creating unanticipated and undesirable consequences. Furthermore, technological consequences may become irreversible if the magnitude and speed of change is greater than the adaptive capacity of the environment, ourselves or other species. Finally, modern science, because of its foundation of mechanistic reductionism, is intrinsically unable to predict all deleterious side effects.

Chapter 2: When Things Bite Back explores, in depth, some of the many unintended environmental and social consequences of modern technologies, ranging from environmental pollution, global warming, species extinction, topsoil loss and ecological disruptions by genetically engineered organisms to social alienation; death and destruction brought about by chemical, nuclear and other high-tech weaponry; antibiotic resistance; human overpopulation and the decline in biological fitness.

Chapter 3: Technology, Exploitation and Fairness advances the thesis that many technologies are regrettably used for control and exploitation of both humans and the environment, leading inevitably to detrimental consequences for those exploited.

Chapter 4: In Search of Solutions I: Counter-Technologies and Social Fixes discusses the limitations of technologies that attempt to counter the negative effects of previous technologies and also shows why technological solutions to social, economic, political and psychological problems are often ineffective because they generally address symptoms rather than causes.

Chapter 5: In Search of Solutions II: Efficiency Improvements analyzes a wide range of historical data to demonstrate that most efficiency improvements have not been able to halt or reverse the growth in the use of limited resources but instead accelerate their consumption, a phenomenon called the rebound effect or the Jevons paradox.

Chapter 6: Sustainability or Collapse? argues that there are at least three critical technological challenges that must be met in order to produce long-term sustainability: avoiding serious environmental impacts associated with the large-scale generation of renewable energy, replacing non-renewable materials with renewable substitutes, and completely recycling non-renewable materials and wastes.

Part II: The Uncritical Acceptance of Technology addresses key questions relating to the naïve acceptance of new technologies despite the many negative consequences and limitations discussed in Part I. Why do we believe in technological progress? Is the current exuberant technological optimism justified by the evidence? Are technology assessments and cost-benefit analyses really objective and unbiased? Why do we still believe that increasing material affluence will increase happiness despite evidence to the contrary? Is technology value neutral and autonomous, as is often claimed? Is it prudent to follow the technological imperative "Whatever can be done, will be and should be done"? How democratic is technological decision making? Should profit maximization remain the primary criterion for the selection of new technologies?

Chapter 7: Technological Optimism and Belief in Progress postulates that belief in progress exhibits characteristics similar to those of religious faith and that most techno-optimism is based on ignorance, enabling the corporate-controlled mass media to present new technologies and products in an overly favorable light to a gullible public.

Chapter 8: The Positive Biases of Technology Assessments and Cost-Benefit Analyses demonstrates how each step in the standard cost-benefit analysis procedure has intrinsic problems and ambiguities, some of which are specifically exploited, knowingly or unknowingly, to produce positive recommendations for the development and diffusion of technologies even when they are of marginal or of no benefit.

Chapter 9: Happiness provides extensive evidence that material affluence, consumerism and economic growth brought about by advances in science and technology have failed to improve psychological well-being and, at the same time, have weakened or destroyed many non-materialistic and traditional sources of happiness.

Chapter 10: The Uncritical Acceptance of New Technologies discusses five topics related to the widespread belief in the inevitability of technological change: the myth of value-neutrality, the technological imperative, the loss of freedom and technological dependency, the myth of autonomous technology, and the undemocratic control of technology.

Chapter 11: Profit Motive: The Main Driver of Technological Development demonstrates that profit maximization does not necessarily lead to the development of technologies and products best suited to meet the needs of people in terms of food, health and security.

Part III: The Next Scientific and Technological Revolution poses critical questions about the future of science and technology. Because most problems caused by science and technology in the past were created within the dominant worldview characterized by excessive individualism and the goals of control and exploitation, a paradigm shift to a different view of reality is needed to solve fundamental problems. A more realistic paradigm would lead to a change in the form of economic activities as well as to changes in the practice of science and technology. How could such a paradigm shift be brought about? Can technologies be designed to be environmentally sustainable and socially appropriate while minimizing unintended consequences? What are the characteristics of a self-correcting, critical science? Do science and engineering professionals have social responsibilities?

Chapter 12: The Need for A Different Worldview suggests that a shift is needed to a different view of reality, one that is based on the fact of interconnectedness rather than the illusion of separateness, a view that would result in a change from a growth economy to a steady-state economy and to a change in how science is performed, technology applied and medicine practiced.

Chapter 13: The Design of Environmentally Sustainable and Socially Appropriate Technologies suggests specific environmental and social design criteria for new technologies, the importance of the precautionary principle in preventing unintended consequences, and the need for a more democratic control of technology.

Chapter 14: Critical Science and Social Responsibility outlines ways to increase the awareness of scientists and engineers regarding their social responsibilities as well as ways to transform current science into a critical, self-reflective and self-correcting science.

The arguments advanced in *Techno-Fix* are supported by extensive research, with more than 1,200 footnotes citing at least 600 references, primarily from peer-reviewed academic publications. Key points are also supported by quotations from authorities and original thinkers such as Rachel Carson, Barry Commoner, Herman Daly, Paul and Anne Ehrlich, David Korten, Jerry Mander, Donella Meadows, Jeremy Rifkin, E.F. Schumacher, and E.O. Wilson. In the Appendix, suggestions "For Further Thought" invite readers to engage in critical analyses themselves.

For more information and updates visit technofix.org.

PART I

TECHNOLOGY AND ITS LIMITATIONS

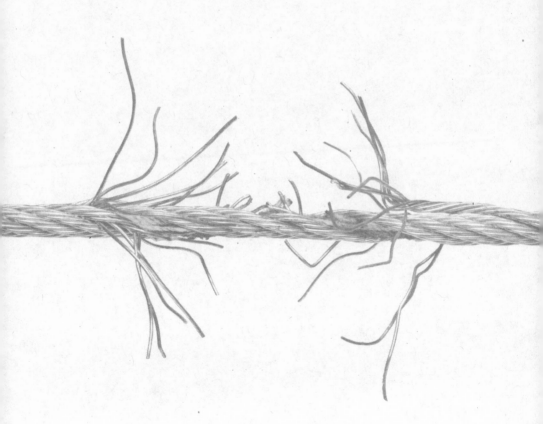

The Inherent Unavoidability and Unpredictability of Unintended Consequences

Interconnectedness

We live in a highly complex and dynamic world where, according to Barry Commoner's insightful first law of ecology, "Everything is connected to everything else."[1] Although we may perceive the natural environment as consisting of many different and isolated components and processes, these are all derivatives of the same cosmos, interrelated and linked together through mutual cause and effect. Science, of course, has been very successful in elucidating some of these causal relationships but, as will be discussed later, only a subset of the totality of such relationships. The fact that "all is connected to all"[2] has profound implications for the application of technology, particularly with respect to unintended consequences.

Interconnectedness in the natural world exists at many different levels, i.e., from the physical, chemical and biological to the sub-atomic. For example, the global cycling of many elements involves physical transport by wind and water over very large distances. Plants and animals depend on properly functioning global water, carbon and nitrogen cycles for their survival. Interconnectedness at the chemical level is even more profound, with thousands of organic and inorganic chemical reactions linking the various parts, which superficially appear unrelated.

Plants and animals are connected through the mutual exchange of oxygen and carbon dioxide. The oxygen produced by plants via photosynthesis is needed by animals for respiration, resulting in the generation of carbon dioxide, which in turn is taken up again by plants and used for their growth. There are thousands of biochemical reactions carried out within living cells, most of which are coordinated through highly complex regulatory networks involving feedback loops and other control mechanisms. Even above the cellular level, there is tight coordination of the functions carried out by the different organs within an organism. Furthermore, there are highly complex interdependencies among different species within a given ecosystem, many of them being part of an intricate food web consisting of elaborate predator-prey relationships.

Naturalist John Muir observed more than a century ago: "When we try to pick out anything by itself, we find it hitched to everything else in the Universe."[3] Because humans are an integral part of nature, whatever they do to nature will ultimately affect them, either positively or negatively. This simple fact has been expressed by many of the world's native people. For example, a Maori proverb cautioned, "Destroy nature, destroy yourself."[4] Chief Seattle warned, "All things are connected. Whatever befalls the Earth befalls the sons of the Earth. Man did not weave the web of life, he is merely a strand in it. Whatever he does to the web, he does to himself."[5]

The obvious truth regarding humans as part of nature escaped the philosophers of the Enlightenment who espoused a conceptual separation between humans and the environment, between observer and observed, thereby paving the way for a mechanistic reductionist science, which, in turn, yielded powerful knowledge on how to dominate, control and exploit the environment. However, according to Eugene Schwartz, commenting on the limitations of science and technology, "the concept of harnessing nature through conquest was in error because it failed to recognize that man was a part of nature and that what happened to nature would in turn rebound upon man."[6] Modern technology aggravates this conceptual error by creating an even greater illusion of separateness from the natural environment:

> All this leads us to believe that we have made our own environment and no longer depend on the one provided by nature. In the eager search for the benefits of modern science and technology we have become enticed into a nearly fatal illusion: that through

our machines we have at last escaped from dependence on the natural environment.[7]

It is perhaps ironic that the initial success of science, based as it was on the conceptual separation between man and nature, is finally, after more than 300 years, demonstrating in disciplines ranging from chemistry to ecology to quantum physics that there is, in fact, no such separation.

Human Improvement upon Nature

One assumption that underlies a substantial number of technological applications is the belief that nature can be improved upon or perfected for the benefit of mankind. Indeed, the whole idea of progress, which was introduced during the Enlightenment, is based on the faith that both human societies and nature can be perpetually improved through the power of reason. Unfortunately, the belief that humans can improve upon nature is outdated and has been shown to be false by science itself, specifically by the discovery of biological evolution. Originally conceived by Charles Darwin but since confirmed by different, often unrelated, scientific disciplines ranging from paleontology and geology to ecology and molecular biology, the evolution of species involves two processes, random mutation and natural selection.[8] The genetic blueprint of life constantly changes in a random fashion as a result of both mutation and recombination. Individuals having a genetically-based phenotype best suited for survival in a given environment will also most likely procreate successfully, thereby out-competing, in terms of number and fitness of progeny, less fit individuals. As a result, the selective pressure which is constantly exerted by nature ensures that over the long run most populations will be at least adequately adapted to their immediate physical and biological environments, including their interaction with other species. In short, the process of evolution guarantees that, within a given environment, species function and interact in a changing but largely optimized fashion. This concept of balanced, optimized adaptation was described by distinguished biologist Barry Commoner in his "third law of ecology:"[9]

> In my experience, this principle ["nature knows best"] is likely to encounter considerable resistance, for it appears to contradict a

deeply held idea about the unique competence of human beings. One of the most pervasive features of modern technology is the notion that it is intended to "improve on nature"—to provide food, clothing, shelter, and means of communication and expression which are superior to those available to man in nature. Stated baldly, the third law of ecology holds that any major man-made change in a natural system is likely to be detrimental to that system.[10]

For example, the environmental pollution caused by thousands of synthetic organic chemicals that do not naturally occur anywhere in nature is likely to have severe negative effects on plants, animals and ecosystems. These artificial compounds, unlike those found in nature, have not been subjected to natural selection over billions of years of evolution to ensure their adaptive fit and coordination. For example, only compounds that can be biodegraded by microorganisms should be released into the environment, ensuring the continued recycling of elements. By contrast, many synthetic chlorinated organic compounds, such as the insecticide DDT, are highly resistant to biodegradation, thereby persisting in the environment and bio-accumulating in the fatty tissues of many animals, including humans. Barry Commoner continues by explaining how evolution, like research and development (R&D), has attempted to optimize the overall performance and coordination of living organisms:

> In effect there are some two to three billion years of "R&D" behind every living thing. In that time, a staggering number of new individual living things have been produced, affording in each case the opportunity to try out the suitability of some random genetic change. If the change damages the viability of the organism, it is likely to kill it before the change can be passed on to future generations. In this way, living things accumulate a complex organization of compatible parts; those possible arrangements that are not compatible with the whole are screened out over the long course of evolution. Thus, the structure of a present living thing or the organization of a current natural ecosystem is likely to be "best" in the sense that it has been so heavily screened for

disadvantageous components that any new one is very likely to be worse than the present one.[11]

In summary, natural selection operating on genetic variability continuously optimizes the balanced functioning of all species with respect to each other within given ecosystems. Therefore, when humans, using science and technology, attempt to optimize nature for their own purposes, they immediately disturb the natural balance. As a result of human intervention, natural processes will function in less than optimal ways, which will have negative repercussions for humans who are also a part of the natural world. In the words of conservation biologist David Ehrenfeld,

> There is the limit, an especially frustrating one, that is described by the maximization theory of von Neumann and Morgenstern, which says in effect that in a complex world we cannot work everything out for the best simultaneously. This limit is why evolution has proven more reliable than our substitutes for it. Evolution is slow and wasteful, but it has resulted in an infinity of working, flexible compromises, whose success is constantly tested by life itself. Evolution is in large measure cumulative, and has been running three billion years longer than our current efforts. Our most glittering improvements over Nature are often a fool's solution to a problem that has been isolated from context, a transient, local maximization that is bound to be followed by mostly undesirable counter-adjustments throughout the system.[12]

Unavoidable Negative Effects of Technology

Because the negative consequences of science and technology often occur in unanticipated forms and in distant locations, and sometimes after significant time intervals, they are often not perceived as related to their causes. Nevertheless, technology will necessarily produce both positive and negative effects.[13] This character of technology creates a serious intellectual challenge for technological optimists, who exclusively focus on the positive aspects of technology while ignoring the, often enormous, negatives.

As Barry Commoner states in his fourth law of ecology "There is no such thing as a free lunch:"

> In ecology, as in economics, the law is intended to warn that every gain is won at some cost. In a way, this ecological law embodies the previous three laws. Because the global eco-system is a connected whole, in which nothing can be gained or lost, and which is not subject to over-all improvement, anything extracted from it by human effort must be replaced. Payment of this price cannot be avoided. It can only be delayed. The present environmental crisis is a warning that we have delayed nearly too long.[14]

All technological manipulations amount, at best, to zero sum games in which the costs balance the derived benefits. It is a mistake to believe that any benefits of technology can be obtained without cost. As Jacques Ellul wrote 35 years ago in *The Technological Society*,

> The technical phenomenon cannot be broken down in such a way as to retain the good and reject the bad.... It is an illusion, a perfectly understandable one, to hope to be able to suppress the "bad" side of technique and preserve the "good." This belief means that the essence of the technical phenomenon has not been grasped.[15]

Irreversible Consequences

The extent of negative effects of modern technology is directly related to the scale of exploitation of nature, which depends not only on the magnitude of human activities but also on the speed at which they are carried out. If both the magnitude and speed of human actions are greater than the adaptive capacity of nature, certain natural processes may cease to function entirely, resulting in irreversible consequences. For example, global climate change has the potential to precipitate the irreversible global collapse of planetary ecosystems and human civilization, destroying much that has been created through millions of years of evolution and thousands of years of cultural development.

As Robert Sinsheimer points out, the resilience of complex systems, such as those of nature and even social institutions, is becoming increasingly undermined by the speed of technological change:

Most states of nature are quasi-equilibria, the outcome of competing forces. Small deviations from equilibrium, the result of natural processes or human intervention, are most often countered by an opposing force and the equilibrium is restored, at some rate dependent upon the kinetics of the processes, the size of the relevant natural pools of components, and other factors. Although we may therefore speak of the resilience of nature, this restorative capacity is finite and is limited in rate.... Because human beings (and most creatures) are adapted by evolution to the near equilibrium states, the resilience provided by the restorative forces of nature has appeared to us to be not only benevolent, but unalterable.... The fragility of the equilibria underlying social institutions is even more apparent than of the equilibria of nature.... Our faith in the resilience of both natural and man-made phenomena is increasingly strained by the acceleration of technical change and the magnitude of the powers deployed.[16]

The resilience of both environment and human societies is limited. The extinction of thousands of species as well as many indigenous human cultures is an example of the irreversible changes brought about by the current pace of technological development and the enormous magnitude of technological exploitation.

Global climate change could well cause irreversible changes to life on Earth. The planet is a self-regulating open system whose complex order and proper functioning are maintained by the constant inflow of solar energy. According to John Peet, open systems such as Planet Earth are particularly vulnerable to perturbations:

Open systems (dissipative structures) tend to maintain a metastable state, often called dynamic instability. This reflects the fact that they are far from a state of equilibrium with their environment and depend on inflows, especially of energy, to maintain their state.... In self-regulation, the system preserves its stability by adapting and adjusting.... A system that is far from equilibrium can reach a point at which it can either dissolve into disorder or evolve to a new, more complex level of organization.[17]

Either of these two adaptive strategies by Earth in response to global climate change, disorderly collapse or the emergence of a new complex system will result in innumerable irreversible changes that could severely threaten the survival of humans and many other species.

Biological evolution is a very slow process. It has taken more than 3 billion years for life to evolve from single-cell organisms to the myriad complex, multi-cellular plant and animal species found today. Somewhat poetically we may say that the human race spent 100 million years as a mammal, 45 million years as a primate and over 15 million years as an ape.[18] Human cultures also evolve, and traditional cultures evolve very slowly. For example, the Desert Culture of the Tohono O'odham (Papago Indians) of the American Southwest is believed to have remained substantially unchanged for many thousands of years.[19]

In contrast to these slow rates of biological and cultural evolution, the rate of current technological change is orders of magnitude greater, thereby posing a potentially insurmountable challenge to environmental and cultural adaptation. According to Chauncey Starr, the speed of technology diffusion is so fast that social or environmental impacts often cannot be assessed and addressed in time to avoid serious negative consequences:

> The bulk of evidence indicates that the time from conception to first application (or demonstration) has been roughly unchanged by modern management, and depends chiefly on the complexity of the development. However, what has been reduced substantially in the past century is the time from first use to widespread integration into our social system. The techniques for societal diffusion of a new technology and its subsequent exploitation are now highly developed. Our ability to organize resources of money, men, and materials to focus on new technological programs has reduced the diffusion-exploitation time by roughly an order of magnitude in the past century. Thus, we now face a general situation in which widespread use of a new technological development may occur before its social impact can be properly assessed, and before any empirical adjustment of the benefit-versus-cost relation is obviously indicated.[20]

Governmental controls, such as environmental regulations and laws designed to protect both the public and the environment from the negative effects of innovative technologies, often lag behind the hasty introduction of technology. At some point, science and technology may be employed to weaken negative feedback cycles. However, the weakening of feedback and the removal of other natural checks and balances may temporarily protect us from the negative effects of new technologies but most likely will result in the delayed appearance of even more serious consequences later. For example, an animal population is generally kept in balance by the limited availability of food and the presence of predators. Humans have used powerful technologies to escape these natural constraints, first by using weapons to eliminate large predators, then by inventing agriculture to increase food supplies and finally by employing sanitation and medical technologies to increase their chances for survival. As a result, human numbers have increased to almost 7 billion today, a number that is completely unsustainable over the long term even with the application of advanced science and technology (see Chapter 6). Attempts to maintain such large human populations over time will cause a devastating series of widespread and unparalleled severe and irreversible consequences, such as habitat loss and species extinction as well as the collapse of human civilization (see Chapter 6).

To summarize, given that nature is a unified whole whose balance is maintained by evolution, it is intrinsically impossible for humans to "improve" upon nature. In fact, temporarily gained benefits will always be followed by corresponding costs. The negative consequences of science and technology at best may be delayed but never avoided. One could, in principle, argue that this is acceptable as long as one could predict the negative effects and assess not only all benefits and costs but also determine who will gain and who will lose, with the goal of properly compensating all parties that might be adversely affected by the introduction of new technologies. Unfortunately (as will be shown in Chapter 8, which examines the limitations of cost-benefit analyses), the accurate prediction of all possible consequences following the application of innovative technologies is inherently impossible, because science itself cannot provide the necessary information. The reason for this shortcoming is that the scientific method, from its beginning until today, is based on mechanistic reductionism, the limitations of which will now be explored.

The Limitations of Reductionism

Prior to the Enlightenment, knowledge was considered valid only when derived from authoritative sources such as holy scriptures or respected classical authors (e.g., Plato, Aristotle, Cicero or St. Augustine). In retrospect, it is amazing that for thousands of years, the validity of this authoritative knowledge was rarely questioned by demanding factual evidence. This changed profoundly during the Enlightenment when Francis Bacon conceived a new method of knowledge acquisition and validation, published in his *Novum Organum* in 1620.[21] In this revolutionary treatise on science, he proposed the testing of hypotheses by the meticulous observation of natural phenomena in carefully designed experiments. At the same time, René Descartes promoted the idea of scientific reductionism, which assumes that an adequate understanding of a complex system (e.g., nature) can be achieved by investigating the properties of its isolated parts.[22]

Mechanistic reductionism soon became, and remains to this day, the foundation of modern science. After all, human senses and even the most technologically advanced scientific instruments can observe only very isolated and specific aspects of natural phenomena. For example, the human eye can process only a very small fraction of the electromagnetic spectrum. Similarly, the ear can process pressure waves occurring only within a relatively small frequency range. Most scientific instruments are designed primarily as extensions of the human senses, and they accurately measure only selected properties of natural phenomena.

Science has been successful because experiments are carefully designed to observe the effects of changing only one selected variable at a time, while other conditions are held constant. By explicitly focusing on the elucidation of one or a few sequential cause-effect relationships, using specifically selected measurements, it has been possible to determine the many physical, chemical and biological mechanisms underlying natural phenomena. At the same time, however, because experiments always must be designed in a way to eliminate the effects of confounding variables, other important interactions within complex systems are necessarily ignored, if identified at all. In short, it is not possible to observe everything, and mechanistic reductionist science reflects this very basic limitation. In the words of Barry Commoner,

> Confronted by a situation as complex as the environment and its vast array of living inhabitants, we are likely—some more than others—to attempt to reduce it in our minds to a set of separate, simple events, in the hope that their sum will somehow picture the whole.... Each of these separate views of the environmental system is only a narrow slice through the complex whole. While each can illuminate some features of the whole system, the picture it yields is necessarily false to a degree. For in looking at one set of relationships we inevitably ignore a good deal of the rest....[23]

Scientific reductionism is based not only on the assumption that highly complex, integrated systems can be understood by subdividing them into separate, discrete and often functionally unrelated parts but also on the belief that the physical world is orderly and that human reason is able to discover this order and thereby to harness nature in the service of mankind. It is not surprising that the early Enlightenment philosophers and scientists had faith in the existence of an orderly and deterministic universe. After all, many of them were devoutly religious, believing that God had created a perfect world and that it was their personal mission to discover the underlying design of God's creation. For example, Isaac Newton's quest to understand the movements of celestial bodies, which ultimately resulted in his three laws of motion, was motivated by his desire to unveil the Creator's laws governing the clock-like universe. Unfortunately, the belief that nature is orderly and deterministic has now been proven false by science itself, most lately by quantum physics, and it is therefore highly questionable whether natural systems can ever be sufficiently well understood to be manipulated without creating unpredictable and often highly undesirable consequences.

One of the great dangers of scientific reductionism is the generation of "half knowledge," that is, knowledge of very specific cause-and-effect mechanisms without understanding or even being aware of all the complex relationships within the entire system.[24] As the Taoists of China knew more than 2,000 years ago, ignorance is particularly pernicious if one does not perceive that ignorance: "To know that one does not know—that is high wisdom. The fault of those who make mistakes is that they think they know when they do not know."[25] Ignorance of ignorance and lack of wisdom is the state of affairs in most of science and technology today, resulting in so-called "myopic engineering"[26] and all its unintended, negative consequences:

> The technologist's success is undone. The reason for this failure is clear: the technologist defined his problem too narrowly, taking into his field of vision only one segment of what in nature is an endless cycle that will collapse if stressed anywhere. This same fault lies behind every ecological failure of modern technology: attention to a single facet of what in nature is a complex whole.... Now the reason for the ecological failure of technology becomes clear: Unlike the automobile, the ecosystem cannot be subdivided into manageable parts, for its properties reside in the whole, in the connections between the parts. A process that insists on dealing only with the separated parts is bound to fail.[27]

What is even worse, science—by relying almost exclusively on mechanistic reductionism—not only makes us ignore the complexity of natural systems but in effect generates, according to Jeremy Ravetz, man-made scientific ignorance:

> Through all the centuries when progress became an increasingly strong theme of educated common sense, science could be seen as steadily advancing the boundaries of knowledge. There seemed no limit in principle to the extent of this conquest, and so the areas of ignorance remaining at any time were not held against science—they too would fall under the sway of human knowledge at the appropriate time. Now we face the paradox that while our knowledge continues to increase exponentially, our relevant ignorance does so even more rapidly. And this is ignorance generated by science![28]

Indeed, the generation of knowledge cannot be separated from the generation of ignorance. It may be that even the second law of thermodynamics applies here as well, indicating that knowledge, which can be considered a highly ordered form of information, can be generated only at the expense of creating more disorder, that is, more ignorance. In fact, a brief survey of major problems facing humanity today suggests that most of them were created by the application of science and technology, and that the greatest challenge will be to overcome our ignorance in dealing with these self-created disasters. For example, it will be an enormous, if not impossible, task to determine all of the possible negative consequences of global climate change and to find solutions to them. Our collective ignorance about how to address these derivative problems is far greater than the scientific knowledge and related technologies that created them.

In summary, the negative and sometimes irreversible consequences brought about by the application of science and technology are not only inherently unavoidable but also intrinsically unpredictable. Consequently, we have to be continuously vigilant for the possible occurrence of negative side effects of innovative technologies, which could occur at any time and any place in unpredictable ways. As Langdon Winner recognizes in *Autonomous Technology*: "Eternal vigilance is the price of artificial complexity."[29]

In the next chapter, we explore examples of the many environmental and social consequences of advanced technologies.

CHAPTER 2

When Things Bite Back: Some Unintended Consequences of Modern Technology

Unintended Environmental Consequences

As discussed in Chapter 1, many negative environmental consequences resulting from the technological exploitation, control and modification of nature are inherently unavoidable because human actions cannot really "improve" nature, a complex interconnected system that is continually adapting to change through the process of evolution. In addition, the conservation of mass principle as well as the first and second laws of thermodynamics can be invoked to demonstrate that it is impossible to escape the negative environmental effects of newly introduced technologies. To review, the first law of thermodynamics states that energy within a closed system may change in form but overall must remain constant; thus any gain in energy must be balanced by an equal loss of energy somewhere else. The second law of thermodynamics maintains that chaos or disorder (entropy) within closed systems must increase with time. However, by means of energy inputs, it is possible to increase order within a subsystem at the expense of creating more disorder elsewhere in the system.[30] To understand the implications of these basic laws of nature, it is useful to briefly review economic and industrial activities in terms of the conservation of mass/energy principles and the law of entropy.

The survival of individual organisms depends on the intake of high-energy, low-entropy foods and the excretion of low-energy, high-entropy wastes. Similarly, the successful functioning of industrial and economic systems requires the extraction of low-entropy (i.e., highly ordered) and high-energy matter from the environment and its conversion into useful products and services with the concomitant and unavoidable production of low-energy, high-entropy (i.e., highly disordered) wastes that, whether treated or not, re-enter the environment. As will be discussed in Chapter 6, it is impossible to recycle wastes without the input of additional energy, the generation of which increases entropy. In short, industrial and economic activities consist of extracting highly ordered matter from the environment and returning highly disordered wastes. For example, energy generation is possible only by converting low-entropy materials such as coal, petroleum, natural gas and uranium into high-entropy wastes such as carbon dioxide, sulfur dioxide, nitrous oxide, radioactive materials and waste heat. As a more specific example, the compounds that are present in coal consist of highly compact, ordered molecular structures (reflecting a low-entropy state) that, after combustion, are converted into carbon dioxide, which becomes highly dispersed after mixing with other gases in the atmosphere (reflecting a high-entropy state). Consequently, according to the physical laws of nature, it is impossible to avoid negative environmental impacts of human economic and industrial activities, a fact that Herman Daly summarizes by providing the following two reasons:

> The first [reason] is the first law of thermodynamics (conservation of matter/energy). The taking of matter and energy out of the ecosystem must disrupt the functioning of that system even if nothing is done to the matter and energy so removed. Its mere absence must have an effect. Likewise, the mere insertion of matter and energy into an ecosystem must disrupt the system into which it is added.... The second reason is the second law of thermodynamics, which guarantees that the matter/energy extracted is qualitatively different from the matter/energy inserted. Low-entropy raw materials are taken out, high-entropy wastes are returned.[31]

Jeremy Rifkin summarizes the second law of thermodynamics this way: "Each technology always creates a temporary island of order at the expense of greater disorder in the surroundings."[32] It has been suggested by a number of scientists that the increase in entropy (disorder) in the environment is directly related to environmental damage and ecosystem disruption.[33] Thus, given that the second law of thermodynamics guarantees that for each unit of "order" (neg-entropy) created in the human-based economy, more than one unit of "disorder" (entropy) is created in the surrounding environment,[34] it follows that all industrial activities must lead to unavoidable environmental disruptions. This was recognized almost three decades ago by N. Georgescu-Roegen, an economist who was the first to incorporate entropy considerations into economic theory.[35] He states, "No one has realized that we cannot produce 'better and bigger' refrigerators, automobiles, or jet planes without producing also 'better and bigger' waste."[36] As will be explored further in Chapter 6, the second law of thermodynamics also has serious implications for the pursuit of so-called "sustainable development" because any additional economic growth most likely will result in further environmental degradation.

Given the rapid and more than 100-fold expansion of world industrial output during the past 250 years,[37] most of it occurring during the past 50 years, it should not be surprising that we are now confronted with a large array of environmental problems: enormous amounts of garbage; toxic and nuclear wastes; air, groundwater, soil and sediment pollution; oil spills; loss of topsoil, acid rain; photochemical smog; reduced visibility; global climate change; disruption of global biogeochemical cycles; rainforest destruction; habitat loss; and species extinction. Hundreds of books have already been written on these and other environmental problems caused by the widespread applications of science and technology. We will focus here only briefly on three representative examples: pollution by synthetic organic chemicals, global climate change and species extinction.

Every year, the chemical industry produces over 100 million tons of synthetic organic chemicals, representing over 65,000 different compounds in regular commercial use.[38] The problem is, of course, that these chemicals are synthetic, few being naturally occurring, and therefore are, in Rachel Carson's words, "totally outside the limits of biological experiences."[39] It would require an indeterminate number of generations for humans and other animals to adapt to these novel compounds through the process of biological evolution. Because there are no naturally occurring degradative pathways for many of these synthetic organic chemicals, these alien substances bioaccumulate in the fatty tissues of animals, including humans. Organic chemicals produced by a relatively small number of manufacturers and widely dispersed around the world enter into the tissues of billions of humans and animals. The laws of physics, in particular the second law of thermodynamics, guarantee increasing entropy and disorder in the environment.

Unfortunately, synthetic organic chemicals are not rigorously tested for health and environmental effects prior to widespread use. Although there has been some progress during the past three decades in testing organics for acute toxicity (i.e., lethality) and for a few selected toxicological endpoints such as mutagenicity or carcinogenicity, our ignorance of the potential health and environmental effects is immense and will remain so, given the inherent limitations of scientific reductionism (see Chapter 1). There will never be sufficient research funding to test the tens of thousands of organic chemicals for all foreseeable side effects, many of the effects being not yet even conceivable. Consider, for example, the recent discovery that many organic chemicals such as DDT, PCBs, dioxins and plasticizers, at exceedingly low concentrations, mimick hormones, thereby affecting the sexual development of animals, including humans.[40] Paul and Anne Ehrlich describe the subtle effects of these endocrine disruptors as follows:

> Sensitivity to these compounds is especially high in early stages of development, and reproductive systems, both male and female, seem to be partially affected. Among the potentially disastrous consequences that may result from prenatal exposure to female hormone-mimicking chemicals are marked declines in male fertility, including very low sperm counts, and abnormal development of sexual organs in either sex.... Some other compounds, by contrast, seem to have an opposite effect: they block the action of female hormones, in effect masculinizing the individual. Prenatal exposure to dioxin...can exert both feminizing and masculinizing effects.[41]

Some have suggested that sexual orientation may not be a free personal choice, as is sometimes claimed, but rather a function of the types of contaminants to which one has been exposed, either pre- or postnatally.[42] As Robert Glennon, professor of law and specialist on water resource issues, writes, "I am all for sexual freedom and cross-dressing, but I think it should be a matter of personal choice rather than of having drunk the wrong water."[43] Clearly, significantly more research is needed to determine whether there is a link between environmental contaminants, such as endocrine disruptors, and sexual orientation in mammals, including humans.

We now turn to another unintended consequence of modern technology. Global climate change is essentially caused by the increase in atmospheric carbon dioxide concentrations (i.e., from pre-industrial levels of 280 ppm to more than 390 ppm today[44]) because of fossil-fuel combustion at the rate of approximately 7 billion metric tons of carbon per year. For more than a century, humans have been burning fossil fuels from deposits that required millions of years to accumulate, creating a severe imbalance in the global carbon cycle. Consider, for example, that the rate of fossil carbon release into the atmosphere is about 400 times the net primary productivity of the planet's current biota.[45] Clearly, the rate of increase in atmospheric CO_2 concentration is much greater than the rate at which the planetary climate system can adapt. As a result, many negative consequences have already begun to appear. The Intergovernmental Panel on Climate Change is predicting an increase in global mean temperature of about 2 to 5°C (3.6 to 9°F) within the next hundred years.[46] Greater average temperatures will result in an altered hydrogeological cycle, which translates into higher frequencies of severe droughts and floods. Furthermore, the predicted temperature increase will cause the thermal expansion of the world's oceans and the partial melting of polar ice, thereby causing a substantial rise in sea levels.[47] More than a billion people living in coastal regions, many of them living in large metropolises such as New York City, Bangkok and Shanghai, will likely be affected. The response to sea-level rise will either be flight from coastal regions, resulting in hundreds of millions of environmental refugees, or construction and continuous monitoring of many thousands of miles of seawalls and dikes, an example of how humans have created a hostile world in which constant vigilance is needed for survival.[48] Also possible is irreversible "abrupt" climate change, which may be triggered by the slowing of the thermohaline ocean circulation, which is predicted to occur if atmospheric CO_2 concentrations exceed

550 ppm.[49] This is likely to happen within the next 50 to 100 years unless drastic measures are taken to reduce global carbon emissions.[50] Abrupt climate change would cause such severe weather changes that adaptation by humans and many other species would be extremely difficult, specifically in most of Europe, where temperatures could become too low for the practice of agriculture. As a result of these and other uncertainties, the total costs of global climate change could amount to 50 percent of the world gross product, indicating that a significant fraction of economic activities will have to be redirected toward the effective mitigation of and adaptation to climate change.[51]

Probably the worst environmental crime of the present and looming future is the extinction of myriad plant and animal species that came into existence over millions of years of evolution. A species that becomes extinct is gone forever. The damage is irreversible and, if carried to extremes, the web of life that has evolved over the past three billion years will be destroyed. Should humans claim the right to destroy life forms they never created? Such disrespect for life will have many negative repercussions, such as the disruption or collapse of entire ecosystems, which will certainly be followed by a sharp reduction in the many "free" services nature provides for humans.[52] Unfortunately, given the inherent limits of scientific reductionism (see Chapter 1), it is unknown how many of the approximately 7 million plant and animal species can be removed "safely" before ecosystems cease to function.[53]

The primary causes of species extinction are habitat loss and chemical pollution. For example, a 90 percent reduction in habitat area will eventually result in a 50 percent decrease in the number of species.[54] While rates of extinction and rates of speciation (i.e., the formation of new species) have been more or less in balance during most of the Earth's history, present rates of extinction are about 1 million times greater than current rates of speciation,[55] resulting in a massive decline of biological diversity. The present rate of species extinction is alarming: according to various estimates, ranging from best- to worst-case scenarios, between 1,000 and 100,000 plant and animal species disappear each year, which translates into 2.7 to 270 irreversible extinctions every day.[56] About one quarter of all bird species have been driven to extinction during the past two millennia.[57] It is highly unlikely that this unraveling of the web of life can continue much longer without severe repercussions to humans. As Donella Meadows concludes,

> The wildly varied stock of DNA, evolved and accumulated over billions of years...is the source of evolutionary potential, just as science libraries and labs and scientists are the source of technological potential. Allowing species to go extinct is a systems crime, just as randomly eliminating all copies of particular science journals, or particular kinds of scientists, would be.[58]

Unintended Consequences of Industrialized Agriculture

Since the invention of agriculture, humans have used animal labor and waste to produce plants whose edible parts contain the solar energy captured through the process of photosynthesis. These sustainable farming practices came to an end with the industrialization of agriculture approximately a century ago, when powerful farm machinery replaced human and animal labor, and non-renewable fossil-fuel inputs were used to increase crop productivities through the application of synthetic fertilizers, herbicides and pesticides. As a result of these technological innovations, agricultural productivity has increased significantly. For example, 100 years ago, an American farmer needed 150 minutes to produce one bushel of corn; in 1955 it took him just 16 minutes; and today it takes less than 3 minutes.[59] In 1950, about 15 people were fed per farm worker; today one farmer feeds more than 70 people.[60] Unfortunately, the benefits of this more efficient food production are associated with a wide range of environmental and social costs.

Initially, and almost immediately, the industrialization of agriculture led to many painful social consequences as mechanization caused massive rural unemployment of farm workers, resulting in a mass exodus of farmers from the countryside in search of jobs in the city. In the United States, millions of family farms and thousands of rural communities were destroyed, leaving behind a depopulated industrialized farmland operated by a few large agribusinesses.[61]

In addition to social consequences, the industrialization of agriculture has caused, over time, numerous negative environmental impacts. First, the plowing up of hundreds of millions of acres of grasslands and forests, using powerful farm machinery, led to the loss of enormous quantities of fertile topsoil through erosion by wind and water. As a result, the average depth of topsoil has fallen from 21 inches 200 years ago to no more than 6 inches today.[62] Soils, which are produced naturally on a time scale of inches per thousand years, are now being lost through erosion on a scale of inches per decade.[63] Indeed, about 90 percent of US cropland is losing soil far faster than it is being formed.[64] Worldwide, the situation is even worse. During the past 40 years, nearly one third of the world's arable land has been lost due to erosion and continues to be lost at a rate of more than ten million hectares per year.[65] Clearly, any form of agriculture that depends on the irreversible exploitation of soil fertility—"soil mining"—cannot be sustainable.

Another serious problem with industrialized agriculture is its dependence on large-scale irrigation. Currently, approximately 40 percent of the world's food supply is produced on irrigated cropland.[66] Worldwide, humans now appropriate more than half of the Earth's accessible fresh water for their own use, most of it for irrigation.[67] As a result, lakes, rivers and creeks are running dry, contributing to the extinction of many aquatic species.[68] As the collapse of earlier hydraulic societies in the Fertile Crescent of Mesopotamia should have taught us, irrigation agriculture is not sustainable because the buildup of salts and other toxic contaminants will ultimately make the soil unsuitable for crops and will leave behind a desert wasteland.[69]

A virulent series of environmental problems has been created by the use of artificial herbicides and pesticides, some of which were originally developed as chemical warfare agents by Nazi scientists during World War II but were later marketed by US chemical companies to farmers for pest and weed control.[70] The use of insecticides and herbicides increased in the US from only 200,000 pounds in the 1950s to an astounding 1,100,000,000 (1.1 billion) pounds in 1993, which translates into four pounds of poison for every person living in North America.[71] Worldwide use of pesticides is around 4.5 billion pounds per year. In addition to the dramatic increase in the amount of pesticides used in the past 50 years, the "killing power" per pound of chemical has increased tenfold during this period as well.[72] There are about 21,000 different commercial pesticide and herbicide products that contain various mixtures of at least 860 active toxic ingredients.[73] For about half of them, the residues remaining on food products cannot be determined by current analytical methods.[74]

Considering these facts, we should ask the obvious question that was posed more than 45 years ago by Rachel Carson in *Silent Spring*: "Can anyone believe it is possible to lay down such a barrage of poisons on the surface of the Earth without making it unfit for all life?"[75] The answer, of course, is that it is not possible to attack nature with biological warfare agents without seriously damaging both human health and the environment. Worldwide, about 1 million pesticide poisonings occur each year, resulting in 20,000 deaths.[76] In addition, pesticides kill about 67 million birds and between 6 and 14 million fish each year in the United States alone.[77] Finally, as was mentioned earlier, many of these synthetic organic chemicals bioaccumulate in the fatty tissues of all animals, including humans, where they may cause a number of acute or chronic health effects, many of them yet unknown. Known effects range from cancer and birth defects to immune dysfunction and impairment of sexual reproduction.[78]

Unfortunately, the industrialization of agriculture is not yet complete, as the recent promotion of genetically engineered crops by so-called life-sciences companies has shown. However, the large-scale application of genetic engineering technologies is expected to lead to many irreversible and disastrous consequences.

Unintended Side Effects of Genetic Engineering

Genetic engineering involves the removal, insertion or modification of specific DNA sequences (i.e., genes) within an organism's genetic material with the goal of removing undesirable traits, introducing new traits or enhancing existing characteristics of microorganisms, plants or animals. While traditional breeding techniques have been used for thousands of years to improve particular plants and animals to better meet human needs, the new gene-splicing biotechnologies are intrinsically different in that they enable the transfer of genetic traits *across* species boundaries. Plants have been engineered to become resistant to specific herbicides, enabling farmers to apply massive quantities of these toxins without affecting crops; but these toxins potentially endanger human health and the environment. The ability to recombine genes from completely unrelated species in myriad ways has increased our control over nature by many orders of magnitude compared to traditional breeding practices. Since more powerful control of nature generally translates into greater exploitation for monetary gain, we find "life-sciences" companies viewing these innovative biotechnologies as a way to dramatically increase profits. As is the case with the introduction of any new technology, the list of envisioned benefits of genetic engineering is long and deceptive, ranging from solving the problem of world hunger to curing, or even preventing, human diseases.

Genetic engineering presents one of the most extreme examples of the belief that nature can be "improved" through human ingenuity— this, despite the fact that we barely know enough to carry out the necessary techniques of molecular biology and have little or no knowledge about the long-term ramifications. Indeed, it is surprising that molecular biologists, who more than anyone else should have an understanding of the biological complexity of nature and the adaptive role of evolution, can honestly believe that genetic modification of plants and animals will not have major negative consequences. This lack of concern for potential system-level consequences may very well be the result of the extreme technical specialization that is a hallmark of reductionism as it currently exists in science and engineering, but it may also indicate a lack of social and professional responsibility, a topic discussed further in Chapter 14. Whatever the reasons, this is an example in which scientific reductionism provides limited knowledge on how to manipulate nature but at the same time creates even greater ignorance in the prediction and prevention of negative side effects (see Chapter 1). As Jeremy Rifkin points out in *Biotech Century*,

> The reseeding of the planet with a laboratory-conceived second Genesis is likely to enjoy some enviable short-term market successes, only to ultimately fail at the hands of an unpredictable and noncompliant nature. While the genetic technologies we have invented to recolonize the biology of the planet are formidable, our utter lack of knowledge of the intricate workings of the biosphere we are experimenting on poses an even more formidable constraint. The introduction of new genetic engineering tools and the opening up of global commerce allow an emerging "life industry" to "reinvent" nature and manage it on a worldwide

scale. The new colonization, however, is without a compass. There is no predictive ecology to help guide this journey and likely never will be, as nature is far too alive, complex, and variable to ever be predictably modeled by scientists.[79]

Even the conservative US Office of Technology Assessment reported to Congress that "in the long term (10-50 years), unforeseen ecological consequences of using recombinant organisms in agriculture are not only likely, they are probably inevitable."[80] The lack of knowledge of hazards should not be considered as safety.[81] Rifkin continues:

> There is not a single instance in history in which the introduction of a major technological innovation has had only benign consequences for the natural world.... Whenever a genetically engineered organism is released, there is always a small chance that it too will run amok because, like non-indigenous species, it has been artificially introduced into a complex environment that has developed a web of highly integrated relationships over long periods of evolutionary history. Each new synthetic introduction is tantamount to playing ecological roulette.[82]

What is of particular concern is that genetically modified organisms, specifically microbes and plants whose wind-blown pollen is essentially impossible to contain, will continue to reproduce and will colonize terrestrial and aquatic habitats, thereby making ecological disruptions ubiquitous and irreversible.

Beyond the use of genetically modified crops, there are many other potential applications of recombinant DNA technologies, many of them holding great promise but having an equally great potential to cause harm. Recombinant DNA technologies have already been used and will continue to be employed in the production of pharmaceuticals (e.g., insulin and growth hormone). This is likely to be safe unless some of these genetically engineered microbes escape into the environment, where they could propagate and cause irreversible damage. Despite serious ethical concerns, genetic engineering may also be used to create new patentable animals, specifically with the goal of improving meat production and profits for agribusiness. Furthermore, many inherited diseases may be prevented by population-wide genetic screening, but

the resulting private medical information could be used to discrimi-
nate against persons who are found to have higher susceptibilities to
certain disorders. There may also be a future campaign to create the
perfect human (i.e., "designer babies") through germ-line modification
but, as Robert Sinsheimer observes, "it seems paradoxical that a living
organism emergent from the evolutionary process after billions of years
of blind circumstance should undertake to determine its own future
evolution."[83] Finally, as is the case with many technological innovations,
genetic engineering biotechnologies can and probably will be applied in
warfare. For example, following the completion of the human genome
project, it may be possible for military scientists to design biological
warfare agents that are "ethnic weapons," that is, which infect only people carrying certain genes, thereby allowing the targeted elimination of
"undesirable" racial groups.[84]

According to Rifkin, the future of genetic engineering is uncertain,
leaving many important questions unanswered:

> In reprogramming the genetic code of life, do we risk a fatal
> interruption of millions of years of evolutionary development?
> Might not the artificial creation of life spell the end of the natu-
> ral world? Do we face becoming aliens in a world populated by
> cloned, chimeric, and transgenic creatures? Will creation of mass
> production, and wholesale release of thousands of genetically
> engineered life-forms into the environment cause irreversible
> damage to the biosphere, making genetic pollution an even
> greater threat to the planet than nuclear and petrochemical pol-
> lution?[85]

Unintended Consequences of the Automobile

Prior to the invention of the internal combustion engine, everyday
life was arranged in such a way that most locations related to work and
family activities could easily be reached by foot, while horses and, later,
trains facilitated occasional long-distance travel. When the first automobiles became widely available about a century ago, they had instant
appeal because they had the very obvious benefit of profoundly increasing freedom and mobility. The early benefits of the automobile could

have been preserved if its use had remained limited. Unfortunately, this has not been the case, and the number of automobiles has grown exponentially. In 1900, there were only 8,000 cars in the United States. Today, there are over 160 million. Worldwide, the automobile industry produces 40 million new cars each year, with 530 million already in use. The number of automobiles is expected to double to more than a billion within the next 20 years, the increase being greatest in developing countries.[86] Such large-scale use of the automobile will inevitably result in significant social and environmental costs.

One obvious negative consequence of the car culture is the high number of traffic deaths and injuries: between 40,000 and 55,000 people have been killed every year in the US since 1962. In 1998, there were 41,200 deaths and almost 3,200,000 injuries, which translates into a motor-vehicle-related death every 13 minutes and an injury every 10 seconds.[87] Worldwide, some 885,000 people are killed on the road each year,[88] which is equivalent to five fatal jumbo jet crashes every day.[89] Automobiles have killed over 30 million people since they were first introduced a century ago, more than the total number of military deaths suffered in World War I and World War II combined.[90]

In addition to killing people, cars also destroy the environment. Significant environmental impacts occur during the three life-cycle stages of the automobile: manufacture, use and disposal. Cars are responsible for smog and air pollution (particulate matter, nitrous oxides, volatile organic compounds and carbon monoxide) and are indirectly responsible for pollution occurring during the production and distribution of transportation fuels: oil spills at sea such as the recent Deepwater Horizon oil well explosion in the Gulf of Mexico, refinery emissions and the contamination of the nation's aquifers by several hundred thousand underground tanks that leak gasoline and fuel oil. The widespread use of tetraethyl lead as an octane enhancer in gasoline results in the irreversible dispersal of this toxic metal in the environment.[91] Transportation by automobile requires a vast network of well-maintained paved roads, many of which allow human access to remote areas, which results in the disturbance or destruction of sensitive habitats and wildlife. There are more than 3.5 million miles of roads (i.e., about 1,250 times the distance from New York City to Los Angeles) in the United States,[92] and nationwide, pavement covers an area larger than the State of Georgia.[93] In addition, an estimated 1 million vertebrate animals are killed by automobiles each day, which translates into 11 deaths per second.[94] Finally, cars also contribute significantly to the problem of climate change. In the United States, road transportation uses about 66 percent of all oil consumed and accounts for 28 percent of all carbon dioxide emissions.[95] Globally, cars discharge 1 billion tons of carbon dioxide each year, approximately 16 percent of all anthropogenic CO_2 emissions.[96] Given the projected growth of the world's vehicle fleet, automobiles may become the major contributor to climate change.

The history of the automobile provides an example of how a newly introduced technology may initially hold great promise, such as freedom and mobility, only to deliver exactly the opposite after its widespread adoption. In the United States, more than 6 billion hours are lost each year as a result of traffic congestion, costing the economy at least $48 billion. In 39 metropolitan areas with populations greater than 1 million, one third of all automobile travel takes place under congested conditions. In major cities such as London, New York, Paris and Tokyo, the average speed during rush hour traffic is often less than 10 km per hour (6 miles per hour), less than the velocity of a horse-drawn carriage a century ago.[97] Given the projected growth in car ownership, time lost to traffic congestion is expected to substantially increase.[98] Unfortunately, as will be discussed in Chapter 5, the common approach to solving problems created by technology via the application of more technology has failed in the case of traffic congestion: roads that are specifically built to reduce congestion fill up as soon as they are opened, making the original congestion problem even more intractable.[99]

As with many other technologies, the automobile has had a profound impact on social interactions among people. It is not surprising that the automobile was so readily adopted in the United States, because it embodies this nation's most cherished values of freedom and individuality. The automobile allows individuals to have maximum freedom of mobility (assuming no traffic jams), but at the same time separates them physically from other people as a result of being literally locked up in a steel and glass box, thereby eliminating any possibility of direct personal contact. As David Korten points out, cars have almost completely destroyed the quality of city life:

Cities once consisted primarily of exchange spaces for people—places such as shops, schools, residences, and public buildings. The pathways that connected exchange spaces were also places to meet and reaffirm relationships with neighbors. The automobile has changed our cities in fundamental ways.... Many of the spaces that once brought us together have been converted into noisy, congested, polluting places that isolate us from one another and destroy the quality of city life.[100]

After the car ruined city life by destroying public places and creating intolerable noise and air pollution, people fled with their automobiles to the suburbs, where they began living in low-density developments that could not sustain a mass-transit system.[101] As a result, those living in suburbs have become completely dependent on their automobiles not only to get to work but also to carry out all other routine daily activities away from home, such as shopping, transporting children to school and recreational events, and visiting friends. Thus, the automobile has led to urban sprawl and the creation of highly dispersed settlements that not only aggravate feelings of social alienation, isolation and placelessness but also make reliance on car transport absolute. Here again we have the paradoxical situation in which the initially promised freedom that cars were supposed to deliver turns into its opposite: total dependence on the automobile and whatever is needed to sustain it, such as imported fossil fuels and a vast network of pipelines, oil tankers, refineries, gasoline stations, car repair shops and millions of miles of roads. Given that during the past century most of the building and transportation infrastructure in the United States has been designed around the private automobile, it would currently be cost-prohibitive to eliminate it, although that might become necessary at some point due to oil shortages, limited renewable fuel supplies (see Chapter 6) and climate change concerns.[102]

Intended and Unintended Consequences of High-Technology Warfare

Much of the massive suffering endured during the major wars of the past hundred years is directly related to the destructiveness of innovative and powerful military technologies. World military expenditures in 2005 exceeded a trillion dollars, with the United States spending approximately 48 percent of the world total, followed by the United Kingdom, France, Japan and China with 4 to 5 percent each.[103] In 2005, 57 percent of the $131-billion federal R&D budget in the United States was allocated to warfare-related projects, confirming the importance of science and technology in maintaining the lead in weapons development.[104]

Military technologies, like most other technologies developed since the Industrial Revolution, are essentially labor-saving devices, i.e., they increase efficiency, in this case the killing efficiency of soldiers. The substantial investment in military research and development by industrialized nations during the past hundred years has resulted in the generation of ever-more-powerful and lethal weapons systems, which after deployment have caused millions of deaths and enormous destruction. In addition, because many military technologies such as bombs, missiles, nuclear weapons and chemical warfare agents kill indiscriminately, the proportion of civilians killed in wars has steadily increased.[105]

Two technological innovations, the machine gun and nerve gas, unexpectedly turned World War I into a protracted and barbarous killing match. The total number of World War I casualties amounted to approximately 21 million wounded and 19 million killed, half of them civilians.[106] In World War II, two key technological inventions, the bomber and the atomic bomb, not only increased the destructiveness of military weaponry by orders of magnitude but also caused incredible numbers of civilian deaths. Approximately 72 million people died in World War II, about 46 million civilians and 26 million soldiers.[107] The application of new technologies often leads to irreversible consequences—and there is nothing more irreversible than death.

Following the development and use of the atomic bomb during World War II, a nuclear arms race ensued between the United States and the Soviet Union.[108] During the Cold War, which lasted more than 40 years, the two super powers competed with each other in the development and production of nuclear weaponry to guarantee their mutually assured destruction. There was a constant threat of nuclear annihilation, not only of targeted cities or even countries, but even of the entire planet if the smoke, dust and radioactive fallout following worldwide atomic warfare were to cause a so-called nuclear winter, completely eliminating agricultural food production in the northern hemisphere, leading to the deaths of billions of people from starvation.[109] Although the threat of nuclear annihilation has greatly diminished since the breakup of the Soviet Union in 1991, it has not been eliminated, because large stockpiles of nuclear weapons and fissile materials still exist, and it is difficult, if not impossible, to prevent this material from falling into the hands of terrorists, blackmailers, political desperados or suicidal maniacs.[110] Thus, continual vigilance will be needed to protect ourselves from our own inventions.

Chemical and biological weapons have also been developed and deployed, such as Agent Orange and napalm. Here is an eye-witness report of the effects of napalm by a BBC journalist:

> In front of us a curious figure was standing, a little crouched, legs straddled, arms held out from his sides. He had no eyes, and the whole of his body, nearly all of which was visible through tatters of burnt rags, was covered with a hard black crust speckled with yellow pus.... He had to stand because he was no longer covered with a skin, but with a crust-like crackling which broke easily.[111]

The deployment of chemical defoliants, such as Agent Orange contaminated with dioxins, resulted in severe environmental and health effects that last to this day. Dioxin, according to Blum, is

> a nearly indestructible pollutant that is regarded as one of the most toxic substances in the world; at least as toxic as nerve gas, and highly carcinogenic. Amongst other health effects associated with exposure to dioxin are metabolic disorders, immunological abnormalities, reproductive abnormalities, and neuropsychiatric disorders.[112]

Unintended Consequences of High-Tech Medicine

The human body is highly complex, consisting of many specialized organs in which thousands of coordinated biochemical reactions occur simultaneously at the cellular level. The functioning of the body is not only dependent on the proper operation of a great number of interrelated physical, chemical and biological processes but is also affected by thoughts and emotions as well as interactions with the environment.

The processes operating within the human body and its interaction with the environment have evolved over hundreds of millions of years. One may define health as that state in which all of these interactions (i.e., those within the body, and between body and environment) are balanced in such a way to guarantee the optimum functioning of the organism. The process of evolution has ensured that most surviving organisms within a population function adequately because the population is adapted to the immediate environment. If environmental or social conditions change too rapidly and exceed the organism's ability to adjust, health deteriorates and diseases develop.

Indeed, for thousands of years prior to the Enlightenment, it was believed that harmony and balance were the sources of health while disharmony and imbalance were the causes of human disease.[113] Early medical practitioners, from Aristotle and Hippocrates to Galen and Ptolemy, believed in the power of natural healing, which generally involved restoring a harmonious balance between body, mind and environment. These earlier concepts of health and disease changed radically 400 years ago when the Enlightenment philosophers, specifically Francis Bacon and René Descartes (see Chapter 1) developed the "New Science" (*Novum Organum*), which was founded upon a passionate faith in both reductionism and a mechanistic interpretation of the world. Reductionism, as mentioned above, is the belief that complex systems such as the natural world or the human body can be best understood by investigating the properties of their isolated parts.[114] Mechanistic theories assume that the entire cosmos, nature in general and organisms in particular, are all essentially machines. Thus, the universe was perceived as a gigantic clock; animals were seen as "beast-machines;" and humans were seen as "perpendicularly crawling machines."[115] This new, mechanical view of the human body had a profound effect on the definition of health and disease, and therefore on the entire practice of modern medicine. In the words of Ivan Illich,

> Descartes' description effectively turned the human body into a clock-work and placed a new distance, not only between soul and body, but also between the patient's complaints and the physician's eye. Within this mechanized framework, pain turned into a red light and sickness into mechanical trouble. A new kind of

taxonomy of diseases became possible. As minerals and plants could be classified, so diseases could be isolated and put into their place by the doctor-taxonomist. ... Sickness was placed in the center of the medical system, a sickness that could be subjected to (a) operational verification by measurement, (b) clinical study and experiment, and (c) evaluation according to engineering norms.[116]

As a result of this new worldview, health was no longer seen as a harmonious balance. Instead, according to Fries and Crapo, the medical model now "defines health as the absence of disease and seeks to improve health by understanding and eradicating disease."[117] Given the machine paradigm, diseases are understood in terms of their physical, chemical or biological mechanisms, often at the molecular level. Detailed mechanistic knowledge of phenomena occurring in nature in general, and within the body in particular, is used for intervention and control. Thus, modern medical practice involves attempts to eradicate diseases through various physical, chemical and biological manipulations.[118] Like a machine, the body is assumed to be near optimum health if the functions of all its isolated parts (organs) have been controlled and improved by medical interventions.[119]

The bias of this mechanistic and reductionist engineering approach to human health is also reflected in the belief that most diseases have clearly identifiable causative agents (specific etiology) and therefore can be eradicated by "magic bullets" that counteract them.[120] An example of a "magic bullet" is the antibiotic designed to kill pathogenic microorganisms, thereby eradicating infectious diseases. The quest for magic bullets continues unabated and is the driving force behind much current medical and pharmacological research, as is apparent in the high-intensity "warfare" against cancer, AIDS and other challenging diseases. There is no doubt that this type of high-technology medicine has been impressively successful in specific cases, but it is also obvious that this mechanistic and reductionist approach to health and disease is very limited and is, therefore, responsible for a number of unforeseen consequences.

One unfortunate and unintended result of the engineering approach to healing is that potential environmental, economic, social, dietary and psychological factors that may be involved in disease causation are often ignored. This is not surprising because scientific reductionism makes it inherently difficult to perform a holistic analysis of diseases, a situation that is further complicated by the increasing fragmentation of scientific knowledge and specialization of medical practice. Consequently, the ultimate causes of a disease may never be addressed and no efforts made to prevent its recurrence. While high-tech medicine yields impressive results when addressing a specific disease in an individual patient, it is an extremely ineffective and expensive way to improve the health of whole populations. If, similar to Jeremy Bentham's utilitarian principle, the goal is optimum health for the maximum number of people, simple preventative measures that address the many social and environmental causes of disease would be much more cost-effective than high-tech medical interventions. Here we may very well have a case of technological determinism where the allure of groundbreaking scientific research and high-tech medical therapies not only shapes our very definition of health but also favors the expensive treatment of diseases instead of their simple, inexpensive prevention. Norman Temple explains how even fundamental medical research is biased by our fascination with developing high-technology cures rather than identifying preventative strategies:

> Complex research, in the main, is targeted at finding new treatments. Despite receiving the majority of resources expended on research, it has only produced a minor share of the useful information.... If complex research is such an inferior strategy, then why are most medical scientists seemingly mesmerized by it? Two explanations present themselves. First, medicine is oriented toward therapy rather than prevention.... This in turn leads to an emphasis on complex research. Second, medicine has for decades had a fatal attraction for high tech science. Indeed, this also helps explain why medicine concentrates on therapy and shuns prevention. It is therapy not prevention that utilizes high tech science.... Unfortunately, there is little attempt to critically evaluate the grand strategy.[121]

Another problem with the current engineering approach to eradicating disease is that many medical interventions focus on an isolated problem while ignoring the repercussions of these manipulations within the highly complex system consisting of body, mind and environment. As most environmental problems are brought about by manipulating nature without considering its interconnectedness, many negative side effects of high-tech medicine are caused by ignoring not only the intricate balance among the physical, chemical and biological reactions occurring within the body but also by disregarding the complex external interactions. It is therefore not surprising that high-tech medicine, because of its emphasis on mechanistic reductionism, has been the source of many unintended negative consequences. In the remainder of this section we will focus on problems such as physician-caused (iatrogenic) diseases, the widespread and persistent problem of antibiotic resistance and the inhumane conditions that patients endure when their lives are artificially prolonged by high-tech medical interventions.

Iatrogenic diseases are the result of the correct application of conventional diagnostic and therapeutic procedures and therefore are distinct from harm caused by malpractice.[122] Given the enormous complexity of the human body, the fragmentation of scientific knowledge, the rapid growth of high-technology procedures and the specialization of medical practice, it should be obvious that there are significant risks when diagnostic tests are performed, potentially toxic drugs administered or complicated surgeries carried out. In a recent analysis of the number of deaths from iatrogenic causes, Dr. Barbara Starfield of Johns Hopkins University reported that each year in the United States 12,000 patients die as a result of unnecessary surgery, 7,000 from medication errors, 20,000 from other mistakes made in hospitals, 80,000 from hospital infections that are caused primarily by antibiotic resistant pathogens (see discussion below) and 106,000 from the negative side effects of drugs.[123] Thus, approximately 225,000 patients die each year of iatrogenic causes, making medical practice the third leading cause of death in the United States, after deaths from heart disease and cancer. In addition, millions of non-fatal injuries and diseases, some of them permanently disabling, are also caused each year by medical diagnostic procedures and therapies. It appears that physicians who practice modern high-technology medicine frequently and inadvertently violate the Hippocratic Oath which admonishes "First, do no harm."[124]

It is interesting to note that the majority of deaths caused by the negative side effects of drugs are the direct result of mechanistic, reductionist thinking. Indeed, as Dr. Melvin Konner points out, the lack of specificity of many drugs inherently predisposes them toward undesirable side effects:

> In the end, what we are searching for is a self-contradiction: a chemical agent which acts powerfully on an illness—a living thing, whether microbe, tumour or failing organ—but which is completely without other, unwanted actions in the body. The trouble is that an agent that has a powerful biological effect in one area has powerful biological effects in others.[125]

Chemotherapy, for the treatment of cancer, provides an excellent example of negative drug side effects. Chemotherapy can be devastating because the drugs not only interfere with the growth of malignant cells but also with the functioning of normal cells. It may ultimately be possible to design chemical and biological agents that target only tumor cells, thereby eliminating unwanted side effects. But even this and other similar successes would have the additional unintended consequence of discouraging the implementation of simple, inexpensive, preventative measures resulting from the recognition of environmental, social and dietary causes of diseases.

The increasing death toll from hospital infections caused by pathogens that have become resistant to antibiotic drugs is another example of how the limited focus on manipulating infectious diseases, without understanding or considering the process of pathogenic evolutionary adaptation, is not only reversing earlier medical progress but is also creating serious difficulties for future treatment of infections. The search for antibiotics started shortly after Louis Pasteur and Charles Sedillot suggested in 1878 that infectious microorganisms are the cause of many well-known diseases. This so-called "germ theory" prompted physicians and scientists to search for "magic bullets," i.e., drugs that would kill the disease-causing agent, bacterium or parasite. In 1928, Alexander Fleming discovered the first true antibiotic, penicillin, which became available for widespread use in 1944.[126] As early as 1945, Fleming alerted the public that misuse of penicillin could lead to the selection and propagation of bacterial mutants resistant to the drug.[127] As Dr. Melvin Konner explains in The Trouble with Medicine,

Microbes, like all forms of life, are constantly evolving, adapting
to new conditions that the environment presents to them. Penicillin did not spell doom for the *streptococcus* and *gonococcus*; it
merely presented them with an unusual evolutionary challenge.
In the end the microbes could almost be seen to be thumbing
their noses at us; the more pharmacological bullets were fired at
them, the more the bacteria adapted—a classic case of an "evolutionary arms race" between a predator and its prey.[128]

The phenomenon of antibiotic resistance provides an example of how
nature responds to an imbalance in which one species (*homo sapiens*)
tries to eradicate others (pathogenic microorganisms) but inadvertently makes the target more resistant to attack, thereby restoring the
previously existing balance. The occurrence of antibiotic resistance is
another reminder that technological progress is often short-lived and
that initial benefits are frequently followed by rising costs and failures.
Indeed, soon after the introduction of penicillin in the 1940s, a number
of pathogens such as certain strains of *gonorrhea* and *staphylococcus* were
found to have become resistant to this antibiotic.[129] It was soon discovered that these pathogens acquire their resistance not only through
chromosomal mutation but also by the acquisition of new resistance
genes via plasmids (i.e., small fragments of DNA found in the cytoplasm
of cells), which are readily transferred from one microorganism to another, thereby enabling antibiotic resistance to spread rapidly.[130] Ever
since the first appearance of antibiotic resistance, pharmaceutical company scientists have been in an "arms race" against bacterial evolution,
creating new arsenals of antibiotic drugs, only to find that the pathogens
become resistant to them within a few years.

A major reason for the rapid appearance of resistant pathogens has
been the widespread overuse and abuse of antibiotics worldwide. Because antibiotics were initially available in the United States without a
doctor's prescription, a situation which is unfortunately still the case in
many developing countries, many patients used antibiotics inappropriately. For example, self-medication often results in the administration of
antibiotics for medical conditions not caused by bacteria, such as viral
cold and influenza. In addition, as soon as disease symptoms subside,
many patients stop taking the drug before it has had a chance to totally
eradicate all pathogens, leaving those resistant to antibiotics surviving,

multiplying and spreading in the environment. It is estimated that in the United States, about 50 percent of all antibiotic use is unnecessary or inappropriate.[131] In addition, most of the 6 billion farm animals in the United States receive antibiotic treatments during their short lives, often to compensate for the crowded and unsanitary conditions in which they live. About 24 million pounds of antibiotics are given to these animals at small, sub-therapeutic levels for growth promotion, a practice that was banned by the European Union in 2006.[132] Since pathogens are not eradicated at these low doses, this method of fattening animals for slaughter is an excellent way to select for antibiotic resistant microbes, which then spread with the animal's fecal matter through soil, water and air.[133]

As a result of this massive overuse and abuse of antibiotics, the occurrence of infectious diseases that have become untreatable by antibiotics is rising, indicating that we are rapidly losing the war against microorganisms. Nearly 2 million patients in the United States acquire hospital infections each year, with more than 70 percent of the disease-causing pathogens resistant to at least one commonly used antibiotic; and about 90,000 of these infected patients die each year, an increase from 13,300 in 1992.[134] According to a recent article in *Science*,

> Tuberculosis (TB) is back with a vengeance. Once nearly vanquished by antibiotics, at least in the developed world, tuberculosis resurged in the late 1980s and now kills more than 2 million people a year—second only to AIDS among infectious diseases. Especially frightening is the emergence of drug-resistant strains.[135]

As Stuart Levy concludes in *The Antibiotic Paradox—How Miracle Drugs Are Destroying the Miracle*, "This situation raises the staggering possibility that a time will come when antibiotics as a mode of therapy will be only a fact of historic interest."[136]

The invention of new surgical procedures, the application of powerful drugs and the use of various high-tech life-support systems has enabled physicians to save the lives of many seriously wounded and diseased patients. While in many cases the practice of high-technology medicine has yielded impressive results, it also has created a number of unexpected negative consequences.

Many terminal patients are so miserable after receiving heroic medical interventions that legitimate questions arise as to whether it would not have been better for them to have refused treatment and to have passed away in peace and dignity. High-tech medicine, because it is based on objective science and quantification, is clearly biased toward increasing the "quantity of life," generally measured in years of survival, rather than the "quality of life," which is much more difficult to ascertain. In addition, modern medicine is very effective in treating acute conditions while being less successful with the chronic diseases that often follow "successful" initial treatment. For example, thousands of head-injury victims, who would have previously died within a few days, are now being saved and continue to live under terrible conditions with chronic pain and serious disabilities.[137] Similarly, a recent study of the resuscitation of elderly patients following cardiac arrest yielded the depressing result that of the few (8 percent) who could be successfully resuscitated, more than half had suffered irreversible hypoxic damage to the brain and other organs and had to be committed to chronic care facilities, with little chance of long-term improvement.[138] Even the use of advanced diagnostic procedures can psychologically devastate the lives of patients and their families when, for example, they result in the early detection of an illness for which there is no cure, greatly increasing the psychological suffering from "death expectancy."[139]

The very concept of death has been redefined by the availability of life-saving technologies, a typical example of the so-called technological imperative that states, "If it can be done, it should be done." Thus, the very availability of medical technologies often results in their use to prolong life at any cost, thereby bypassing any personal deliberate, ethical decision-making process to determine whether this is the right thing to do. As Richard Taylor observes in *Medicine Out of Control,*

> It highlights the problems of defining death and dying in a technological age in which various body functions can be taken over by machines for prolonged periods. It demonstrates clearly how the medical establishment, by and large, has shrunk from coping with the new dilemmas and how its inaction has allowed modern technology to dictate its actions.... The use of highly sophisticated life support systems carries with it the responsibility

to define death and dying in the context of its application. The medical establishment has evaded this responsibility: they are no longer professionals but technocrats.[140]

Because physicians and families may be the decision-makers regarding the employing or withholding of life-saving technologies, they are in fact determining when death will occur.[141] It is understandable that most doctors want to avoid the uncomfortable situation in which they may feel themselves to be causally or morally responsible for a patient's death.[142] Consequently, they may apply any and all treatments available to buy a few more months, even if it leads to prolonged suffering and a slow painful death without peace or dignity. Consider spending your last days in an intensive care unit:

> Rows of physiological preparations (also known as human beings) lie surrounded by an astounding array of mechanical and electronic gadgetry. A tube or catheter of some description violates every natural orifice, and perforations in various parts of the body are made especially for the placement of others. Multicolored fluid is pumped in, similar fluid drains out, respirators sigh, dialysers hum, monitors twitch, oxygen bubbles through the humidifiers. The unfortunate hostages, mercifully unresponsive to their environment (either through natural causes or drugs) lie silent while this ritual of desecration takes place.[143]

When machines have taken control of the patient's bodily functions, there is very little peace, dignity or humanity left in the process of dying. Here we have an example of the inseparability of positive and negative effects caused by the unquestioned application of advanced science and technology (see Chapter 1). Is this really progress?

Finally, because of our desire to use high-tech medicine in the pursuit of inherently unachievable goals such as the total eradication of disease and death, there will be an unending demand for innovative technologies and treatments, thereby interminably escalating health care costs. As David Callahan observes,

> The contemporary medical enterprise has increasingly become one that considers the triumph of illness and the persistence of

death both a human failure and a supreme challenge still to be
overcome. It is an enterprise that feeds on hope, that constantly
tells itself how much farther it has to go, that takes all progress
to date as simply a prologue to the further progress that can be
achieved. Nothing less than total control of human nature, the
banishment of its illnesses and diseases, seems to be the implicit
ultimate goal....While most people, moreover, seem to agree with
the principle that treatment should be stopped when it does no
more good (a hard notion to oppose), almost everyone might
choose more treatment if there was some promise of some success. But that is just what medical progress always seems to offer,
some promise of some success; and that is why, when there is
doubt, treatment continues.[144]

The open-ended goal of defending ourselves against disease and death
has caused costs to escalate at rates greater than inflation, thereby consuming an increasing proportion of the nation's GDP. For example, total
health care spending in the United States was $2 trillion, or 16 percent
of GDP, in 2005 and is expected to reach $4 trillion, or 20 percent of
GDP, by 2015.[145] Given that funds for medical care will always be limited,
efforts should be made to address the causes of cost escalation, some of
which are unrealistic goals and the application of expensive technologies to achieve them. As will be discussed further in Part III, there are at
least two ways to address the fundamental problem of health care cost
escalation. First, greater efforts can be made to prevent disease through
changes in lifestyle, diet and environment, which is a very cost-effective
way to increase public health. Second, given that no one can live forever, it would be more prudent to accept old age and death as something
natural, not something to be opposed to at any physical, emotional and
financial cost.

Unintended Consequences of Technological Revolutions

Human numbers increased from an estimated 10,000 at about 130,000
to 150,000 years ago to nearly 7,000,000,000 today.[146] While this population increase appears to be exponential, a closer analysis shows that
the expansion of human populations has actually occurred in several
distinct stages in response to changes in climate and a series of three

major technological revolutions: the tool-making, agricultural and industrial revolutions. The invention of advanced stone tools during the Upper Paleolithic period enabled hunter-gatherers to expand from their origins in Africa into Eurasia approximately 40,000 to 50,000 years ago, with the total population increasing to as much as 10 million on the eve of the agricultural revolution. The invention of agriculture about 10,000 years ago greatly multiplied the carrying capacity of the land and resulted in a stabilization and expansion of the food supply. As a result, the human population increased to more than 100 million by about 2000 years ago.[147]

The rapid and synergistic development of science and engineering, which began more than 300 years ago, led to a proliferation of technologies, most importantly the steam engine, the internal combustion engine and the electric generator. A major effect of the Industrial Revolution was an enormous increase in food production brought about by the industrialization of agriculture, first through the introduction of fossil-fuel-powered farm machinery, which facilitated the rapid conversion of forests and prairies into farmland, and later through an increase in crop productivities through the application of fossil-fuel-based fertilizers, pesticides and herbicides. Essentially, non-renewable fossil energy was substituted for land and labor. As a result of the Industrial Revolution in agriculture, the carrying capacity of the land expanded at least tenfold, with human numbers increasing from approximately 545 million 350 years ago to about 7 billion today.[148]

The Industrial Revolution began in Europe about 300 years ago and subsequently has spread throughout the world via intensive technology transfer, initially during colonization by Europeans and more recently as part of attempts to "develop" poor nations. The "Green Revolution" is one of the best examples of such technology transfer. As technological development in Europe led to rapid population increases there during the 19th and 20th centuries, technology transfer to "developing" nations has been primarily responsible for the current worldwide population explosion.[149]

There are many negative consequences of human overpopulation. More people generally translate into more problems.[150] Increasing competition over scarce resources such as land, water, food, minerals and energy resources often results in poverty, famine, war and genocide (see Chapter 3). As more humans control and exploit nature, they create or aggravate existing environmental problems such as chemical pollution, species extinction and global climate change. Finally, quality of life decreases as more people move into urban centers, where they often live under crowded, unnatural, unhealthy, noisy and stressful conditions as, for example, in the world's "mega-cities."

It is important to recognize that the three major increases in world population were based on improved access to energy. When, through technological innovation, people gained access to more energy than previously had been available, food production increased and populations grew in direct proportion.[151] As early hunter-gatherers expanded their territories, more solar energy was captured in the form of additional plants and animals. The subsequent development of agriculture intensified the photosynthetic conversion of solar energy into food biomass with an even greater effect on population size. The current access to fossil energy has led to increased crop productivities both in terms of land and labor, fueling the most recent and largest increase in human population.

The fact that fossil energy stocks are non-renewable and likely to be depleted in the near future[152] indicates that the current world population, much of which depends on fossil-fuel inputs for food production, cannot be sustained indefinitely. With the anticipated decline in the availability of cheap fossil fuels, world agricultural production will ultimately have to revert to the more labor-intensive and less productive farming methods of earlier times. It is highly unlikely that enough food can be provided without fossil energy inputs for the approximate 9.5 billion people that are expected by 2050,[153] given that the very stability and relatively low population density of traditional agricultural societies was determined by their dependence on the limited but steady inflow of solar energy, which could never be increased but only allocated to arable acreage, pasture and forest to provide crops, animal energy and fuel in the desired proportions.[154]

It is clear that the very success of the Industrial Revolution has placed humans in a very precarious position: unless there are concerted efforts to voluntarily and significantly reduce the size of the human population within the next hundred years, a global population collapse is inevitable. This would lead to massive human suffering on a scale never before witnessed.

The Decline in Fitness of Future Generations

For many thousands of generations, the process of biological evolution guaranteed that humans were well adapted to their natural environment, thereby maximizing their chances of survival. Until very recently, on average seven out of ten children died before reaching reproductive age, assuring that only the most vigorous, those who were best adapted to the relatively harsh environment, passed on their genes to future generations.[155] The human environment, however, changed drastically following the Industrial Revolution, when better nutrition and improved sanitation increased childhood survival rates. Furthermore, because of various medical interventions such as immunization, antibiotics and more recently post- and even prenatal surgeries, more than 95 percent of newborn children in developed nations survive to reproductive age.[156]

Several hundred years of industrialization have surrounded many humans with an artificial environment very different from that in which their ancestors evolved.[157] Mutations that were lethal or deleterious under natural conditions are no longer eliminated by early death.[158] The direction of natural selection has been altered by modern living conditions. Formerly lethal and deleterious mutations accumulate in the population, thereby slowly reducing the population's ability to survive under more rigorous environmental conditions.[159] For example, since the invention of spectacles, people with poor eyesight have been able to greatly reduce their risk of death from predators and accidents, thereby passing on this visual deficiency to future generations.[160] If technological society were to collapse and spectacles were no longer readily available, millions of people would suddenly find themselves seriously handicapped. Professor James Crow summarizes the consequences:

> However efficient natural selection was in eliminating harmful
> mutations in the past, it is no longer so in much of the world....
> It seems clear that for the past few centuries harmful mutations
> have been accumulating. Why don't we notice this? If we are like

Drosophila [a fruit fly], the decrease in viability from mutation accumulation is some 1 or 2% per generation. This is more than compensated for by much more rapid environmental improvements.... How long can we keep this up? Perhaps for a long time, but only if there remains a social order that permits steady environmental improvements. If war or famine force our descendents to return to a stone-age life they will have to contend with all the problems that their stone-age ancestors had plus mutations that have accumulated in the meantime.[161]

Here we encounter another manifestation of the dual nature of technological "progress": as people attempt to benefit from life in a high-technology society, their descendants are decreasing, generation after generation, in fitness for survival under more natural conditions. When future generations are forced by either temporary or permanent societal collapse to live "closer" to nature, they will suffer a substantially elevated mortality rate. Given that complex civilizations do not persist indefinitely, it is only a matter of time until our descendants will be forced to live again without current interventions. At that point, millions of people will die, in a sense, because of our present lifestyle.

CHAPTER 3

Technology, Exploitation and Fairness

TECHNOLOGY IS OFTEN USED to control and exploit. The terms "control" and "exploitation" are used so frequently in everyday language that they may evoke different meanings based on context or circumstances. To clarify, according to Webster's dictionary, exploitation is defined as an act involving "(a) utilization or working of a natural resource, (b) an unjust or improper use of another person for one's own profit or advantage, and (c) co-action between organisms in which one is benefited at the expense of the other."[162] In short, exploitation involves giving less than the value of that which was received, thereby creating imbalance and injustice. Exploitation generates a skewed cost-benefit balance in which excess benefits accrue to one party and excess costs are externalized onto one or more other parties.

Whenever technology is used to control and exploit, the exploited will by definition suffer negative consequences. Since the application of technological power for control and exploitation is based on ignorance of the fact that "everything is connected to everything else" (Chapter 1) and also violates basic principles of reciprocity and fairness, one could postulate that any type of technological control and exploitation will inevitably lead to negative outcomes and suffering for all parties. Thus, as long as technology is used for control and exploitation, negative social and environmental effects are inherently unavoidable.

Technology and Exploitation

Power, including technological power, is often used to exploit others. However, many individuals, even if they possess the power to exploit, have ethical reservations against exploiting others if they directly witness the suffering of the exploited and can personally identify with the victims. Thus, individuals will generally be less likely to exploit those close to them (those who belong to the same "in group") such as family and friends, but may be unconcerned about the exploitation of those who are unknown, different or far away (those who belong to the "out group"). Many technologies facilitate exploitation by creating a safe distance between exploiter and exploited. This distance is generally physical (between two locations) but can also be temporal (between the present and the future).

Prior to the Enlightenment, most Europeans lived directly from the land and therefore had a very close relationship with their immediate environment. Their feelings of connectedness with nature precluded them from exploiting their environment wantonly. This more or less harmonious co-existence between humans and nature explains, for example, why farming practices were sustainable for hundreds, if not thousands of years. However, with the advent of the "Age of Reason," a conceptual separation occurred between humans and nature, and people began to objectify their environment.[163] Indeed, as was discussed in Chapter 1, the scientific method and its success are based on the separation between observer (human) and observed (nature). This separation between humans and nature created a safe psychological distance between people and their environment, the latter changing from "in group" to "out group." This, in turn, enabled the extensive exploitation of natural resources to an extent that would never have been acceptable earlier and is still not acceptable in many of those "primitive" cultures that have somehow escaped "enlightenment."

Exploitation of nature not only consists of the extraction of raw materials and energy but also involves the creation of all kinds of wastes and pollution. The invention of powerful fossil-fuel-based transportation technologies has made it possible to locate environmentally destructive industrial processes far away from the consumers of the final products and services. In fact, almost everything we use on a daily basis has been manufactured in places we have never seen and probably never will see. Therefore, if the environment is destroyed or harmed in these distant places, we are rarely aware of it and even if we are, it is difficult to become sufficiently concerned about it to move us to action.

As a result of our evolutionary trajectory, our ethical sensitivities are generally limited to people and situations that are in close proximity. Therefore, if technologies enable us to be isolated from exploitative situations in which we unknowingly participate, these technologies actually encourage unethical behavior. For example, if we consume products that are made in the "developing" world by strip mining land or destroying forests or introducing toxic waste into the soil, air or water, we are responsible for these destructive activities, even if transportation technologies that facilitate international trade conceal our involvement. Similarly, the use of electricity, considered by many to be "clean" energy, is generally associated with unsustainable fossil energy use and air pollution at power plants far away and out of sight. Even hydroelectric energy generation is problematic as it destroys terrestrial and aquatic wildlife habitats, in many cases leading directly or indirectly to the extinction of many species.

Advanced technologies may not only spatially separate the exploiter from the exploited but may also temporally separate them, as when the costs and consequences of activities are transferred to future generations. This is particularly the case with technologies that generate negative environmental consequences that future generations will suffer, doubtless against their will. For example, the exploitation of renewable resources such as forests and fisheries beyond their sustainable yields will significantly reduce or destroy their future productive potential, depriving future generations of their fair share of these natural resources. Similarly, the radiation and poisons emanating from nuclear wastes will have to be confronted by hundreds of future generations, who will be forced to deal with the hazards without having received the benefits of nuclear power. The massive use of fossil energy, which ignited the Industrial Revolution and continues to fuel present industrial economies, is responsible for the increase in atmospheric carbon dioxide concentrations, which in turn is triggering global climate change. Again, most of the negative effects, such as sea-level rise, severe weather events and the creation of hundreds of millions of environmental refugees, will be borne by future generations. In summary, modern technologies enable us to derive various benefits while simultaneously transferring unwanted costs to future generations without their consent or ability to prevent us from doing so. This intergenerational externalization of costs is clearly exploitative and unethical.

Factories may also provide physical barriers between workers and owners, enabling the latter to exploit the former without reservation because they may not be sufficiently aware of poor or hazardous working conditions. This was particularly true in Europe and the United States during the early phases of the Industrial Revolution, when workers had to labor for long hours in dark, smoky and stinking factories while the owners lived in luxurious houses at a psychologically safe distance. Transportation and communication technologies have permitted the globalization of markets and have thus separated both consumers and owners in developed nations from the millions of poor factory workers, who may be toiling under unbearable conditions in "developing" world sweatshops thousands of miles away.

Factories are also used to exploit animals. Consider, for example that approximately 90 percent of the meat eaten by Americans today comes from factory farms and feedlots where animals are crowded in concentration-camp-like conditions for the sole purpose of gaining weight for slaughter.[164] In the United States alone, 7,000 calves, 130,000 cows, 360,000 pigs and 24 million chickens are killed every day to support a very ethically questionable meat-based diet. The average American consumes approximately 2,400 animals in a lifetime.[165] The majority of these animals are kept in crowded factory farms where these sentient creatures can barely move and never enjoy life in natural surroundings, only to be slaughtered as soon as feasible in high-tech disassembly lines in slaughter houses. This massive cruelty to animals could never be carried out if people were not physically separated from these hideous acts of exploitation. In fact, many of us would probably become vegetarians immediately if we had to witness the animal suffering on factory farms and in slaughter houses. But modern technologies, by physically separating the rearing and killing of animals from the final consumer of meat, creates a safe psychological distance between exploiter and exploited, thereby promoting animal cruelty.

Another example of how modern technologies create physical and emotional barriers that promote immoral actions is the use of advanced weapons to kill and destroy. Ever since the invention of gunpowder and firearms, it has been possible to kill another person at a distance, often allowing the killer to escape witnessing the intense suffering and death. By enabling killing from an even greater distance, modern warfare technologies allow combatants to easily ignore their natural moral inhibitions as well as international law. This technological suppression of ethical sensitivities can pave the way for massive violence and atrocities.

The psychological distance created by modern weapons technologies is expressed succinctly by Konrad Lorenz who writes,

> The man who presses the releasing button [of modern remote control weapons] is so completely screened against seeing, hearing, or otherwise emotionally realizing the consequences of his action, that he can commit it with impunity.... Only thus can it be explained that perfectly good-natured men, who would not even smack a naughty child, proved to be perfectly able to release rockets or lay carpets of incendiary bombs on sleeping cities, thereby committing hundreds and thousands of children to a horrible death in the flames. The fact that it is good, normal men who do this, is as eerie as any fiendish atrocity of war.[166]

Psychological distancing involving the development of warfare technologies was also involved in the nuclear arms race during the Cold War when, with the push of a button, a worldwide nuclear war could have been precipitated, potentially resulting in mutually assured destruction and a global "nuclear winter."

In summary, many modern technologies have the capacity to severely blunt the normal ethical sensitivities of humans by creating safe distances, physical or temporal, thereby promoting immoral acts such as wanton exploitation, destruction and murder.

Technological Exploitation of Nature

Human have always depended on nature for their sustenance, controlling and sometimes exploiting natural resources to provide food and shelter. Although early hunter-gatherers may have contributed to the extinction of Pleistocene fauna, the degree of control and exploitation of nature increased with the domestication of plants and animals about 10,000 years ago, as humans began modifying and managing their local environments to improve and stabilize food production through animal husbandry, horticulture and later, farming.[167] However, until the Enlightenment and subsequent Industrial Revolution, the extent of exploitation and control of nature was limited, most likely because humans were still very few in number and lacked the technological power to create major global disasters. In addition, prior to industrialization, many of the world's cultures had a profound reverence for nature, considering themselves to be a part of it, often prohibiting or discouraging the wanton abuse of animals, plants and the land. Some of today's indigenous peoples, including the First Nations of North America, still maintain a respectful view of nature.[168] Some, however, have now fully adopted the Western exploitative and destructive orientation.

The concept of harmony with nature was found not only among "primitive" peoples but was also a view common in early Eastern as well as Western civilizations. For example, the Hebrews, as expressed in the Psalms of the Old Testament, celebrated the order and beauty of nature and its intrinsic value quite separate from its usefulness to themselves.[169] Similarly, in ancient China and Greece, according to Alan Drengson, "Humans were seen as ignorant and finite; nature as mysterious and infinitely powerful. It was believed that it is futile for humans to try to control the natural world, or even to drastically alter it."[170] What is particularly interesting is that the Greeks, despite their great interest in advancing scientific knowledge, did not use science to develop useful technologies, as was done following the Enlightenment in Europe, because this would have clashed with their worldview of universal harmony and balance. As Jacques Ellul writes,

> In their golden age of science, the Greeks could have deduced the technical consequences of their scientific activity. But they did not.... The Greeks were suspicious of technical activity because it represented an aspect of brute force and implied a want of moderation.... Here we find the supreme Greek virtue, self control. The rejection of technique was a deliberate, positive activity involving self-mastery, recognition of destiny, and the application of a given conception of life.... In Greece, a conscious effort was made to economize on means and to reduce the sphere of influence of technique. No one sought to apply scientific thought

technically, because scientific thought corresponded to a conception of life, to wisdom. The great preoccupation of the Greeks was balance, harmony, and moderation, hence, they fiercely resisted the unrestrained force inherent in technique, and rejected it because of it potentialities.[171]

Of course, this changed dramatically with the Enlightenment, when scientific knowledge was explicitly applied via innovative technologies to control and exploit nature. As Eugene Schwartz comments,

The advancement of science was accompanied by a growth of knowledge that purported to be moving forward from "lower" to "higher" levels, onward and upward toward the improvement and betterment of mankind. Advancement lay in man's mastery over nature, his control over and freedom from the environment, and his understanding of the "laws" of nature so that he could dominate and exploit it. Reason, science, and progress were a triad that at once became the ends and means of modern scientific civilization.[172]

Clearly, the leading scientists of the Enlightenment, such as Francis Bacon and René Descartes, understood very well that the new science, the *Novum Organum*, was to be used for the domination and exploitation of nature. As Daniel Sarewitz writes,

Francis Bacon's famous dictum "Nature, to be commanded, must be obeyed" grew from the Christian desire both to understand God's laws and to exercise dominion over nature. Bacon postulated an explicit linkage between understanding of nature through scientific inquiry and making use of that understanding to subdue nature.... The urge to explore and the urge to conquer are not so distinct, and each facilitates and motivates the other, as Francis Bacon understood when he claimed, with great foresight, that the understanding of nature was a necessary prerequisite of its conquest.[173]

The rest is history. The new science, relying on a strict separation between observer (the scientist) and observed (nature) as well as on the rigorous testing of hypotheses through experimentation and other forms of data collection, has been extremely effective during the past 300 years in elucidating the laws of nature. As soon as new scientific knowledge was gained, it was almost immediately applied to the development of innovative technologies, first to generate new sources of power (the steam and combustion engines) and then to use this power to extract a wide range of resources from the environment. The combination of (a) the availability of cheap sources of fossil energy, (b) the continuous development of new technologies for exploiting resources and manufacturing products, (c) a culture that promotes a never-ending quest for greater material affluence and (d) the explosion of human populations has produced a worldwide exploitation and control of nature unparalleled in magnitude and extent.

Human Domination of Nature

It is a well-established fact that we have serious environmental problems. However, instead of repeating here a litany of various local environmental disasters reported on a daily basis in the media, we will focus our attention on a global perspective, i.e., the overall scale and dimension of human domination of nature and the recognition of "planetary boundaries."[174] Specifically, we will look at the following three indicators of human "planetary" domination: (a) land use, (b) alteration of the biogeochemical cycles of carbon, water, nitrogen and other elements and (c) the release of anthropogenic organic chemicals.[175]

The following is an estimate of human domination in terms of global land use: Approximately 10-15 percent of the Earth's land surface is occupied by row-crop agriculture or by urban-industrial areas, while another 6-8 percent has been converted to pasture land.[176] Overall, humans appropriate between 30-40 percent of all terrestrial primary productivity (photosynthetically fixed carbon), indicating that two fifths of the land's productive capacity is tightly controlled and managed in order to supply food, fiber and energy.[177] Because many other species depend for their survival not only on habitat but also on food sources from photosynthetically produced biomass, it is clear that land transformation and degradation is a leading cause of species extinction (see Chapter 2).

Unfortunately, with continuing global human population growth and worldwide economic development, environmental pressures related to land use will further increase in the foreseeable future. Consider, for example, that the ecological footprint—the average amount of productive land and shallow sea appropriated by each person for food, water, housing, energy, transportation, commerce and waste absorption—is about one hectare (2.5 acres) in developing nations but about 9.6 hectares (24 acres) in the United States.[178] If every person in the world enjoyed the same level of material consumption as is common in the US, three to four additional Earths would be required, and even more (up to 12) if future population and economic growth is taken into account.[179]

All major biogeochemical cycles, particularly those of carbon, water and nitrogen, have been profoundly impacted by human action. The massive extraction and burning of fossil fuels over the past 200 years has caused a significant rise in the levels of atmospheric carbon dioxide, which is expected to result in global climate change with all its impending negative consequences. Fossil-fuel combustion currently adds about 7 billion metric tons of carbon to the atmosphere each year.[180] To better visualize the magnitude, consider that each year every US citizen, on average, uses about 8,000 pounds of petroleum, 5,150 pounds of coal and 4,700 pounds of natural gas.[181] The dumping of massive amounts of carbon dioxide (CO_2) into the atmosphere has resulted in an increase of CO_2 concentrations from pre-industrial levels of 280 ppm to more than 380 ppm today. Given that today's atmospheric CO_2 concentrations are greater than at any time in the past 420,000 years and probably have not been exceeded during the past 20 million years, this more than 30 percent increase in atmospheric CO_2 levels clearly represents a profound human-caused alteration of the planetary carbon cycle.[182] Even with the most stringent climate change mitigation efforts imaginable, CO_2 concentrations will continue to rise to at least 550 ppm within the next 50 to 100 years.

Humans, because of their need for water, particularly for irrigation agriculture,[183] globally appropriate approximately half of all available fresh water on Earth, leaving millions of non-human animal and plant species to compete for the remaining water for their survival.[184] The extent of the control and management of water resources is astounding: there are currently 40,000 large dams (over 50 feet high) and approximately 800,000 small dams blocking the world's rivers and creeks.[185]

In the United States, only 2 percent of all rivers run unimpeded, and the flow of approximately two thirds of the world's rivers is regulated.[186] Where water from freshwater lakes and rivers is unavailable, humans use groundwater, in many cases non-renewable "fossil" water. For example, one fifth of irrigated cropland in the US is dependent on groundwater pumped, or rather "mined," at an unsustainable rate, from the Ogallala aquifer, which spans portions of eight US states and covers 453,000 square kilometers.[187] In addition, 10 percent of the global grain harvest is currently produced by permanently depleting groundwater supplies.[188] Growing human populations and an increasing need for food production through irrigation agriculture will most certainly put additional pressures on already strained fresh water resources. Worldwide, many major rivers no longer reach the sea. Drawdown now exceeds recharge in many of the worlds' major aquifers.[189]

With the invention of the industrial Haber-Bosch process more than 100 years ago, it became possible to use fossil energy to convert nitrogen from the air into ammonia fertilizer. As a result, farmers no longer depended on fertilizers naturally produced through biological nitrogen fixation but began using synthetic fertilizers to increase crop productivities. Worldwide, the use of synthetic nitrogen fertilizer reached approximately 80 million tons during the 1990s,[190] and is expected to increase to more than 135 million tons by 2030.[191] Overall, humans add as much industrially produced nitrogen to terrestrial ecosystems as do all natural sources combined, thereby severely altering the global nitrogen cycle.[192]

According to a recent analysis of the extent of mobilization, through human action, of all major elements listed in the periodic table, it was found that humans dominate the cycling of most elements, particularly those that are insoluble in their natural state.[193] This is particularly disturbing if this mobilization involves highly toxic metals such as lead, cadmium and mercury.[194] To provide a concrete example of the extent of mineral extraction, each US citizen, on average, uses annually 8,440 lbs (3,828 kg) of stone, 8,250 lbs (3,742 kg) of sand and gravel, 720 lbs (327 kg) of cement, 500 lbs (227 kg) of clays, 430 lbs (195 kg) of salt, 338 lbs (153 kg) of phosphate rock, 1,140 lbs (517 kg) of iron and steel, 49 lbs (22 kg) of aluminum, 21 lbs (9.5 kg) of copper, 14 lbs (6.3 kg) of lead, 12 lbs (5.4 kg) of zinc, 12 lbs (5.4 kg) of manganese and 21 lbs

(9.5 kg) of other metals.[195] Finally, each year the chemical industry produces more 100 million tons of organic chemicals, representing about 70,000 different compounds—most of which have never been tested for health hazards and environmental impacts.[196] Most of these chemicals, after their use, will ultimately end up in the environment, having the potential for yet unknown and serious environmental and health consequences (Chapter 2), as was the case with DDT and many other bioaccumulative organic compounds.[197]

Given this already extensive human domination of nature, which will most certainly increase even further with continued global population growth and economic development, one would think that we will very soon have reached the limits of environmental exploitation and control. Unfortunately, this is not the case. In fact, we are currently in the process of exploring completely new frontiers by attempting, using recent insights in molecular biology and newly developed genetic engineering technologies, to control and exploit life at the molecular level. Jeremy Rifkin explains the implications of these new genetic technologies:

> The new tools are the ultimate expression of human control—helping us shape and define the way we would like to be and the way we would like the rest of living nature to be. Biotechnologies are "dream tools", giving us the power to create a new vision of ourselves, our heirs, and our living world and the power to act on it.... With the newfound ability to identify, store, and manipulate the very chemical blueprints of living organisms, we assume a new role in the natural scheme of things. For the first time in history we become the engineers of life itself. We begin to reprogram the genetic codes of living things to suit our own cultural and economic needs and desires.[198]

In summary, to support billions of people at a high material standard of living, humans currently appropriate a large part of the planet's land surface, interfere significantly with the functioning of numerous natural biogeochemical cycles, introduce into the environment tens of thousands of organic chemicals whose fate and effects are unknown, and attempt to control life at the molecular level via genetic engineering. As was discussed in Chapter 1, such massive intervention in natural processes cannot be carried out without creating a vast number of negative consequences, particularly in view of the fact that current scientific methods, which are based on mechanistic reductionism, are inherently unable to provide the necessary knowledge to address the negative side effects of this exploitation. As Rachel Carson, author of *Silent Spring*, which launched the environmental movement more than 40 years ago, stated, "Control of nature is a phrase conceived in arrogance."[199]

Machines and the Control and Exploitation of Workers

The production of goods and services in free-market economies is carried out through mutual cooperation between the owners of income-producing assets such as factories and machines, and the workers who operate them. There are several reasons why labor-saving machinery was and continues to be employed. First, it greatly increases productivity, thereby reducing the costs of goods and services, which generally translates into greater average material affluence. Second, by eliminating costly labor, it provides a stable and often increasing source of profits for owners. Finally, machines and factories are used to manage and control workers, who might prefer greater autonomy and might rather work on their own family farms or as independent craftsmen and artisans, as was the case before the Industrial Revolution. After all, in the words of David Dickson, "there is little point in introducing machines capable of increasing the efficiency of production without sufficient control over the workforce to ensure that the machines will be operated to their maximum capacity."[200] Consequently, control of the labor force is a necessary prerequisite for exploitation, and most production technologies lend themselves to this purpose. For example, during the early stages of the Industrial Revolution, the removal of weavers from their homes and their placement in factories built to accommodate the new, larger machines, also served to control the workers:

> It seems possible to identify four main reasons for the setting up of factories. The merchants wanted to control and market the total production of the weavers so as to minimize embezzlement, to maximize the input of work by forcing the weavers to work longer hours at greater speeds, to take control of all technical innovation so that it could be applied solely for capital accumula-

tion, and generally to organize production so that the role of the capitalist became indispensible. Factories provided the organizational framework within which each of these could be achieved. Thus, although machines were present in the early factories, they were seldom the reason for setting up a particular factory. The factory was a managerial rather than a technical necessity. It imposed a new discipline on the whole production process, and was described by Charles Fourier as a "mitigated form of convict prison."[201]

By devising a production process that required the continuous cooperation between workers and machines, often linking them together as a single unit, the owners of machines and factories gained control over the labor force. The workers had little choice because they were entirely dependent on operating the machines for their livelihood, particularly after escape routes back to the family farms and small workshops had been eliminated. Even now, ongoing technological innovation continuously threatens workers with replacement by more advanced machinery such as robots and computers, thereby effectively diminishing their power to resist. The end result is the presence of a rather docile, highly disciplined workforce, which is advantageous in generating profits for owners.

Control of human work by machines probably reached its height with the invention of the assembly line, which forced workers to perform isolated repetitive tasks hundreds if not thousands of times each day at a speed dictated by the manufacturing equipment. This mechanized production process was made even more efficient by optimizing all aspects of the worker's bodily movements,[202] as described by historian Howard Zinn:

> One way [to cut costs and increase profits] was Taylorism. Frederick W. Taylor had been a steel company foreman who closely analyzed every job in the mill, and worked out a system of finely detailed division of labor, increased mechanization, and piecework wage systems, to increase production and profits. In 1911, he published a book on "scientific management" that became powerfully influential in the business world. Now management

could control every detail of the worker's energy and time in the factory. As Harry Braverman said, the purpose of Taylorism was to make workers interchangeable, able to do the simple tasks that the new division of labor required—like standard parts divested of individuality and humanity, bought and sold as commodities. It was a system well fitted for the new auto industry.[203]

Henry Ford was the first entrepreneur to apply the concepts of division of labor, assembly-line production and scientific management on a large scale for the mass-manufacture of millions of identical automobiles:

"Fordism", the technology of mass assembly-line production... was based on machinery, and on the "rational" reorganization of work to fit the rhythm of the new machinery.... All possible skills were transferred from human workers to machines, from people to automatic tools and to the moving assembly line.... In the new organization of auto production, unskilled workers were seen by management as rather simple machines which happened to be alive—Henry Ford expressed this attitude most clearly in his description of factory life: "A business is men and machines united in the production of a commodity and both the men and the machines need repairs and replacement.... Machinery wears out and needs to be restored. Men grow uppish, lazy, or careless."[204]

Given the gradual but continuous improvements in factory working conditions during the 19th and 20th centuries, one might be inclined to dismiss the earlier exploitation and control of workers as a historical phenomenon. Unfortunately, while working life has improved, the basic situation is similar. Industrial machines, advanced equipment and technologies, including most recently the computer, are all used to manage and control the labor force. In fact, the working lives of most people in industrialized nations are dictated on a daily basis by the requirements of technologies, whether machinery in factories or computers in offices. Most recently, the computer and associated electronic communication technologies such as the Internet have been used to exert control over workers through surveillance of their email correspondence and website access. Thus, control has been expanded from the simple physical domain. Even away from work, life is often controlled by the dictates of technology.

The question arises as to why people are so willingly and easily managed and controlled by manufacturing and office technologies. The simple answer is that most people, with few exceptions such as the machine-smashing Luddites of early 19th-century England, appear to accept being controlled if in return they gain greater material afflu- ence. For example, Henry Ford paid assembly-line workers sufficiently high wages that they could afford to buy the cars they produced. But, as Abraham Maslow has shown, many people have a strong need for personal autonomy and therefore resent being managed and controlled by others. [205] Individuals may not be aware of the control because it has been disguised and legitimized by linking it to the perceived need for greater technological productivity and efficiency. This ideology of in- dustrialization is described by David Dickson:

> What the ideology disguises is the degree of political exploita- tion and manipulation that has, in almost all cases, accompanied the industrialization process and hence the development of contemporary technology. Industrialization has appeared to ne- cessitate, and has hence legitimated, man's exploitation of both fellow men and the natural environment. The apparent need for authoritarian discipline and hierarchical organization of the factory required to operate complex production-line equipment, for example, is held to justify the accompanying relationships between management and workers.[206]

In summary, many manufacturing and office technologies are employed to manage and control the labor force to efficiently generate products, services and profits. Technologies, which serve as intermediaries be- tween owners and workers, between exploiters and exploited, help to disguise and even legitimize this control. This legitimization often involves invoking some type of technological determinism, the belief that technologies evolve of their own accord to be ever more productive and efficient and that technological progress is in fact inevitable. This is often expressed in the statement "You can't stop progress," implying that it is futile to even question technological developments. Thus, we passively accept being managed and controlled by technologies, unaware that, as described in greater detail in Part II, technological change is neither inevitable, nor automatic, nor necessarily desirable but is instead consciously directed by those who benefit most from technological innovations. We now turn to another more insidious form of control by technology: the control of mind and the modification of culture by television and related video technologies.

Television: A Powerful Tool for Social Control and Manipulation

The powerful electronic media of television and related video/film technologies have been used with great success for the manipulation of the masses. The very act of spending countless hours in front of the television or computer screen separates individuals from one another and from the natural environment, thereby eliminating points of comparison with the real world. Previous personal experiences of the world by individuals are replaced by artificial images of experiences by strangers in distant places. The ever-increasing isolation from reality makes it extremely difficult to know what is actually true and what is not. According to Jerry Mander, who was for many years a TV advertising executive,

> Living within the artificial, reconstructed, arbitrary environments that are strictly the products of human conception, we have no way to be sure that we know what is true and what is not. We have lost context and perspective.... Whoever controls the processes of re-creation, effectively redefines reality for everyone else, and creates the entire world of human experience, our field of knowledge. We become subject to them. The confinement of our experience becomes the basis of their control of us. The role of the media in all this is to confirm the validity of the arbitrary world in which we live. The role of television is to project that world, via images, into our heads, all of us at the same time.[207]

By sitting passively in dark rooms watching TV or computer screens, we may experience a state of sensory deprivation and exhibit alpha brain waves similar to those that appear during hypnosis. In this state, we are highly receptive to the images and messages provided through the electronic medium. We may let thousands if not millions of alien images

and messages enter our minds, unaware of their power and nearly de-
fenseless against their effects. Television and video are the perfect media
to fill our minds with pre-designed experiences and thoughts, and the
high speed at which different images and messages appear keep our attention on the medium, thereby keeping us constantly distracted and
preventing us from sensibly evaluating the content and thinking for
ourselves.

Television images and messages are broadcast simultaneously from
centralized locations to millions of viewers throughout the nation. The
content is controlled by a handful of people, primarily those who are in
a position to pay for extremely limited, and therefore extremely expensive, broadcasting time. Many large corporations are able to use their
vast wealth to control advertising and influence program content, which
most frequently results in redefining happiness and the meaning of life
in terms of material consumption. In summary, television and other related electronic media are powerful tools for manipulating the attitudes
and behavior of millions of unsuspecting viewers who generally turn
to these media solely for relaxation and entertainment. John Cavanagh
and Jerry Mander conclude,

> Television is the original "virtual reality". The situation verges
> on the bizarre, the stuff of science fiction.... "They're sitting
> night after night in dark rooms. They're staring at a light. Their
> eyes are not moving. They're not thinking. Their brains are in a
> passive-receptive state and nonstop imagery is pouring into their
> brains, images from someplace where they are not, thousands of
> miles away. The images are being sent by a very small number of
> people.... The whole thing seems to be some kind of experiment
> in mind control." And it just may be.[208]

It is also interesting to recognize that the current system of industrial
mass production in effect necessitates a system of mass consumption requiring millions of people to buy identical products, things they would
not necessarily choose on their own. Television and other mass media
are the perfect tools for stimulating this mass consumption through
highly manipulative and incessant advertising. The primary purpose of
TV appears to be that of providing an outlet for corporate advertising,
with movies and news serving as entertaining fillers to keep the viewer

in a receptive state of mind.[209] This is aptly summarized by Todd Gitlin: "The commercial is the purpose, the essence; the program is the package."[210] In the United States, corporations spent approximately $150 billion in 2006 on advertising,[211] an amount that by far exceeds the annual US federal budget of about $90 billion for elementary, secondary and university education.[212]

Corporations have been highly successful in reaching consumers through TV advertising. The following TV viewing statistics are alarming and confirm the highly addictive character of the medium: Approximately 99.5 percent of all American homes have television sets, and 95 percent of the US population watches television regularly. On average, adults watch TV four hours per day, or 1,460 hours per year, while children spend even more time watching, on average five hours per day.[213] Over an entire lifetime (80 years), people in the United States spend about 20 years of their waking life watching TV, being manipulated by those who control TV content and wantonly wasting their limited time, thereby depriving themselves of engaging in other activities that are known to result in greater happiness and sense of well-being (see Chapter 9). In the United States, citizens now spend more time watching television than anything else besides sleeping, working and attending school.[214] In addition, total annual media consumption in the US (i.e., TV, radio, newspapers, magazines, news from the Internet, etc.) is more than 3,000 hours per person,[215] which translates, on average, into more than eight hours per day.

Internationally, the situation is little different, with about 80 percent of the global population having access to television.[216] The intensity with which corporations use television advertising and movies to seduce unsuspecting viewers to a materialistic and energy-intensive consumer lifestyle has been pointed out by Cavanagh and Mander:

> In the United States, the average viewer of television sees about twenty-eight thousand commercials every year. That is twenty-eight thousand times that they are hit by extremely invasive imagery saying virtually the same thing. One may be about tooth paste, and another about cars, cosmetics, and or drugs. But the intent of each of these commercials is identical: to persuade people to view life as a nonstop stream of commodity satisfactions. Cumulatively, globally, the commercialization effect is immense.[217]

Cavanagh and Mander then summarize:

> We have the most powerful and pervasive communications
> systems in history, dominated by a tiny handful of corporate
> people, describing how life should be lived. Is this good? Is it o.k.
> for billions of people to be receiving nonstop doses of powerful
> images and information controlled by such few sources, essen-
> tially telling them to be unhappy about their own cultures and
> values—how they live and who they are—to get onto the com-
> modity treadmill, to put their trust in corporations, and to em-
> brace a global homogenization of Western values? Will this bring
> a sustainable, equitable society? Many think not.[218]

In addition to being used to promote a materialistic culture and lifestyle
worldwide, electronic media, particularly TV, have been employed to
reinforce dominant ideologies and disseminate political propaganda.
In fact, as is pointed out by distinguished linguist and political analyst
Noam Chomsky in *Propaganda and the Public Mind*, democracies require
propaganda in order to function smoothly:

> It should be expected that it's in the democracies that these ideas
> [of propaganda] would develop. Because in a democracy you have
> to control people's minds. You can't control them by force. There's
> a limited capacity to control them by force, and since they have
> to be controlled and marginalized, be "spectators of action," not
> "participants,"…you have to resort to propaganda.[219]

Television, because of the centralized control of its content, is well suited
to the broadcasting of political propaganda.

Finally, the addiction to TV and other electronic mass media sup-
presses citizen political and environmental activism, which is another
form of subtle social control. After wasting more than 3,000 hours per
year on the mass media, there is simply very little time and energy left
for people to engage in any kind of citizen activism. TV, indeed, serves
as a weapon of "mass distraction." Because watching TV isolates us from
other like-minded citizens and transforms us into passive viewers, we
become far less likely to exchange ideas with others on how to improve
social, economic, political or environmental conditions. By broadcasting a "virtual reality," TV isolates us from reality. If we do not know what

is true, how can we contribute in a positive and effective way to society? If we do not get out into nature and experience the natural world for ourselves but only view landscapes on a TV screen, how informed or concerned will we be about environmental issues?[220]

In summary, modern electronic mass media, particularly TV, are extremely powerful mind-control tools that are used by a relatively small number of influential entities to redefine reality for hundreds of millions of unsuspecting viewers, to promote a materialistic consumer culture throughout the world, to reinforce dominant ideologies, to disseminate political propaganda and to suppress social and environmental activism.

Military Technologies

As we have seen, advanced technologies are employed to extract a wide variety of material and energy resources from the environment; these are then converted into various products and services for mass consumption, the latter of which, as mentioned above, is promoted by advertising via modern communications technologies such as television, radio and the Internet. Thus, advances in science and technology during the past 200 years have played a key role not only in raising the material standard of living for millions of people but also in creating an ever-increasing need for various natural resources such as minerals and fossil fuels. Unfortunately, many key renewable resources (e.g., water, timber, fisheries, arable soil) and non-renewable resources (e.g., oil, coal, gas, uranium, strategic metals) are not only limited in quantity but are frequently distributed unevenly among nations, which leads to global competition for certain energy and mineral supplies.[221]

This worldwide competition for scarce resources is generally mediated through free-market trade mechanisms. However, if these fail or are found unfavorable, military force, or the threat of it, is often used to guarantee continued access to the limited resources which are considered vitally important to industrial economies.[222] Thus, high-tech military technology plays a key role in ensuring the continued exploitation and control of natural resources that are essential to maintaining the materialistic consumer lifestyle.

A brief historical review of the major military conflicts of the 20th century indicates that many of them were caused, at least in part, by competition for scarce resources, particularly petroleum. No highly industrialized nation can survive without a steady supply of oil. As early as the beginning of the 20th century, the industrializing nations of Europe turned to their colonies for supplies of fossil fuels and minerals. Since Germany had no resource-rich colonies, one of its objectives during World War I was to seize colonies from its competitors, France and England. In fact, it has been suggested that World War I was primarily motivated by competition for exploitable colonies in the "developing" world.[223]

Gaining and defending access to petroleum also played a key role in World War II. For example, Germany's lack of access to Middle Eastern oil supplies was one of the factors leading to the aggression against Russia, which possessed productive oil fields.[224]

The rapid and continuous development of new and increasingly powerful military technologies has allowed the industrialized nations to dominate and exploit "developing" nations, which do not have the wealth or expertise to develop the high-tech weaponry necessary to defend themselves and their resources. The importance of military supremacy for colonial and imperial conquests is summarized by Michael Parenti:

> We need to be reminded that only by establishing military supremacy were...colonizers able to eliminate the crafts and industries of Third World peoples, control their markets, extort tribute, undermine their cultures, destroy their villages, steal their lands and natural resources, enslave their labor, and accumulate vast wealth. Military supremacy was usually achieved after repeated and unspeakable brutal applications of armed violence.[225]

To summarize, many nations over the centuries have used their access to military technologies, some of which were advanced for their time, to exploit the peoples and resources of other nations. It has been said that "the Third World is not 'underdeveloped' but overexploited."[226] Given that the affluence of industrialized nations could not be maintained without exploiting "developing" nations, massive military power is required to enforce and maintain the unjust, exploitative global economic order. The imperial quest for foreign markets, natural resources and cheap labor is the primary reason for the many violent conquests and invasions witnessed during the past two centuries. Many modern technologies are employed to control and exploit nature, people and even future generations, and this is being done on a grand scale. Because the very act of exploitation violates the principles of reciprocity and fairness, the cost-benefit balance is skewed, often generating massive destruction of nature and unthinkable human suffering.

In summary, technological control and exploitation essentially ignores the fact that "everything is connected to everything else."[227] Unfortunately, technology increases the distance, both spatial and temporal, between exploiter and exploited, thereby blunting normal ethical sensitivities and allowing the perpetration of hideous atrocities. Since separateness between exploiter and exploited is an illusion, the exploiters ultimately cannot escape the negative consequences of their actions. Thus, in the long run, exploitation is counterproductive. For example, while the citizens of developed nations have benefited for more than 100 years from the exploitation and use of fossil fuels, none are likely to escape the destructive consequences of global climate change. As long as advanced technologies are used for exploitation, worldwide negative effects are inherently unavoidable.

There have been major attempts to solve the problems created by science and technology by applying more of the same. In the next chapter, we will discuss the failures of these counter-technologies and social fixes to mitigate the unintended consequences of earlier technologies.

CHAPTER 4

In Search of Solutions I:
Counter-Technologies
and Social Fixes

THE UNINTENDED CONSEQUENCES brought about by the application of science and technology have introduced entirely new and complicated problems requiring our attention and ingenuity to address effectively. As Alan Drengson points out, it appears that humans are better at causing problems than solving them:

> Too much time has been taken up in dealing not with genuine problems but with those that should not have been created in the first place.... The majority of the problems we now face are human-caused.... We fail to see that problems caused by our activities often are a result of those very activities.... Consequently, we redouble our efforts to solve "new" problems, but use the same methods that caused the problems in the first place.[228]

Although many contemporary problems were created by earlier applications of science and technology, there is nevertheless a very strong belief that more science and technology will be the solution.[229] Faith in the effectiveness of techno-fixes has been repeatedly expressed as well as occasionally questioned by various writers:

> "Technology creates problems, which technology can solve."[230]

> "Many are confident that the problems created in part by science and technology will be solved by more science and technology."[231]

71

"Where one technique has failed, another is called to its rescue; where one engineer has goofed, another—or several more—are summoned to pick up the pieces."[232]

"We still hear that new generations of machines will solve the problems left by prior generations of machines."[233]

"Society continues to throw science and technology at its accelerating scientific and technological problems."[234]

"It compounds the crises by attempting to solve them by the same measures and means that generated the problems in the first place."[235]

"To maintain technology, technology itself must be increasingly devoted to the production of counter-technologies which are attempts to mitigate the more serious and harmful effects that technology brought about in the first place."[236]

"Technology today is offering solutions to everything in every sphere. You can hardly think of one for which it does not come up with the answer. But it would like us to forget that in virtually every case, it has created the problem in the first place that it comes round to say that it will transcend. Just a little more technology. That's what it always says."[237]

Some techno-optimists even claim that technology is an autonomous self-correcting system, which is expressed in the faith that technological innovations will automatically, without human guidance, solve problems created by previous technologies.[238] This claim is highly unrealistic because, as will be shown throughout this chapter, most techno-fixes have very limited capacity to solve either technical or social problems.[239]

Technological fixes can be loosely grouped into three different, sometimes overlapping, categories. First, there are so-called counter-technologies, which are technologies specifically developed to oppose and neutralize the negative effects created by other technologies. For example, environmental remediation technologies are counter-technologies because they attempt to remove the pollution that was generated by previous technologies. Second, there are so-called social fixes that involve the use of science and technology to solve social, economic, political or cultural problems. A good example of a social fix is the attempt to solve the persistent problem of world hunger by increasing food production via the application of industrialized agriculture despite the fact that hunger and starvation may not be due to food shortages but rather the result of various economic and political factors.[240] Third, efficiency improvements serve as techno-fixes by increasing the effectiveness of technological processes in an incremental fashion, thereby attempting to gradually increase the benefits while reducing the costs of a particular technology. For example, an increase in the fuel efficiency of cars will not only increase benefits (i.e., the distance that can be driven per volume of fuel used) but will also decrease costs (i.e., expenses for fuels and environmental pollution). We discuss the first two types of technological fixes, counter-technologies and social fixes, in this chapter and will focus on the effectiveness of efficiency improvements in Chapter 5.

Counter-Technologies

Counter-technologies are technologies that attempt to neutralize the negative effects caused by other technologies. Before analyzing the effectiveness of counter-technologies, it is instructive to briefly review the causes of unintended consequences that result from the application of science and technology. As was mentioned in Chapter 1, both positive and negative effects are necessarily produced by any technology. It is impossible for technologies to provide only benefits without any costs in a world in which everything is connected and in which evolution has operated over millennia. Also, as discussed in Chapter 1, unavoidable negative consequences are intrinsically unpredictable because reductionist science cannot provide the necessary information on all possible interactions and cause-effect relationships within complex systems.

These two general characteristics, unavoidable negative consequences and their inherent unpredictability, apply to counter-technologies as well. Thus, while counter-technologies may be successful in ameliorating the undesirable effects caused by previous technologies, they will, in turn, generate another sequence of unintended and unpredictable consequences. Consequently, counter-technologies are inherently incapable of solving technology-created problems without leaving in their wake "residue problems."[241] If these residue problems were to be addressed by further counter-technologies, second-generation residue problems would be generated. Thus, the application of counter-technologies leads to incomplete "quasi solutions."[242] The negative effects of previous technologies can never be completely neutralized without creating new problems.

It is instructive to evaluate the manner in which counter-technologies address the typical unintended physical, chemical or biological side effects that were caused by earlier applications of other technologies. There is a wide range of undesirable physical consequences that may be caused by the deployment of technologies, most of them involving the addition or removal of mass and/or energy. For example, surface mining may remove large quantities of minerals, rock, soil and vegetation from the land. Filling may remediate this situation, but the material must be obtained from elsewhere, thereby creating deficits and associated environmental problems in other locations. When a power plant releases waste heat into a river, thereby potentially damaging fish by decreasing dissolved oxygen levels, it is theoretically possible to remove this excess heat from the river, but, in accordance with the first law of thermodynamics, this can be done only by transferring it to another location, where hopefully it will have fewer negative effects (i.e., total energy is always conserved and never destroyed). Finally, when technologies disperse undesirable matter over a large area, as is the case for metal-containing pesticides sprayed on farmland, lead-containing dust from car exhausts or zinc-oxide-containing dust from automobile tires, it may be theoretically possible but is completely impractical to gather up and concentrate these widely dispersed materials. If this were attempted, the second law of thermodynamics ensures that it would require extremely large amounts of energy, the generation of which would have, in turn, significant negative effects.[243] In summary, counter-technologies may partially or even completely reverse the negative physical consequences brought about by previous technologies but only by creating matter and energy imbalances elsewhere. Like all technologies, counter-technologies are initially considered successful because focus is placed, following the tradition of scientific reductionism, primarily on their desirable effects while ignoring their undesirable ones, i.e., ill effects that may be further removed in time and space or completely unknown.

Technologies may also create unintended negative chemical and biological consequences. Such effects generally result from interference with naturally occurring chemical or biochemical reactions. For example, the combustion of fossil fuels to generate energy has the undesirable side effect of releasing CO_2 into the atmosphere, which disturbs the global carbon cycle, thereby causing climate change. One could in principle reverse this fuel oxidation reaction by reducing the atmospheric CO_2 back to carbon. However, given that no industrial process is 100 percent efficient, it would take more energy to reduce CO_2 to elemental carbon than was initially released during the combustion of fossil-fuel carbon to CO_2. Thus, significant amounts of (net) energy would be needed to reverse the chemical reaction that initially created CO_2. Since any type of energy generation, including renewable energy production, has negative environmental impacts (see Chapter 6), it is clear that counter-technologies that attempt to reverse undesirable chemical reactions will have negative environmental effects associated with the generation of the energy required to implement the "fix." Finally, in cases where technologies cause irreversible biological effects such as species extinction, there are no counter-technologies that can remedy such permanent loss.

Social Fixes

While counter-technologies attempt to address problems created by the application of previous technologies, social fixes involve the use of technologies to solve social, cultural, economic and political problems. Because many social problems manifest themselves in the form of symptoms that appear to be treatable by science and technology, it has been tempting to redefine these complex social problems as simple technical challenges. This is another example of the limited focus of scientific reductionism in which complexity is reduced to simple and isolated cause-effect relationships that are more amenable to technological manipulation.

The use of social fixes is widespread in industrialized nations. For example, instead of recognizing that the environmental crisis is caused by overpopulation and overconsumption, it is redefined as a scientific and technical problem, requiring the remediation of past pollution and the development of cleaner technologies "to pave the way for a more sustainable future" (see Chapter 6). Instead of recognizing that traffic congestion is the result of a pernicious car "culture" that needs to be diminished if not abolished, the problem is addressed by building more roads, thereby inviting more traffic and aggravating the problem. Instead of realizing that many psychological conditions such as depression, restlessness and anxiety may be the result of living in a "sick" society characterized by high levels of stress and alienation, these problems are solved by medical professionals through the prescription of powerful antidepressants and tranquilizers that mask the patient's symptoms as well as the social antecedents. Instead of acknowledging that the leading causes of death in industrialized societies—coronary heart disease and cancer—are largely the result of poor lifestyle choices such as smoking, high-fat diets and lack of exercise, these diseases are treated by doctors employing a large arsenal of expensive, invasive, high-tech procedures. Instead of resolving economic, political or cultural differences and tensions with other nations through negotiation and diplomacy, lethal high-tech weaponry is used to bully, control and exploit (see Chapter 3).

It is tempting for politicians and policy-makers to rely on science and technology to mask complicated social problems because it seems easier to invent and deploy new technologies than to change people.[244] As Kenneth Stunkel and Saliba Sarsar remark,

> The physicist Alvin Weinberg has argued that we can get around social problems by "reducing them to technological problems."... It is easier to change technology than to change the way people behave. Technological engineering is far more expedient and effective in the short term than social engineering.[245]

For example, it is far easier to promote the development of clean technologies to reduce environmental problems than to change the materialistic consumer lifestyle. Similarly, it is much less controversial to build more roads to ease traffic congestion than to mandate that commuters car pool, bike or walk. Likewise, it is much simpler to provide ready access to contraceptives than to change people's sexual mores. It also much easier for doctors to prescribe antidepressants and tranquilizers than to promote economic and workplace policies that would make everyday life less competitive, stressful and alienating. Similarly, many people apparently prefer invasive surgery over changing their daily exercise and eating habits. Finally, despite the astronomically high number of lives lost in the wars of the last century (see Chapters 2 and 3), many politicians around the world still seem to believe that it is more expedient to spend astronomical amounts of money developing and deploying high-tech weapons to resolve conflicts instead of avoiding conflict through diplomacy.

The reliance on technological solutions to social problems has a number of unfortunate side effects. First, like all counter-technologies, social fixes have unintended physical, chemical and biological consequences. In addition, as Eugene Skolnikoff points out, "technology alone cannot be expected to solve an important societal problem without creating new social, economic, and political problems along the way."[246] Finally, since social fixes deal only with symptoms, the underlying causes of many social, economic, political and psychological problems are ignored. As a result of this limited "myopic engineering" approach, the search for truly effective and lasting solutions to difficult social problems is delayed, often until a situation has become almost unmanageable.

Environmental Counter-Technologies

All environmental technologies can be considered counter-technologies because they have been specifically developed to address the pollution created by other technologies. This is reflected by the hopeful but unrealistic proclamation that "the deterioration of the environment produced by technology is a technological problem for which technology has found, is finding, and will continue to find solutions."[247] Despite this enthusiastic assurance by a true believer in the effectiveness of techno-fixes, we will show that while most environmental remediation technologies may be able to address the pollution generated by previous technologies, they often create side effects that are worse than the original problem.

As discussed in Chapter 1, the conservation of mass principle is succinctly expressed in Barry Commoner's second law of ecology: "Everything must go somewhere."[248] Thus, a pollutant, if not transformed naturally by either chemical or biological reactions, will remain indefinitely in the environment, and its total mass will not change with time. Many physical treatment technologies attempt to reduce the risk posed by a pollutant by using various strategies such as (a) limiting the dispersal of the contaminant (e.g., land-filling), (b) reducing toxic effects by dilution (e.g., smokestacks) or (c) transferring contaminants from one medium to another (e.g., air-stripping of contaminated water).

Since the total mass of a pollutant is not reduced by physical treatment technologies, the effectiveness of these technologies in solving environmental problems is highly questionable. It is clear that such treatment only transfers risk either from one place to another or from the present to the future. For example, land-filling limits the dispersal of contaminants only at the present time. Since landfill integrity cannot be maintained indefinitely, contaminants can be expected to re-enter the environment and cause negative effects in the future.

The rationale for diluting pollutants or transferring contaminants from one medium to another is based on the assumption that these treatment strategies render the pollutants less harmful. As was pointed out in Chapter 1, this assumption is very questionable considering that reductionist science is inherently unable to elucidate all negative environmental consequences. For example, dilution will certainly reduce the risk of localized and obvious acute toxicity but will likely increase the probability of more widespread subtle chronic effects, which are currently unknown or difficult to monitor. Similarly, it is uncertain whether transfer of a pollutant from one medium to another will actually reduce the overall risks. More likely, this type of treatment strategy is chosen because the hazards of a pollutant in one medium are well known (i.e., chlorinated solvents in groundwater used for drinking) while very little data are available about the same compound in the medium to which it is transferred (i.e., chlorinated solvents dispersed in air).

In summary, physical treatment technologies are unable to provide effective solutions to environmental pollution because they transfer risks either from one place to another or from the present to the future. This transfer of risk has aptly been termed the "Rearranging Effect."[249] Physical treatment technologies are often considered effective only because spatially and temporally transferred risks are not known and because there is little concern about what happens elsewhere or at a later time.

In addition to being limited by the conservation of mass principle, the effectiveness of environmental technologies is also constrained by the second law of thermodynamics. As was mentioned earlier, the second law of thermodynamics states that order (neg-entropy) can be created in one part of a (closed) system only at the expense of creating more disorder (entropy) elsewhere in the system.[250] Since entropy increase is correlated with and can be considered a measure of environmental pollution,[251] the second law of thermodynamics ensures that it is impossible to remediate dispersed contaminants without creating as great or greater environmental problems elsewhere.

Consider, for example, the common treatment of groundwater that has been polluted by gasoline or other fuels leaking from underground storage tanks, of which there are about 250,000 in the United States alone.[252] When contaminating hydrocarbons reach groundwater, they dissolve and disperse, thereby potentially polluting millions of gallons of drinking water. A common way to remediate this problem is to pump the polluted groundwater from the aquifer, treat it to remove the contaminants, often by trapping them onto activated carbon, and reinject the cleaned water back into the ground. From a thermodynamic perspective, this "pump and treat" operation reduces "disorder" (entropy) and pollution in the aquifer by concentrating the highly dispersed contaminants onto activated carbon. But this reduction in entropy in the aquifer can be achieved only by increasing entropy elsewhere in the environment. In this case, entropy is increased by the combustion of fossil fuels, which is required to operate the groundwater pumps. The high-entropy wastes resulting from fossil-fuel combustion consist of highly dispersed air pollutants such as particulates, sulfur dioxide, nitrous oxides and carbon dioxide, the latter of which is a cause of global climate change. In short, the remediation of highly dispersed groundwater contaminants by the common "pump and treat" operation is achieved by creating highly dispersed air pollutants. Because of a single-minded focus on reducing the well-known risks of contaminated drinking water, the lesser known but significant risks of air pollution and climate change are ignored. In general, remediation technologies that attempt to reverse pollutant dispersion appear to be effective only when viewed from a very limited perspective. However, when one considers all side effects, both spatial and temporal, it is extremely questionable whether these remediation technologies can truly improve the overall quality of the environment.

It is also important to recognize that from a practical perspective most highly dispersed persistent contaminants are unlikely to ever be removed from the environment. Again, the second law of thermodynamics ensures that it will require tremendous amounts of energy to concentrate highly dispersed materials.[253] Consequently, no serious attempts have been made to remediate persistent chlorinated pesticides such as DDT, now found in the fatty tissue of almost every organism on Earth, or lead contamination along US roads and freeways, which is the result of the previous use of leaded gasoline, or the widespread metal and hydrocarbon contamination found in many marine and freshwater sediments. As soon as contaminants have dispersed into the environment, remediation technologies are extremely ineffective in addressing the problem—in fact, overzealous attempts might well be counterproductive. It is therefore extremely important to prevent pollution, a topic which will be discussed in more detail in Chapter 6.

Pollution can be prevented either by abstaining from manufacturing harmful products and their associated wastes altogether or by "neutralizing" hazardous compounds before they disperse into the environment. A simple example is the neutralization of acid waste with alkali to create a mixture of neutral pH, which is much less harmful than the original waste. There are many technologies that use either chemical or biological reactions to render a waste stream much less reactive prior to release into the environment. Examples of chemical and biological treatment technologies are incineration, neutralization and bioremediation.

While chemical and biological treatment technologies are certainly more effective than physical remediation efforts, which simply attempt to limit or reverse contaminant dispersal, it is important to realize that many of these technologies also have undesirable side effects.[254] For example, incineration may generate flue gases containing minute quantities of dioxins that are highly poisonous. In addition, it generates ashes containing toxic metals, which must be land-filled, thereby transferring environmental problems to future generations. Even bioremediation, a low-impact technology relying primarily on native bacteria to metabolize pollutants, can cause the formation of intermediates that are more toxic than the original contaminant (e.g., carcinogenic vinyl chloride is produced during chlorinated solvent bioremediation).[255] Also, the addition of certain process chemicals used to stimulate bioremediation may have a negative impact on local soil or sediment ecology. Consider finally the energy requirements and resulting pollution by major waste-water treatment plants. For example, the wastewater treatment facilities at the BASF chemical complex in Germany use as much energy as a city of 50,000 inhabitants.[256] Given that the generation of any type of energy, non-renewable as well as renewable, is always associated with the creation of various negative environmental impacts (see Chapter 6 for a more detailed analysis), it is clear that even wastewater treatment plants, despite their ability to successfully remove many hazardous chemicals from factory effluents, will cause unintended environmental problems elsewhere. One can only hope that there is an overall benefit to the environment, i.e., that the reduction of risk through the removal of toxics from wastewater is greater than the increase in risk caused by air pollution and carbon dioxide emissions during energy generation by fossil-fuel combustion. Not only is science inherently limited in correctly quantifying the risks, there is also the difficulty of defining risk (see Chapter 8). Should we consider risks to humans only or to specific animals, to all animals, to plants, to the entire environment? In summary, even effective chemical and biological treatment technologies are likely to have negative environmental consequences, many of which may not be known or even envisioned because of the narrow focus of reductionist science.

There are already efforts to develop the "next generation" of environmental counter-technologies to address the impending threat of global climate change. Instead of reducing fossil energy use and encouraging the development of renewable energy, there are plans to separate carbon dioxide from the flue gases of coal-fired power plants and sequester it either in deep geologic reservoirs (e.g., depleted oil fields, coal seams or saline aquifers) or in the oceans. These strategies would allow for the continued use of fossil fuels without further increasing atmospheric carbon dioxide concentrations, but it is highly likely that they will result in negative side effects. The problem with geologic sequestration is that carbon dioxide storage is not permanent, given that it is impossible to avoid slow but continuous leakage. The leaked carbon dioxide could cause a number of harmful environmental consequences such as the acidification of groundwater and associated undesirable changes in geochemistry (e.g., mobilization of toxic metals), water quality (e.g., leaching of nutrients) and ecosystem health (e.g., pH impacts on organisms). In addition, a sudden catastrophic release of large amounts of CO_2, as a result of either reservoir fracturing by earthquakes or pipeline failures, could result in the immediate death of both people and animals, particularly as CO_2 is odorless, colorless and tasteless and thus likely to escape detection.[257]

The problem with ocean disposal of CO_2 is that it will acidify seawater, which will lead not only to the dissolution of exoskeletal components of corals and bivalve mollusks but also to metabolic suppression in fish, causing retarded growth and reproduction, reduced activity, loss of consciousness due to disruption of oxygen transport mechanisms (deep sea fish hemoglobins are extremely sensitive to pH) and, if persistent, death.[258] Even minor disturbances of the pH are likely to have significant impacts, given that the deep sea environment has been remarkably stable for millennia and that the organisms there have evolved to be specifically adapted to their unique ecosystem. In addition to detrimental effects on deep sea biota, CO_2 disposal may also negatively affect microbial populations and thus cause changes or disruptions in marine biochemical cycles. The main concern is that even small changes in the biochemical cycling of essential elements (i.e., carbon, nitrogen, phosphorus, silicon and sulfur) may have very large consequences, many of them secondary and difficult to predict.[259]

Finally, several rather bizarre geo-engineering proposals for combating global climate change deserve mention, if for no other reason than to demonstrate the persistent belief that counter-technologies, rather than prevention, are the solution to human-caused problems.[260] The geo-engineering concept of ocean fertilization involves the dispersal of iron, normally a limiting nutrient, over hundreds of millions of square miles of ocean surface to stimulate the growth of carbon-rich phytoplankton, some of which will sink to the ocean floor, thereby temporarily removing carbon from the global carbon cycle. As in the case of CO_2 injection, the main concern is whether large-scale ocean fertilization will result in negative unexpected consequences to marine ecosystems and biogeochemical cycles. For example, large-scale eutrophication could result in the depletion of oxygen, leading to deep ocean anoxia, which in turn would shift the microbial community structure toward organisms that produce methane and nitrous oxide, i.e., greenhouse gases with much higher climate-change potential than CO_2. It is difficult if not impossible to predict all secondary and higher-order effects of ocean fertilization on the ocean food web structure and dynamics, including changes in the biogeochemical cycling of important elements such as carbon, nitrogen, phosphorus, silicon and sulfur.[261]

Another geo-engineering proposal involves the deflection of sunlight from the Earth using a giant space mirror spanning 600,000 square miles.[262] Clearly, this idea verges on science fiction, given that it would be an incredible challenge to build a mirror four times the size of California, let alone catapult it into space.

Finally, the "father" of the hydrogen bomb, nuclear physicist Edward Teller, has proposed another scheme to deflect sunlight. Probably inspired by earlier concerns about a "nuclear winter" (see Chapter 2) that would follow a nuclear holocaust that had caused dust and smoke to obscure the sun, Teller proposes propelling tens of thousands, if not millions, of tons of metal or mineral particles into the atmosphere to scatter and thus reduce sunlight reaching the Earth by about 1 percent, which is deemed just sufficient to cool the Earth and compensate for global warming.[263] As expected, Teller, an expert in not considering the consequences of his inventions, did not even speculate about the unintended consequences of such an approach.

In the final analysis, environmental counter-technologies may also be considered social fixes because, as will be discussed in more detail in Chapter 6, the causes of environmental problems are not only polluting technologies but, more fundamentally, human overpopulation and continued economic growth. Consequently, unless the relevant sociocultural issues are addressed and the size of the human population stabilized and reduced, and the materialistic consumer lifestyle largely abandoned, there is little chance that our environmental problems will be solved or that we will ever achieve sustainability.

Military Counter-Technologies

Most nations claim that military technologies are needed for defense against international threats, so that one nation's weapons systems serve as counter-technologies to another nation's weapons systems. Thus, anti-tank missiles are developed to defend against tanks, surface-to-air missiles against aircraft, mine-sweepers against mines, etc. The use of military technologies for defensive purposes is clearly justified, both ethically and legally, following the long Western tradition of the "just war doctrine,"[264] which is also reflected in the UN Charter that sanctions self-defense in response to an enemy invasion.[265] Unfortunately, there is often no sharp distinction between defensive and offensive technologies. As a result of this ambiguity, nations may perceive defensive technologies in other nations as offensive threats, and this perception may fuel the development of the respective counter-technologies. As a result, an arms race may be precipitated in which successive cycles of counter-technologies are designed and produced.

The situation becomes even more complicated when there are weapons, such as thermonuclear bombs, against which there is no effective defense in the form of a true "physical shield."[266] Under these circumstances, atomic weapons are developed to counter the enemy's nuclear threat by deterrence, implying that these weapons will never actually be used because the fear of nuclear retaliation will freeze the opponent into inaction.[267] As Robert Holmes describes in *On War and Morality*,

> Consider that the other side of the deterrent coin is that whatever deterrent value nuclear weapons are assumed to have, they also serve as a provocation, increasing the fear that their possessor will use them to initiate an attack, and thereby increasing the likelihood that an adversary will seek to counteract that threat by acquiring a deterrent of the same or greater magnitude.... Weapons designed to deter, if perceived by an adversary not as a deterrent but as a threat to his own security, will provoke him to increase his armaments and thus intensify suspicions that he is indeed providing himself with the first-strike capability that the weapons are meant to deter.[268]

However, there is always the risk that military counter-technologies will be used, intentionally or accidentally. Indeed, history has shown that weapons believed to be ideal deterrents because of their horrendous destructiveness were sooner or later used, making wars more protracted and barbarous. For example, after inventing dynamite in the 1860s, Alfred Nobel was convinced that this powerful explosive would prevent wars in the future. "Perhaps my factories," he later told the Baroness von Suttner, "will put an end to war even sooner than your congresses:

on the day that two army corps can mutually annihilate each other in a second, all civilized nations will surely recoil with horror and disband their troops."[269] Unfortunately, he was badly mistaken, as dynamite was extensively used a few years later in World War I and in wars thereafter. The inventor of the machine gun also optimistically believed that his invention would save the world from the horrors of war. As James Turner Johnson remarks in *Can Modern War be Just,*

> Similarly, Maxim, the inventor of the first reliable machine gun, believed it was a humanitarian weapon that would, because of its destructiveness, cause wars to be much shorter and might even lead to the disappearance of war from history. World War I gave the lie to this expectation; ironically, it was the machine gun that kept another effort at a quick victory, the charge across no-man's land, from succeeding,... creating human carnage unlike anything before in military history.[270]

It is therefore questionable whether the development of counter-technologies to earlier counter-technologies can permanently put an end to war. It is more likely that it will instead lead to a situation in which sooner or later the weapons and counter-weapons will be used. In addition, as Ernest Braun points out, the rapid obsolescence of military hardware worldwide avoids the problem of market saturation, thereby ensuring continued arms sales and profits for weapons manufacturers:

> Military technology is a prime example of a technology that cannot stand still, where the impetus for rapid innovation stems from the need to keep abreast and ahead of all potential enemies. In peacetime, or at times of limited warfare, the rapid obsolescence of weapon systems is also the only means of keeping orders flowing. If arsenals are not depleted by enemy action, how can the armaments industry keep going without the obsolescence of weapons? As in the civilian industry, obsolescence becomes the main tool to fight market saturation.[271]

As mentioned above, military counter-technologies may be considered a form of social fix because they are believed to effectively resolve international conflicts whose causes may be differences in social, cultural, economic or political values and goals.[272] Thus, while military technologies may be able to suppress conflicts by deterrence for some time, they may also be used in brutal and barbarous wars, as history has repeatedly shown. This is profoundly dysfunctional because the underlying problems might well have been resolved through intelligent diplomacy or other means.

Medical Counter-Technologies

Advanced medical technologies and procedures are often applied to counter the negative effects brought about by other technological innovations. For example, emergency surgical procedures were initially developed in response to injuries suffered in wars but are now applied on a regular basis to deal with the mutilations caused by accidents involving automobiles (see Chapter 2), trains, airplanes and industrial machinery.[273] Similarly, antibiotics are used in the attempt to abolish microbial infections, which were very rare in early hunter-gatherer societies but which became much more frequent as human numbers increased, causing people to live under crowded and often unsanitary conditions that provided favorable environments for pathogens to spread.[274] In addition, problems caused by medical practice (i.e., iatrogenic disease), such as resistance to single antibiotics, are addressed by next-generation medical counter-technologies, such as combination antibiotics. Finally, even contraceptives may be considered a counter-technology because they may be used to reverse human overpopulation caused by earlier technological developments that provided better nutrition, sanitation and medical procedures, and thereby fueled population growth. It has, however, become an essential counter-technology given the present state of world overpopulation.

Medicine is also often used as a social fix to provide biological solutions to social, cultural, psychological, existential or even moral problems. As Richard Taylor explains in *Medicine Out of Control*,

> The response of the medical establishment to the massive social problems associated with life in urbanized, industrialized countries has been to approach these problems, or rather the expression of them, in an individualized and often purely biological fashion. The medical model of disease has been extensively used in dealing with alcoholism, psychoneurosis, narcotic abuse, road

accidents, suicide and attempted suicide, coronary heart disease and overuse of tranquilizers. Doctors, as biological scientists, have, by monopolizing these conditions of man, given the impression that their solution lies in some new technical innovation or a new drug rather than in the changing of the underlying social, environmental, and economic causes.[275]

An example of a social fix is the use of medicine to address diseases that are the result of poor lifestyle choices such as smoking, overeating and lack of exercise. By not addressing the root causes, which are essentially cultural and psychological, the medical establishment encourages behavior that does great harm. As René Dubos observed almost 50 years ago, "It is the function of a doctor to make it possible for his patients to go on doing the pleasant things that are bad for them—smoking too much, eating too much, drinking too much—without killing themselves any sooner than is necessary."[276] As a recent article in *Science* pointed out, pharmaceutical companies see great profit potential in selling drugs for treating obesity, which is caused simply by overeating and lack of exercise and could therefore be easily prevented by changes in lifestyle:

> Drugmakers have been salivating over the prospect of creating anti-obesity medications. Obesity is a rising pandemic that includes 60 million adults in the United States alone, and although most physicians champion diet and exercise as the best way to fight fat, many people are desperate for an easier way to avoid corpulence and consequences such as heart disease, stroke, and diabetes. It's a drugmaker's dream.[277]

Finally, medicine also provides a social fix for the many psychological disorders that are often caused when people attempt to adapt to stress, alienation and unnatural living conditions, which are prevalent in modern industrialized societies.[278] As Richard Taylor observes, the medical establishment often considers psychological problems as diseases of maladaptation and thus prescribes powerful psycho-active drugs instead of addressing the underlying causes:

> Even when social factors are recognized as antecedents of various medical and psychological conditions, it is usually considered

that the problem lies in the maladaptation of the individual to the society in which he lives. The reverse, that is, that there is something drastically wrong with the society that humans are being asked to adapt to, receives scant attention.... The treatment of "behavioural problems" in children, "maladjustment problems" in adolescents or anxiety, depression, or insomnia in adults, on an individual basis, employing psychological and pharmacological therapy, tends to play down the powerful social and economic forces that lead to these problems.[279]

Unintended Consequences of Counter-Technologies and Social Fixes

As demonstrated throughout this chapter, counter-technologies and social fixes attempt to ameliorate undesirable symptoms while ignoring the underlying causes. Instead of recognizing these symptoms as warning signs of deeper problems, techno-fixes are used to mask them. As a result of ignoring systemic causes, the underlying problems continue to grow unchecked, thereby not only delaying but also complicating the implementation of effective and lasting solutions later. As Philip Bereano summarizes,

> The most frequent charge made against the idea of the technological fix is that it consists of stop-gap mechanisms that diffuse efforts towards genuine social change. One has the impression that semantically the metaphor of the "fix" derives more from the drug culture than from the industrial one. As narrow-range, short-sighted, adaptive mechanisms, shortcuts fail to confront the underlying problem, which often may worsen during the euphoric interval.[280]

Technological optimists sometimes defend techno-fixes with the claim that they will "buy time" for addressing the many underlying complex technological, social, cultural, psychological, economic and political problems. While this argument may initially sound convincing, Stunkel and Sarsar point out that there are really no technological shortcuts to the problems facing industrial societies:

The weakness of the technological fix, however, is not its short-term efficacy but its seductiveness, the temptation to see technology as an autonomous reality that can be manipulated outside the system that produced it, making it unnecessary for people and societies to change their behaviour and reorder priorities. Technological fixes can and do encourage undisciplined consumption and waste, putting off a socially and economically disruptive day of reckoning.[281]

Similarly, René Dubos warns,

Developing counter-technologies to correct the new kinds of damage constantly being created by technological innovations is a policy of despair. If we follow this course we shall increasingly behave like hunted creatures, fleeing from one protective device to another, each more costly, more complex, and more undependable than the one before.[282]

Indeed, unless we confront the root causes of our complex technological and social problems, we will, like drug addicts, apply one techno-fix after another, often with increasing potency, only to find that our problems become progressively worse. Humans may indeed be better at creating problems than solving them. As E. F. Schumacher observes, it seems that the net effect of our progress in science and technology has been an ever-increasing number of problems, some of them so complex and large that they may even threaten our survival:

The third illusion, which is still rampant, is that science can solve all problems. I have no doubt that science can solve any individual problem when it is clearly defined. But my experience is that as it solves problem "A" it creates a whole host of new problems. It is quite a thought that there are more scientists alive today than there have been in all previous human history taken together. What do they all do? They solve problems very efficiently. Aren't we running out of problems? No. We have more and more. This seems to be a bottomless pit. They grow faster than we solve them.... What did our forefathers do? They survived, obviously, without all these scientists. And without all these problems. So we have to ask ourselves if something has gone wrong in the development of technology.[283]

In conclusion, the techno-fix offers no lasting solution to the many complex problems confronting industrial societies. As will be discussed in Part III, finding sustainable solutions will require rigorous analyses of the underlying root causes of technological, social, psychological, cultural, economic and political problems. Root cause analyses will often point to the necessity of a paradigm shift, a change in worldview and values. Let us first, however, discuss another approach that is commonly used to ameliorate problems caused by the application of technology: efficiency improvements.

CHAPTER 5

In Search of Solutions II: Efficiency Improvements

Aₛ WAS DEMONSTRATED in the previous chapter, counter-technologies and social fixes are superficial attempts to address the various negative symptoms of more fundamental technological or social problems. Yet, there is another type of technological fix that is applied even more frequently, evoking great hope among techno-optimists and provoking virtually no opposition from the public: efficiency improvements.

It is generally believed that very few problems in modern industrialized society cannot be solved by greater efficiency. It is assumed that dwindling non-renewable fossil energy supplies will be countered by better energy efficiency, world hunger will be solved by more efficient genetically engineered crops, environmental pollution will be addressed by designing more eco-efficient industrial processes, rising health care costs will be curtailed by more efficient medical technologies and sluggish economic growth will be reversed by increasing the efficiency of labor. As Thomas Princen points out in *The Logic of Sufficiency*, the belief in efficiency as a solution to technical and social problems is a relatively recent phenomenon, but it has become so pervasive that few recognize its subtle but powerful influence:

> The rise of efficiency in little more than a century, its ascendance from an intuitive idea and common practice to a technical metric

and now to a broad social organizing principle, is arguably a
product of modernization, maybe a definition of modernization.
It is an outgrowth of the confluence of scientific understanding,
rapid technological change, human population growth, fossil-
fuel use, and advances in communications and transportation.
"Those who favor efficiency as the goal of social policy tend to
think of it as a grand value that takes up, incorporates, and balances all other values," says philosopher Mark Sagoff. That it now
extends from managerial mantra to social policy to personal
choice, from education and health care to shopping and child-
rearing and a host of other "nonmarket" activities, suggests its
power in modern life. That few people stop to take notice suggests
its hegemony, its pervasive and unquestioned value, its universal
application. "So interwoven into the fabric of society is this notion of efficiency that it seems only common sense," writes Mary
E. Clark. "The fact that this belief in the supreme rightness of
efficiency is just that, a belief based on certain arbitrary assumptions, escapes us."[284]

Given that we are indeed living in the "Age of Efficiency,"[285] it will be
useful to understand the concept of efficiency. In the most general terms,
technological efficiency (e) can be defined as the ratio of the "amount of
benefit (B) derived per unit limited resource (R)", i.e.,

$$e = \frac{B}{R} \qquad\qquad (5.1)$$

Common examples of benefits (B) derived from the application of
technology are work done by machines, food produced by industrialized agriculture, years of life extended by high-tech medicine, or in most
general terms, material affluence, which is often quantified as per capita
gross domestic product (GDP/person). Examples of limited resource (R)
inputs are non-renewable or renewable fuels and minerals, arable land,
the waste absorption capacity of the environment, or in the most general
terms, time (labor) and money. There are alternative ways to express efficiency ratios such as "unit output per unit input," "unit benefit per
unit cost," or even "unit good per unit bad."[286] The latter ratio indicates
that the purpose of efficiency improvements is to maximize personal

or collective utility in a world of constraints. As the British economist Stanley Jevons observed more than a century ago, "To satisfy our wants to the utmost with the least effort—to procure the greatest amount of what is desirable at the expense of the least that is undesirable—in other words, to maximize pleasure—is the problem of economics."[287] Innovative and efficient technologies are the tools with which achievement of this economic objective is attempted.

A few commonly used measures of efficiency are energy efficiency of the entire economy (gross domestic product (GDP) per total energy use in a given nation), automobile fuel efficiency (distance driven per unit fuel consumed), material use efficiency of the entire economy (GDP per unit material input), efficiency of carbon use of the entire economy (GDP per ton carbon (CO_2) emitted), eco-efficiency (GDP per unit pollution or GDP per unit energy/material input), efficiency of health care (years of life extended per unit medical cost) and efficiency of labor (economic output per hour worked).

Efficiency improvements are sought whenever difficulties arise in obtaining a limited resource (R). For example, fossil-fuel shortages are addressed by improvements in automobile fuel efficiencies. Chemical pollution that overwhelms the waste absorption capacity of the environment prompts the development of more eco-efficient industrial production processes. Limited health insurance budgets promote greater efficiency of medical technologies. The concept of reducing the use of a limited resource (R) by increasing efficiency (e) becomes evident when rearranging the above equation:

$$R = \frac{B}{e} \qquad (5.2)$$

Resource use (R) declines with time only if efficiency (e) improvements outpace the growth in the demand for benefits (B) (i.e., only if e increases faster than B). This, of course, is the case for the most simple situation, in which the demand for benefits is constant via application of some type of economic policy. An increase in efficiency will then translate into a reduction in resource use. For example, consider a policy requiring that houses be heated below a given temperature (i.e., the benefit of heating comfort would be approximately constant). If homeowners were then to insulate their houses to improve the efficiency of their heating system, total energy use for heating would decline. However, given current cultural attitudes, it would probably be very difficult to convince the majority of the population to forego any growth in benefits such as comfort or affluence (per capita GDP) in order to reduce the use of a limited resource.

There is a general demand for "more of everything" (i.e., benefits), ranging from greater affluence to a longer life. This points to the most critical limitation of technological efficiency improvements: given, as will be demonstrated later, that there are inherent technical and thermodynamic limits to efficiency gains (e_{max} = const) and that the growth in demand for all types of benefits is generally open-ended, the overall use of a particular limited resource will not decline with time but will increase. Thus, efficiency improvements alone will not solve the many problems associated with the over-exploitation of natural resources and the pollution of the environment. Moreover, technological innovation in general and efficiency improvements in particular have been key contributors to the rapid growth in material affluence during the past 200 years. It is therefore highly questionable whether efficiency improvements will be able to reduce the use of limited resources unless specific policies are in place that require a reduction in resource use and a stabilization or decrease in the demand for the various benefits.

Technological Progress and Rising Material Affluence

As a result of advances in science and technology, both population size and material affluence have increased substantially during the past two centuries, as indicated by the 100-fold rise in global industrial output since 1750[288] and the 450-fold increase in gross domestic product (GDP) in the United States since 1820.[289] The fact that economic output grew much faster than the number of people resulted in a steep rise in per capita affluence (e.g., per capita GDP increased on average 14-fold between 1820 and 1989 in industrialized nations).[290] These trends are expected to continue.

Continuous development in science and technology during the past 200 years has played a key role in contributing to rising living standards (i.e., per capita GDP) in the industrialized nations.[291] We will examine three aspects of this development: (a) the general nature and drivers of technological innovation, (b) empirical confirmation of the so-called rebound effect, which is often observed in response to efficiency improvements, and (c) factor analysis from neoclassical growth theory.

Technological innovations have increased economic output and per capita affluence because, most fundamentally, modern technologies are nothing more than highly efficient processes designed to convert large quantities of primarily non-renewable energy and mineral resources into a wide variety of products and services while at the same time minimizing the input of human labor. More specifically, advances in science and technology have increased affluence, first, by substituting capital and energy for labor, thereby significantly increasing labor productivity, which translates into rising per capita production and consumption; second, by creating a very large number of new products and services that never before existed, thereby opening new avenues of consumption and generating large demands; and third, by continuously increasing efficiencies, producing greater output with less input of labor, capital, energy and mineral resources, thereby decreasing the costs of goods and services and thus stimulating their consumption.[292]

The observation that efficiency gains do not necessarily decrease the use of limited resources but may rather stimulate their consumption as a result of efficiency-induced price reductions was first made by Stanley Jevons in 1865: "It is wholly a confusion of ideas to suppose that the economical use of fuel is equivalent to diminished consumption. The very contrary is the truth."[293] The phenomenon of efficiency improvements stimulating the consumption of the very resources that were supposed to be conserved is termed the "rebound effect," or the Jevons paradox.

Consider, for example, the hypothetical case in which fuel efficiency for automobiles increases from 15 to 30 miles per gallon, thereby decreasing the fuel costs per mile traveled by 50 percent, assuming constant gasoline prices. An engineering analysis would predict that the 100 percent increase in efficiency would result in exactly a 50 percent reduction in gasoline use. However, the lowered "cost of service" (i.e., cost per mile traveled) would enable people to use some of the saved fuel expenses to drive more, particularly if they had been restrained from doing so before by a limited fuel budget. As a result, total fuel use would "rebound" and some or all of the originally predicted fuel savings would be eliminated through additional consumption. In some cases, technological efficiency improvements may even "backfire," resulting in consumption that exceeds the level prior to the initiation of efficiency measures.

According to Greening and colleagues,[294] there are four categories of market responses to efficiency improvements, specifically (a) direct rebound effects, which are responsible for the increased use of goods and services affected directly by their efficiency-induced price reduction, (b) secondary rebound effects, which lead to increased consumption of other goods and services as a result of real income increases, particularly if the direct rebound effect is relatively small (i.e., a re-spending effect), (c) economy-wide effects, primarily price and quantity readjustments, and (d) transformational effects such as changes in consumer preferences, social institutions and the organization of production.

There has been intense debate during the past 20 years regarding the effectiveness of energy efficiency improvements in reducing total energy consumption: some energy policy specialists claim that rebound and take-back effects are minimal, while others maintain that most, if not all, efficiency gains are lost because of increased consumption.[295] While neoclassical growth theory (see below) predicts rebound and even backfire under certain conditions,[296] it has been extremely difficult to measure the overall effects of energy efficiency improvements on total energy use, a fact that should not be unexpected given the complex set of possible market responses. However, the direct rebound effect may be estimated in some cases,[297] and there is now consensus that it is usually relatively small. For example, the direct rebound effect for fuel-efficient cars[298] as well as for improved space heating technology[299] is about 10 to 30 percent, indicating that efficiency improvements will be between 70 and 90 percent effective in reducing energy consumption in these specific cases.

Although the direct rebound effect is generally small, the secondary rebound effect is likely to be relatively large. In general, the smaller the direct rebound, the greater the secondary, because any savings not spent on the original service will be spent (sooner or later) on other goods and services. This re-spending effect is a major contributor to rising affluence and at the same time poses serious problems for resource conservation. Consider, for example, an individual switching to a more energy-efficient mode of transportation (e.g., from automobile to bicycle). The transportation money thus saved will be re-spent on other consumptive activities, most of which will require direct or indirect (embodied) energy. As Bruce Hannon has shown, unless a consumer is informed and seriously concerned about the energy content of various products and services, he is unlikely to select those with the lowest energy intensity and therefore could unintentionally contribute to increased total energy consumption.[300]

As discussed next, neoclassical growth theory provides additional macroeconomic evidence that developments in science and technology have played a key role in promoting economic expansion. In traditional agricultural societies, economic output (primarily in the form of food products) depended on only two factors of production: land and labor. The use of "capital" such as simple tools and machinery played only a minor role. With the Industrial Revolution, however, the contribution of land declined and that of capital and energy increased substantially. Consequently, neoclassical growth theory generally utilizes production functions in which economic output (Q) depends on capital (K), labor (L) and (sometimes) energy (E).[301] The critical question is how much is technological progress, compared to increases in capital and labor, responsible for the total growth in economic output?

The following fundamental equation of growth accounting has been used by economists to determine how much technological progress, relative to increases in capital and labor, has been responsible for the total growth in economic output:[302]

$$\% \text{ Q growth} = \% \text{ L growth} + \% \text{ K growth} + \text{TC} \qquad (5.3)$$

where TC represents technological change, also called total factor productivity (TFP). The contribution of technological change cannot be measured directly but is determined indirectly as the residual after all other components of output (Q) and input (L, K) are quantified and accounted for. Technological change not only includes the effects of scientific advances and technological innovation but also improvements in the education of the labor force and economies of scale. A number of studies which estimated the contribution of total factor productivity (TFP) to economic growth in the United States during different time periods, found that during times of rapid industrialization (i.e., from 1870–1960), technological change was responsible for about 80 to 90 percent of the observed economic growth.[303]

Based on these three independent lines of evidence, it can be concluded that advances in science and technology have been the primary drivers of increased affluence in industrialized nations during the past two centuries.

Efficiency Improvements and Limited Resources

Science and technology has caused growth in material affluence (B) as well as continuous improvement in efficiencies (e). According to equation 5.2 (R=B/e), the use of limited resources (R) will decline with time only if technological efficiency improvements (e) occur faster than the technology-induced growth in (material) benefits (B). Thus, from the perspective of reducing the use of limited resources, the critical question is whether efficiency improvements increase faster than the demand for benefits. This question can be addressed by analyzing a wide range of historical data to determine whether efficiency improvements have been able to reduce the use of the limited resources they were intended to conserve.

Energy Efficiency and Total Energy Use

The energy efficiency (e_e) of the total economy is defined as gross domestic product (GDP) per total primary energy use (TPEU), i.e., e_e=GDP/TPEU. According to a recent analysis by the International Energy Agency (IEA), the energy efficiency of the total economy increased by almost 50 percent in all IEA nations between 1973 and 2000. Unfortunately, these significant gains in energy efficiency were insufficient to halt or reverse the continuing rise in total primary energy use, which grew more than 36 percent during this period.[304] The reason for the failure of energy efficiency to reverse the growth in energy consumption is that the economy expanded (i.e., GDP grew by 200 percent, far exceeding the 50 percent improvements in efficiency). If the size of the economy of IEA countries had remained constant, total energy use would have decreased by more than 30 percent, resulting in a significant reduction in the use of scarce non-renewable energy resources.[305]

Automobile Fuel Efficiency and Total Automobile Fuel Use

Whenever there is a shortage of crude oil supplies and a concomitant increase in gasoline prices, demand increases for greater fuel efficiency in automobiles. Indeed, following the oil crisis of the 1970s, there were significant efforts in the United States to improve fuel efficiency, resulting in an increase from an average of 15.8 miles per gallon in 1975 to 27.8 miles per gallon in 1991.[306] Unfortunately, greater automobile fuel efficiencies alone will not reduce the global use of gasoline, for several reasons. First, as mentioned in Chapter 2, the number of automobiles worldwide is expected to double to more than a billion within the next 20 years.[307] Second, there are indications that people will drive more and will drive longer distances.[308] In addition, increased ownership of inefficient sport utility vehicles (SUVs) in the United States has caused a decrease in the average fuel economy of the entire US automobile fleet, resulting in increased overall transportation fuel consumption.[309] One of the leaders in US efficiency research and implementation, Amory Lovins, founder of the Rocky Mountain Institute and promoter of the super-efficient, aerodynamic, light, hybrid Hypercars,™ concedes that "Hypercars cannot solve the problem of too many people driving too many miles in too many cars; indeed, they could intensify it, by making driving even more attractive, cheaper, and nearly free per extra mile driven."[310]

Indeed, an analysis of automobile fuel efficiency and fuel-use statistics confirms that improvements in efficiency in the United States, Europe and Japan (IEA countries) did not decrease overall transportation fuel demand. In fact, the situation for automobile fuel use and efficiency is similar to that observed for total primary energy supply and energy efficiency. The fuel efficiency (e_f) for a given automobile fleet is defined as the distance driven (i.e., total passenger kilometers, TPK) per total fuel energy (TFE) use for automobile transportation (i.e., e_f=TPK/TFE). Overall, automobile fuel efficiency improved by 20 percent from 1974 to 1998 in IEA countries but this was accompanied by a more than 40 percent rise in consumption of automobile fuel.[311] The primary reason for the failure of automobile efficiency improvements to reverse the growth in fuel use is that the number of total passenger kilometers increased by almost 75 percent during this period. If the latter had remained constant, the total fuel energy use by automobiles would have declined by 20 percent and would have resulted in a significant conservation of fossil fuels.[312]

Lighting Efficiency and Total Energy Use for Public Lighting

The lighting efficiency (e_l) for illuminating public roads is defined as the amount of lighting service (LS, in lumen-hr) per total energy use for public lighting (TEUL, in GWhr) (i.e., e_l=LS/TEUL). According to a recent study of public lighting in the United Kingdom,[313] the efficiency of lighting (i.e., the lamp efficiency, in lumens/watt) increased almost 19-fold from 1923 to 1996. However, this large efficiency improvement did not reduce the amount of energy used in public lighting, which increased more than 36-fold during the same period because of the almost 700-fold expansion in public lighting. Thus, lamp efficiency improvements did not decrease energy use but instead facilitated the expansion of public lighting, a classic example of the rebound effect.[314]

Efficiency of Materials Use and Total Material Requirements

The efficiency of materials use (e_m) for the total economy is defined as gross domestic product (GDP, in $US) per total material requirements (TMR, in tons) (i.e., e_m=GDP/TMR). According to a detailed analysis of materials use in the United States economy,[315] the efficiency of total material use (e_m) increased by 60 percent from 1975 to 1993. Despite this significant improvement in materials-use efficiency, the total use of materials did not decline during this 18-year period but remained constant as a result of a concomitant increase in GDP by 60 percent. If there had been no economic growth, total material requirements would have been reduced by almost 40 percent during this period.[316]

In a recent analysis of total materials use in the Finnish economy,[317] it was found that the efficiency of materials use increased by 70 percent from 1960 to 2005. Despite this large gain in efficiency, the total material requirements in Finland during this period more than doubled from 109 to 228 million tons. This was because during this 45-year period, the Finnish economy (GDP) grew much faster (255 percent) than the improvements in materials use efficiency (70 percent). These examples illustrate how the use of limited (material) resources increases with time if economic growth outpaces materials use efficiency improvements.

Efficiency of Carbon Use and
Total Atmospheric Carbon Dioxide Emissions

Carbon dioxide (CO_2), an end-product of fossil-fuel combustion, is currently discharged without treatment into the atmosphere, which serves as a sink for this very large and ubiquitous waste stream. Worldwide emissions of CO_2 are currently approximately 7 gigatons per year[318] and are expected to increase substantially unless strong counter-measures are taken.[319] As a result of these anthropogenic carbon emissions, atmospheric CO_2 concentrations have risen 34 percent from pre-industrial levels of 280 ppm to approximately 393 ppm in 2010[320] and are projected to increase to 550-970 ppm by 2100.[321] As discussed in Chapter 2, increasing atmospheric CO_2 concentrations are already producing global climate change, and many negative effects such as severe weather events, sea-level rise and the collapse of local ecosystems will increase both in frequency and scale unless a serious effort is made to improve the efficiency of carbon use in the global economy.

The efficiency of carbon use (e_c) in the total economy is defined as gross domestic product (GDP) per total carbon emitted (CARBON) (i.e., e_c=GDP/CARBON). Despite the steady improvement in the global efficiency of carbon use at a rate of approximately 1.3 percent per year,[322] global carbon emissions continue to rise because worldwide economic expansion (i.e., GWP (gross world product) growth) continues to outpace improvements in the efficiency of carbon use (e_c). For example, despite a more than 60 percent increase in the efficiency of carbon use in the United States from 1980-2002, total carbon emissions have not decreased but, instead, have increased by 20 percent because of the large increase (200%) in GDP that occurred during the same period. Thus, as long as the rate of economic growth outpaces the rate at which technological innovation can improve the efficiency of carbon use, total carbon dioxide emissions to the atmosphere will increase with time, further aggravating global climate change.

Road Improvements and Traffic Congestion

As was discussed in Chapter 2, one of the unintended consequences of the car culture has been widespread traffic congestion, which is exactly the opposite of the freedom of mobility that had been originally promised for the automobile. The most common approach to address traffic congestion has been to improve the "efficiency" of road networks by building new streets and highways with additional lanes. Unfortunately, road improvements never have and never will solve traffic congestion problems, because new roads and lanes accommodate and thus stimulate increased use, another example of the rebound effect. It is tragic that the construction of new roads is still believed to be a solution to traffic problems because it has been known for more than a century that this techno-fix does not work. As early as 1907, a writer in the *Municipal Journal and Engineer* observed that, even though road-widening projects should have relieved traffic congestion, "the result has appeared to be exactly the opposite."[323] In 1916, US president Woodrow Wilson remarked that automobile drivers "use [roads] up almost as fast as we make them."[324]

As Katie Alvord notes in *Divorce Your Car*, the fact that new road construction has been unable to relieve traffic congestion has not changed since the beginning of the automobile age:

> Road building not only failed to keep up with quick growth in car numbers; as paving accelerated, it encouraged more car-buying and driving. As historian John Rae observed, the building of new streets and highways to accommodate cars touched off "a race between road and vehicle that is still in progress, with the vehicle consistently ahead."[325]

Similarly, Jane Holtz Kay summarizes the situation in *Asphalt Nation*: "Each improvement stimulated traffic. ... and demanded more improvements, which brought more traffic, and so on, down to the present and seemingly on to an indefinite date in the future."[326]

Numerous European studies of road building projects intended to ease congestion have conclusively demonstrated that traffic volumes increased after completion of the new roads.[327] For example, one detailed analysis of the British road-building program found that a 50 percent *reduction* in road construction would in fact result in *less* traffic congestion (i.e., *increase* average traffic speeds by about 9 percent).[328] In conclusion, improvements in the efficiency of road networks will not reduce traffic congestion unless there are policies that specifically limit the use of automobiles, a topic addressed in Part III.

Labor-Saving Technology and Number of Hours Worked

As mentioned above, many modern technologies are fundamentally nothing more than highly efficient processes designed to convert large quantities of energy and mineral resources into a wide variety of products and services while at the same time minimizing the input of human labor. In general, the labor-saving efficiency of a given technology could be defined as the number of products or services that are delivered per unit time. One of the main reasons for the introduction of labor-saving technologies is that they promise to reduce human work time and thereby increase leisure time. Unfortunately, as will be seen in the following examples, the introduction of labor- or time-saving technologies has in most cases not resulted in a reduction of work time and an increase in leisure but, instead, has only brought about a large increase in the number of products and services.

The washing machine provides a good example of a household appliance that was supposed to save time spent on laundry. Unfortunately, the average time devoted to laundering clothes using washing machines has remained basically the same[329] or has even increased slightly,[330] because people have many more clothes as well as clothes that consist of different materials that need to be washed separately.[331] As a result of the widespread use of washing machines, social standards for cleanliness have increased dramatically, which in fact has forced people to wash their clothes much more frequently than they did 50 years ago.[332]

Similarly, the vacuum cleaner did not reduce the time needed for household cleaning but instead was used to increase the quality of life for average people who could now afford to have many large carpets, a luxury that until then had been reserved for the wealthy, who employed servants to keep them clean.[333] The introduction of thousands of processed foods and supermarket chains were supposed to save housewives time in cooking.[334] Although the average time spent in preparing food for an urban household decreased by half an hour from the 1920s to the late 1960s, the complexities of modern life, such as longer driving distances to supermarkets, dealing with larger stores and record keeping, canceled these time-saving gains.[335] Consider, for example, that contemporary women spend about one full working day per week on the road and in stores, compared to less than two hours per week for women in the 1920s.[336]

Overall, despite the introduction of many labor-saving home appliances, the full-time housewife today spends about the same number of hours (i.e., 55–60) on housework per week as did women in the early 1900s.[337] By contrast, women employed outside the home have benefited from labor-saving appliances and prepared foods, spending only 26 hours per week on housework.[338] Unfortunately, for a woman who is employed full time (i.e., 40 hours per week), this translates on average to at least 66 hours of work per week, not counting commuting to and from work. Thus today's working women have at least one hour less leisure time per day, not counting commuting time, than their forebears had in the early 1900s. As E. F. Schumacher remarked, "The amount of genuine leisure available in a society is generally in inverse proportion to the amount of labor saving machinery it employs."[339] In summary, labor-saving household technologies have not increased leisure time for working women but instead have enabled them to earn additional income to increase the family's material standard of living.

The automobile is another example of a time-saving technology that has not decreased the time spent on transportation but has increased travel distances instead. The widespread distribution and acceptance of the car has enabled people to take jobs further from home, to buy from shops that are not local and to take part in new and more distant leisure activities. As was discussed in Chapter 2, the automobile has changed the entire transportation and building infrastructure so that people not only have become almost totally dependent on the automobile but also must travel much greater distances to just meet their daily needs such as going to work, shopping and socializing. The observation that the total time spent for traveling has remained about constant while travel distances have increased has led to the formulation of a "law" regarding the constant amount of time devoted to traveling, called "Zahavi's Law" or "Hupke's Constant."[340] In short, the automobile has failed to reduce the overall time spent in meeting personal transportation needs.

Machines have been used since the Industrial Revolution to improve labor productivity in manufacturing, which can be defined as the GDP output per person-hour worked. Gains in labor productivity brought about by technological innovation and efficiency improvements can be used to either increase per capita income (i.e., the material standard of living) or decrease working time, or both.[341] In the United States, the almost 12-fold increase in labor productivity from 1870 to 1989 was converted into a sixfold increase in affluence (per capita GDP) and a reduction in total hours worked by half (i.e., from 2,964 to 1,604 hours per year).[342] In a perfectly free market, workers could choose to split the benefits derived from labor productivity gains into their preferred mix of income and leisure time. Unfortunately, as Juliet Schor has shown, there is currently a structural bias toward increased working hours, which is due to various employer incentives.[343] Given that US labor productivity has more than doubled since 1948, we could produce today a 1948 standard of living, which was generally acceptable to our parents and grandparents, by working only four hours a day.[344] Instead, because of structural biases in the labor market and the constant promotion of consumption by the mass media, Americans now work 200 hours more per year than 30 years ago.[345] In short, labor- and time-saving technological innovations have not been used to reduce the annual work load or increase leisure time for US citizens, but were employed instead to increase material affluence (per capita GDP). As Lewis Mumford observed more than 60 years ago in *Technics and Civilization*,

> The justification of labor-saving devices was not that they actually saved labor but that they increased consumption. Whereas, plainly, labor-saving can take place only when the standard of consumption remains relatively stable, so that increases in conversion and in productive facility will be realized in the form of actual increments of leisure. Unfortunately, the capitalist industrial system thrives by a denial of this condition. It thrives on stimulating wants rather than by limiting them and satisfying them. To acknowledge a goal of consummation would be to place a brake upon production and to lessen the opportunities for profit.[346]

Medical Progress and Health Care Costs

Although there is no generally accepted definition of medical progress, one could define the overall efficiency of medicine or the health care system as the average number of years that life is prolonged per unit of health care expenses paid. As was mentioned in Chapter 2, total health care spending in the United States was $2 trillion, or 16 percent of GDP, in 2005, and is expected to reach $4 trillion, or 20 percent of GDP, by 2015.[347] A common approach to addressing escalating costs is to call for improvements in efficiency, such as streamlining the health insurance system or developing new technologies that promise to save time, labor and costs. Unfortunately, as has been observed with many other efficiency solutions, technological innovation in general and efficiency improvements in particular are unlikely to reverse the escalation of health care costs and may even aggravate the situation. Consider, for example, that the widespread application of high technology in medicine is believed to be responsible for 50 to 85 percent of the growth in health care costs.[348] There are a number of reasons why progress in medical science and technology will increase rather than decrease health care costs. These are (a) the greater availability and accessibility of medical tests and treatments due to efficiency-induced cost reductions (i.e., the rebound effect), (b) the never-ending hope for new cures, which, if successful, become a permanent need and (c) the goal of prolonging life as long as possible, no matter what the costs.

As is the case with most efficiency improvements, greater efficiency generally translates into reduced costs for various goods or services, thereby stimulating their consumption. Thus, even if technological innovations make certain diagnostic tests or treatment procedures more efficient and cost-effective, the overall costs involved will not necessarily decrease because the tests and procedures may be employed more often because they are now cheaper to administer and therefore more accessible.[349] This is a typical example of the rebound effect, in which efficiency-induced cost reductions stimulate the consumption of those goods and services that are produced more efficiently.

As is the case with most technological innovations, new medical products and treatments are introduced to the market to encourage their "consumption," thereby increasing the income and profits of the medical establishment. Indeed, making new diagnostic tests, drugs and high-tech treatments available to various patient populations is one of the best ways to compete successfully in a nation's health care market. The medical enterprise continuously stimulates the desire for new high-technology cures, thereby increasing demand for them, and after these innovative treatments have become commonplace, the original desire for them becomes transformed into a permanent need. As Daniel Callahan points out in *What Kind of Life: The Limits of Medical Progress*,

> Our aspirations and hopes are at first made credible by medical progress, then, as they move closer to realization, they come to have the status, and the insistence, ordinarily accorded to need. This is the transformation of desire or aspiration into putative need. No one thought, a century ago, that a person suffering from heart disease "needed" a heart transplant. Death was simply accepted. But the advent of heart transplants was stimulated first by hope, and that hope became concrete need as transplantation succeeded. People now "need" heart transplants. A whole new category of need has been created, a need originally thought to encompass only younger people but now coming to include older patients as well.... The line between need and desire or aspiration is, therefore, not only increasingly in principle indistinguishable, but it seems clearly the case that the best way to stimulate more desire is to meet real needs. That makes certain that the desires not yet met will be defined as needs still to be met—and so on, infinitely. Desires beget needs, and needs met beget further needs.[350]

In addition to escalating health care costs stimulated by desires for new cures and desires becoming needs, the situation is further aggravated by the fact that one of the implicit goals of the medical-industrial complex is to conquer death, if not permanently, then as long as is technically possible. This unrealistic, open-ended crusade against death creates a situation in which demand for medical treatment is infinite, in which patients who have been cured of one disease will need treatment for the next one, one that inevitably follows. As Callahan notes,

> As some problems are solved there will then be others to take their place.... The offices of doctors will always be full, no matter how much progress is made. There will always be those on the edge of that progress—an office full of different patients than, say, a century ago, but still, and always, a full office. There will always be death, pain, and suffering, and there will always be a medical frontier.... When will we know when and how to stop?

Not when and how to stop because further progress cannot be made—further progress can always be made; we have no reason to disbelieve that. But knowing how and when to stop because further progress entails either too great an economic or social price or too little likely improvement in the human condition, or both, is a far harder decision.[351]

He concludes that efficiency improvements alone will never solve the problem of rising health care costs as long as there is an open-ended crusade against death and thus an unlimited demand for medical treatment:

> There is a hard philosophical truth at which we have avoided looking, one that must be radically disquieting for any hopeful beliefs about the possibility of some ultimate efficiency. It is simply the burden of mortality: Illness, decline, aging, and death can only be forestalled, kept at bay, never permanently vanquished.... Consider the implications of the reality of mortality for cost containment. One implication is that, as the easier causes of disease and death are understood and managed, the costs of dealing with those that are more complex and intractable will and must grow, at least if we insist upon moving ahead with a continued high sense of urgency to find cures for those that remain. Efficiency can only be relative in that context and costs must increase. At the same time, as we well know, the price of the progress can be an increased prevalence of chronic illness, the result of slowing but not vanquishing decline and death.... The more successful that progress, the more of it we want, and the more of it we want, the more economically insupportable it becomes.[352]

In conclusion, unless old age and death are accepted as natural and not something to be fought with the latest arsenal of high-technology medical treatments, efficiency improvements will not be successful in stemming the tide of rising health care costs. The same overall conclusions apply to all other efficiency improvements presented in this chapter: gains in efficiency have not been able to reverse the use of limited resources because no efforts have been made to limit growth, affluence or demand. As will be discussed in more detail in Part III, technological innovation and efficiency improvements will be successful in reducing the use of limited resources and in promoting sustainability only if they are accompanied by a fundamental change in values and worldview; that is, we must accept that unlimited economic and material growth are neither possible nor desirable.

Inherent Limits to Efficiency Improvements

The techno-optimist will counter the suggestion that economic growth and material affluence should be limited by claiming that technological efficiency improvements only need to be accelerated in order to outpace the growing demand for various economic and material benefits, thereby assuring that the use of limited resources will be reduced. (See equation 5.2.) Historical evidence has clearly demonstrated that in most cases efficiency improvements have not kept pace with economic growth and demand, but rather, as mentioned above, have led to increased use of limited resources. Theoretically, it might be possible, through heavy investment in research and development, to accelerate the rate at which efficiency improvements are made; however, there are inherent thermodynamic and practical limits to all efficiency improvements. Consequently, after maximum efficiency has been achieved, economic and material growth cannot continue without further depleting limited energy, mineral and environmental resources, thereby threatening the sustainability of industrialized societies and our planet.

The conversion of energy and mineral resources into commercial products and services often involves many sequential processes, each of which has a maximum achievable efficiency. For example, petroleum extracted from a reservoir (primary energy) must be converted into gasoline (secondary energy), which is transformed within an automobile's combustion engine to provide useful work (available energy) in the form of locomotion (service). According to the second law of thermodynamic (see below), there will always be energy loss in each conversion step, because achieving 100 percent efficiency is not only practically but theoretically impossible. For example, the refining of crude oil into gasoline requires energy that is irreversibly lost as heat. The conversion of the energy embodied in gasoline into useful work within a combustion engine is always less than 100 percent because of entropy constraints (see below). The conversion of useful work into locomotion is subject to various physical constraints such as friction between tires and the road as well as air resistance, both of which dissipate useful energy as heat.

The second law of thermodynamics can be used to determine the maximum energy conversion efficiency that is theoretically achievable for various processes. According to the second law, it is impossible to have an engine that converts heat energy into useful energy (work) with 100 percent efficiency because some of the energy must be released into the surroundings as waste heat.[353] While the energy conversion efficiency of an idealized heat engine (Carnot cycle) approaches unity at infinitely high engine temperatures, this maximum efficiency cannot be reached in practice because there are no materials that could withstand the necessary high temperatures.

The efficiency of converting primary energy to useful energy, also called supply-side efficiency, is currently around 37 percent at the global level,[354] and the potential for increasing it substantially is very small. For example, according to a theoretical analysis by Eberhardt Jochem,[355] the global average supply-side efficiency could be increased twofold at most. Further improvements are thermodynamically impossible because Carnot efficiency limits would have been reached.[356]

There may be greater potential for increasing so-called end-use efficiencies, which are the efficiencies of processes that convert useful energy into various energy services such as moving a vehicle, operating a machine, warming or cooling a room or providing space illumination. Indeed, some techno-optimists believe that end-use efficiencies could be improved at least 10-fold.[357] While this may be true in selected cases for a single innovation (e.g., LED lighting), it is unlikely that the efficiency of the many energy services that are needed to conduct the majority of economic activities such as transportation, manufacturing, agriculture, construction and heating/cooling can be improved more than two- or threefold.[358] In summary, overall energy conversion efficiencies, which are the product of supply-side and end-use efficiencies, can realistically be increased by at most fivefold, and only if there is massive investment in research and development.[359]

Similarly, there are inherent limits to increasing the efficiency of materials use. While some techno-optimists (e.g., those belonging to the "Factor 10 Club") believe it is possible to achieve the same level of material affluence (per capita GDP) with ten times less material input,[360] there are serious doubts that this can be accomplished for most products and services. For example, there are certain minimum unavoidable material requirements for growing food, and there are limits to making automobiles, houses, furniture and other products lighter and thinner without the risk of losing their necessary functions.[361] Ecological economist Herman Daly points out the obvious: that one cannot indefinitely "angelize" the economy.[362] Similarly, Cutler Cleveland and Matthias Ruth warn,

> There is no compelling macroeconomic evidence that the US economy is "decoupled" from material inputs.... We caution against gross generalizations about material use, particularly the "gut" feeling that technical change, substitution, and a shift to the "information age" inexorably lead to decreased materials intensity and reduced environmental impact....Claims that a substantial decoupling of economic production from material input has occurred or is feasible (Factor 10 Club) should be viewed for what they are: assertions that currently have little convincing empirical support.[363]

As is the case for energy and materials-use efficiencies, there are limits to the efficiency of labor, or labor productivity. While labor productivity has been and probably can be further improved in manufacturing and agriculture through the application of additional automation as well as information technology, it is unlikely that the efficiency of labor can be increased in any profound way for all those jobs that cannot be performed by robots or computers and therefore require the presence of human beings. All jobs in the so-called service sector fall into this category, such as the restaurant and hotel business, transportation, retailing, landscaping, etc. In addition, it will be impossible to replace professionals working in medicine, law, research and education with machines. In conclusion, there are inherent thermodynamic and practical limits to all efficiency improvements. As a result, it is impossible to pursue the goal of unlimited economic growth and affluence without further depleting already limited energy and material resources. The continued pursuit of such an illusory goal may well lead to environmental collapse.

Unintended Consequences of Efficiency Solutions

As mentioned in the discussion of the rebound effect (the Jevons paradox), one unintended consequence of implementing efficiency improvements is that it often stimulates the consumption of the same limited resource that was supposed to be conserved. In addition, negative consequences are likely to arise if the concept of efficiency gains the status of a dominant social paradigm so that efficiency solutions are sought to the exclusion of potential non-technical solutions.

A society in which most processes have become ultra-efficient is much more vulnerable to resource shortages. Given that in such a society, efficiency (e) would be very close to its maximum value and therefore essentially constant, any unexpected change in the supply of energy or material resources (R) would, according to equation (5.1) above, immediately result in a concomitant decrease in benefits (B). After the oil crisis of the 1970s, the response to the sudden decline in crude oil availability was the increase in technological efficiencies to compensate for the otherwise inevitable decline in material benefits (i.e., per capita GDP); that is, the detrimental effects of sudden fuel limitations were buffered by both short- and long-term increases in energy efficiency. This kind of response would not be possible if processes were operating close to maximum efficiency already. Consequently, a maximally efficient society is likely to become very vulnerable to resource shortages because these would translate directly and immediately into painful reductions in material living standards. In addition, because of the increasing need for continued technological innovation to maintain an ever-increasing efficiency, dependence on scientists and engineers would increase but would, at some point, outstrip their ability to produce as well as the society's ability to fund the expanding enterprise.

Another problem is that efficiencies may become ends in themselves. Instrumental values are transformed into ultimate values, and no time is spent critically examining alternative goals and objectives. As Albert Einstein stated, "Perfection of means and confusion of goals seem to characterize our age."[364] Indeed, the confusion is so profound that many readily accept scientific and technical efficiency as a worthwhile ultimate goal, only to discover later—often when it is too late—that it is in serious conflict with many other highly cherished personal and cultural values. As Langdon Winner points out in *Autonomous Technol-*

ogy, we are confronted here with the strange phenomenon of "reverse adaptation," which is

> the adjustment of human ends to match the character of avail-
> able means.... People come to accept the norms and standards
> of technical processes as central to their lives as a whole.... Ef-
> ficiency, speed, precise measurement, rationality, productivity,
> and technical improvements become ends in themselves ap-
> plied obsessively to areas of life in which they would previously
> have been rejected as inappropriate.... Reverse adaptation is an
> interesting situation which is the exact opposite of the idealized
> relationship of means and ends.... Ends are adapted to suit the
> means available (instead of the reverse).... Reverse-adapted sys-
> tems represent the most flagrant violation of rationality.[365]

Widespread acceptance of efficiency as an ultimate value, despite the fact that it may create a society in which previously held values are destroyed, may be due to the fact that many believe that the pursuit of efficiency is a rational, objective and value-neutral enterprise. Unfor-tunately, as will be discussed in Chapter 10, technology in general and efficiency in particular are never value neutral. They embody specific values that are often not perceived by the uncritical observer. The very definition of efficiency is value-laden. Efficiency optimizes that which is considered valuable (benefits) while minimizing that which is consid-ered not valuable (costs). Since the power of technology is based on the manipulation and control of material resources, most technical efficien-cies optimize quantifiable material benefits such as more "miles per gallon," more "lighting per kilowatt-hour electricity," more "consumer products per unit energy and material inputs," more "material products and services per hours worked"—in short, more material affluence with less input of energy, material and labor. Thus, by continually focusing on the optimization of technical efficiencies, materialistic values and goals are strengthened while non-material values of every sort are neglected. Emphasis on the technological achievement of materialistic objectives is in conflict with many cherished personal and cultural values and, in fact, often destroys what is most enjoyable in life.

There are a number of reasons why excessive focus on efficiency im-provements may degrade quality of life. First, technological innovation and increased efficiency may translate directly into the increased exploitation of workers and destruction of the environment, often with disastrous consequences for both. For example, the conditions of factory workers worsened considerably after the invention of the assembly line, which forced workers to perform stupefying, repetitive tasks at high speed under inhumane conditions and which eliminated all traces of individuality and creativity, all in the name of increasing the efficiency of manufacturing.

Second, technological efficiency has a built-in bias toward the quantifiable and material. As a result, anything that is tangible and quantifiable is generally overvalued and subjected to technological improvements, while anything that is intangible and unquantifiable tends to be undervalued and ignored. Thus, previously cherished values such as fairness, equity, freedom, creativity, faith and aesthetics are neglected and wither. As Kenneth Stunkel and Saliba Sarsar observe,

> All technologies, whether structural, mechanical, electrical, or chemical, measure their objects. They seek quantitative standards of efficiency, an unmistakable hallmark of technological rationality. Manipulation and control are easier with quantification, an objective clearly in tension with the non-quantitative standards of ethics, religion, or art, where feeling, tradition, self-understanding, or other kinds of non-objective judgments predominate.... The goal of all technique is to find the most efficient means of dealing with a problem and sweep away all merely "subjective" alternatives.[366]

Third, because efficiency improvements depend heavily on pure logic and rational problem solving, anything related to the subjective is by definition ignored, creating an inhumane world devoid of love and empathy. This sentiment is expressed by Andrew Kimbrell:

> In daily life we treat nothing that we love or care about solely or even primarily on the basis of efficiency, but on the basis of empathy. Whether child, spouse, friend, or pet, love—not efficiency—is the primary mover. Yet this realization has not made its way into our public policies...or towards the rest of the natural world.[367]

Life in industrialized societies is unlikely to be very enjoyable as long efficiency remains a dominant organizing principle. As William McDonough and Michael Braungart observe,

> Efficiency isn't much fun. In a world dominated by efficiency, each development would serve only narrow and practical purposes. Beauty, creativity, fantasy, enjoyment, inspiration, and poetry would fall by the wayside, creating an unappealing world indeed.... An efficient world is not one we envision as delightful.[368]

How can we avoid an unpleasant future where efficiency has become a dominant organizing principle? The obvious solution is to reverse the current situation in which technical means dictate societal ends (i.e., the problem of "reverse adaptation") by first defining and setting goals and objectives (ends) and then employing science and technology (means) to meet them efficiently. It is most important to carefully choose the goals because it is counterproductive and even dangerous, in the words of Paul and Anne Ehrlich, to do "efficiently that which should never be done at all."[369] Similarly, as Herman Daly warns, "Indeed, if our ends are perversely ordered, then it is better that we should be inefficient in allocating means to their service."[370] Or, as McDonough and Braungart point out, the outcome of efficiency "depends on the value of the larger system of which it is a part. An efficient Nazi, for example, is a terrifying thing. If the aims are questionable, efficiency may even make destruction more insidious."[371]

Efficiency improvements, by themselves, can become counterproductive. What is needed is a conscious effort to direct technological innovation toward the achievement of clearly defined societal goals that reflect shared values. For example, if we strive for a humane and sustainable world, we should first clearly define goals such as agreeable working conditions, adequate leisure time, material sufficiency, psychological well-being, minimum pollution, a stabilization or reduction in the use of limited energy and material resources and so forth. Only after specific goals have been set should we employ the most innovative science and technology to meet them in the most efficient manner. As will be discussed in Chapter 12, the critical examination of our core societal values and the setting of specific goals for the achievement of a humane and sustainable world will be quite a challenge to the status quo, which currently employs science and technology to accomplish rather different objectives, such as maximal exploitation of both people and the environment. Indeed, such a change would require a major paradigm shift, a revolution in worldview. Unless we undertake this critical challenge, technological innovation and efficiency improvements will continue to promote unsustainable growth, which will inexorably lead to environmental and societal collapse.

CHAPTER 6

Sustainability or Collapse?

P RIOR TO THE AGRICULTURAL REVOLUTION, humans had lived for
millennia under circumstances approaching steady-state conditions; that is, both population size and material affluence were more
or less constant. As was discussed in Chapter 2, the invention of agriculture led to an increase in food production which, in turn, resulted
in the slow but steady growth of the human population. In addition, because farmers have a much more sedentary lifestyle than their hunter-
gatherer ancestors, it became possible, for the first time, to increase one's
level of material affluence by accumulating property.

Because food production using prehistoric farming methods depended solely on the amount of sunlight that could be captured by
plants cultivated on arable land, the growth of agricultural societies
was ultimately limited by the availability of solar energy. However, this
situation changed radically with the Industrial Revolution, when not
only new sources of fossil energy became available but new machines
(primarily the steam and internal combustion engines) were invented
that could convert fossil fuel energy into useful work. This unique combination of increased energy availability and technological innovation
has been the primary reason for the increase in human population and
material affluence (per capita GDP) during the past 200 years. As industrial ecologist Braden Allenby observes,

It is seldom recognized how substantially economic activity has grown since the industrial revolution. Since 1700, the volume of goods traded internationally has increased some 800 times. In the last 100 years, the world's industrial production has increased more than 100 fold. In the early 1900s, production of synthetic organic chemicals was minimal; today, more than 225 billion pounds of synthetic organic chemicals are produced in the US alone per year.... Since 1900, the consumption of fossil fuel has increased by a factor of 50.[372]

As was discussed in the previous chapter, per capita GDP increased on average 14-fold between 1820 and 1989 in industrialized nations."[373] This trend is expected to continue. According to various development scenarios published in the Third Assessment Report of the Intergovernmental Panel on Climate Change,[374] gross world product (GWP) is projected to increase 12- to 26-fold by 2100 (relative to 1990 values), and per capita affluence (GWP/person) 4- to 19-fold during this period. It is clear that such massive expansion in economic production and consumption will be a serious threat to environmental sustainability. An obvious way to avoid environmental and societal collapse would be to limit economic growth and transition to a steady-state economy (see also Chapter 12). Unfortunately, this does not seem possible in the current political climate, in which the world's nations are committed, often with the fervor of fundamentalists, to economic growth as the solution to most social, political, cultural and even environmental problems. Because of the pervasiveness of this belief, it has become almost heretical to question the doctrine of unlimited economic growth. This situation is further aggravated by the fact that mainstream "neoclassical" economists do not believe that the economy depends on material and energy inputs from the environment. In the words of David Korten,

> Take the case of the obvious reality that the human economy is embedded in and dependent on the natural environment. As far back as 1798, Thomas Robert Malthus suggested that environmental limits might make population growth a problem for the future of humanity. Neoclasscial economists have dealt with this inconvenience by adopting a model of analysis that assumes

that economies consist of isolated, wholly self-contained, circular flows of exchange values (labor, capital, and goods) between firms and households without reference to the environment. In other words, their model assumes away the existence of the environment. Then, in defiance of logic, they use this model to prove that the environment is of little importance to the function of the economy. Those who challenge the possibility of infinite growth on a finite planet are dismissed with the stinging epithet "neo-Malthusianism."[375]

The critical question, in the context of this chapter, is whether continued economic growth is environmentally sustainable and whether science and technology can prevent environmental and societal collapse.

Sustainable Development and Eco-Efficiency

The term "sustainable development" has received unprecedented popularity ever since it was first introduced at the World Commission on Environment and Development more than 20 years ago. The Brundtland report claimed that sustainable development would "ensure that humanity meets the needs of the present without compromising the ability of future generations to meet their own needs."[376] Unfortunately, in order to reach an agreement among widely disparate parties, the concept of sustainable development was kept deliberately vague and inherently self-contradictory, which resulted in a situation such that, in the words of ecologist David Reid, "endless streams of academics and diplomats could spend many comfortable hours trying to define it without success."[377]

Within a few years after publication of the Brundtland report, however, the international business community designed a more concrete definition of the ambiguous term. They envisioned sustainable development as a combination of continued economic growth and environmental protection. For example, the International Chamber of Commerce stated that "sustainable development combines environmental protection with economic growth and development."[378] Similarly, the Business Council for Sustainable Development, under the leadership of the Swiss millionaire industrialist Stephan Schmidheiny, agreed that "sustainable development combines the objectives of growth with environmental protection for a better future."[379] The European Organization for Economic Cooperation and Development (OECD) also issued a major report on eco-efficiency in an effort to promote "sustainable" economic growth.[380]

To ensure that continued economic growth and environmental protection can go hand in hand, business leaders promoted the concept of "eco-efficiency" as the primary tool for achieving industrial sustainability. In fact, the Business Council for Sustainable Development originally coined the term eco-efficiency and defined it as "adding maximum value with minimum resource use and minimum pollution."[381] Or, more specifically, in the words of Stephan Schmidheiny, "Corporations that achieve ever more efficiency while preventing pollution through good housekeeping, materials substitution, cleaner technologies, and cleaner products and that strive for more efficient use and recovery of resources can be called 'eco-efficient'."[382] The Business Council for Sustainable Development also suggested that eco-efficiency should be the main corporate response to the goal of sustainable development:[383] "Economic growth in all parts of the world is essential to improve the livelihood of the poor, to sustain growing populations, and eventually to stabilize population levels. New technologies will be needed to permit growth while using energy and other resources more efficiently and producing less pollution."[384]

In short, the complex societal challenge of sustainable development was reduced to the purely technical problem of improving industrial eco-efficiency or "producing more with less."[385] While this approach clearly results in temporary win-win solutions for both industry and environment, it is highly questionable whether technological solutions alone are sufficient to guarantee the long-term sustainability of industrialized societies.[386]

To establish that technology has indeed the potential to assure sustainability and prevent environmental and societal collapse, it must be shown that progress in science and technology will result in a reduction of total environmental impact (I), as expressed either in terms of resource use or pollution. According to the well-known IPAT equation,[387] total impact may be expressed as

$$I = PAT = \frac{PA}{e_{eco}} = \frac{GDP}{e_{eco}} \qquad (6.1)$$

where P is the population size (number of people, or more correctly, number of consumers), A is affluence (per capita GDP) and T is the impact of technology (resource use or pollution per GDP). Note that the technology factor (T) is the inverse of eco-efficiency (e_{eco} is the GDP per resource use or pollution), indicating that increases in eco-efficiency will result in a reduction of the technology factor (T) and thus total environmental impact (I). As the first textbook on industrial ecology, the new discipline of sustainability science and engineering, states "It is the third term (T) in the equation that offers the greatest hope for a transition to sustainable development, and it is modifying this term that is the central tenet of industrial ecology."[388]

While there is hope in promoting sustainable development with the help of science and technology, there are limits to the degree to which eco-efficiencies can be improved. As was discussed in Chapter 5, because of thermodynamic and practical constraints, overall energy conversion efficiencies can realistically be improved at most by about fivefold, and only if there are massive investments in research and development to achieve this ambitious goal. Similarly, because there are minimum unavoidable material inputs to many necessary products and services, there are limits to the extent to which the economy can be dematerialized.[389] According to an analysis by leading eco-efficiency authorities (i.e., members of the "Factor 4 Club"), it may be possible to reduce the use of materials by a factor of two while at the same time doubling economic growth; that is, eco-efficiencies possibly could be improved fourfold.[390] Finally, because of constraints imposed by the physical law of entropy (i.e., the second law of thermodynamics), it is impossible to have industrial processes that have zero environmental impact; in other words, the T-factor can never become zero.

The fact that there are clear limits to the extent to which eco-efficiencies can be increased, or, equivalently, the T-factor can be decreased, has serious implications for sustainable development, also called "sustainable economic growth." Consider, for example, that if economic growth were to continue for 100 years at a rate of 2 to 3 percent per year, which is the implicit goal of most developed and developing nations and a target recommended earlier by the President's Council on Sustainable Development,[391] the size of the economy (GDP) would increase 12- to 19-fold, respectively. Note that this is also very close to the estimates of the Intergovernmental Panel on Climate Change (IPCC) in which, depending on scenario conditions, gross world product (GWP) is projected to increase 12- to 26-fold by 2100 relative to 1990 values.[392] Under these circumstances, the growth of the economy (GDP) will be much greater than the maximum technically achievable gains in eco-efficiency (e_{eco}), which according to equation (6.1) above, translates into a significant increase in total environmental impact (I). For example, if the size of the world economy (GWP) were to expand 20-fold by 2100 but eco-efficiency increased only fivefold during the same period, then total environmental impact (I), measured as resource use or pollution, would increase fourfold. This would continue the historical trends presented in Chapter 5 in which economic growth has almost always outpaced efficiency gains, resulting in increased energy and materials use and environmental pollution.

In conclusion, because of the inherent thermodynamic and practical limits of any and all eco-efficiency improvements, continued economic growth into the indefinite future will result in increased resource use and pollution, a situation that clearly is not sustainable and, in fact, will accelerate global collapse.

Three Conditions for Long-term Sustainability

If gradual improvements in eco-efficiency are unable to ensure the long-term sustainability of growth-oriented industrial economies, it could be argued that a more radical approach, such as a complete redesign of industrial processes and economic activities, could allow for continued "sustainable economic growth." A closer analysis shows that industrial societies are intrinsically unsustainable at present for at least three major reasons. First, more than 85 percent of all energy used to carry out industrial and economic activities is derived from non-renewable fossil and nuclear sources.[393] Second, many raw materials (e.g., metals, minerals and petroleum) that are crucially essential to the functioning of modern industrial societies are non-renewable. Given the limited supply of non-renewable energy and material resources, it is clear that their continued use is inherently unsustainable. Third, industrial processes discharge wastes at rates exceeding the assimilative capacity of the environment, resulting in various negative unintended health and environmental consequences. In short, industrial societies are unsustainable because they are "linear-flow" economies that convert non-renewable energy and non-renewable materials into products and services that, during production and after use, generate waste that is discharged into the environment.

Industrial agriculture provides a good example of the inherent unsustainability of modern farming practices. Industrial agriculture is almost completely dependent on non-renewable fossil fuels. Prior to the industrialization of agriculture, no external energy sources other than human and animal labor were required for food production. Indeed, the main purpose of agriculture is, or should be, to use plants for the capture of free solar energy and its subsequent conversion into edible foods via the process of photosynthesis. As a result, pre-industrial agriculture was extremely energy efficient: for each calorie of human or animal energy invested, between 5 and 50 food calories were harvested.[394] As fossil fuels became readily available during the past 100 years, machines have replaced human and animal labor; synthetic fertilizers, pesticides and herbicides have been manufactured from petroleum; and highly processed foods are now often transported from farms to consumers over large distances, frequently more than 1,000 miles.[395] As a result, industrialized agriculture has become extremely energy inefficient: it requires about five to ten calories of fossil fuel energy to obtain one food calorie.[396] As E.F. Schumacher commented, "Industrial man no longer eats potatoes made from solar energy, he now eats potatoes made partly of oil."[397] Considering that fossil energy resources are limited and expected to be depleted in the near future,[398] industrial agriculture is clearly not sustainable. In fact, a decline in global oil production could lead to increased food insecurity worldwide, particularly in developed nations.[399]

Furthermore, as described in Chapter 2, industrial farming often depletes renewable resources such as fresh water supplies and topsoil at rates faster than they are regenerated, a condition that is clearly unsustainable. The use of phosphate fertilizer is also unsustainable because it is obtained from limited, non-renewable supplies of phosphate rock.[400] Finally, industrial agriculture depends on the large-scale application of synthetic herbicides and pesticides, chemicals that are discharged into the environment with the sole purpose of killing certain plants and insects, thereby interfering with natural ecosystem functioning.

If current industrial and economic activities are inherently unsustainable, they could in principle be redesigned by discontinuing current practices and by following three basic conditions for achieving "strong" sustainability,[401] which were automatically satisfied in most pre-industrial societies:[402]

1. Sustainable Energy Generation Condition

All energy used for industrial and economic activities must be supplied from renewable sources at rates that do not cause disruptive environmental side effects. If renewable energy is obtained from biomass, its harvest must not exceed the regenerative capacity of the respective ecosystem (sustainable yield criterion).

2. Sustainable Materials Use Condition

a. All raw materials used in industrial processes must be supplied from renewable resources at rates that do not exceed the regenerative capacity of the respective ecosystems (sustainable yield criterion) and do not cause any other environmental disruptions.

b. The dissipative use of non-renewable materials must be discontinued, or at least greatly minimized, either by finding renewable substitutes or by recycling them to the greatest extent possible.

3. Sustainable Waste Discharge Condition

Wastes can be released into the environment only at a rate compatible with the assimilation capacity of the respective ecosystems, and without negative impacts on biodiversity.

If these three sustainability conditions were to be satisfied, the current linear flow economy would become a circular flow economy in which, like the elemental cycles found in nature, most material flows would be driven by renewable solar energy, and wastes would be recycled to become inputs for production processes.[403] Because of this extensive recycling, also called "closing the materials cycle" by industrial ecologists,[404] very few wastes would be released into the environment. In addition, because most raw materials would be derived from renewable biomass-based sources, their disposal would not cause serious environmental consequences as they could readily be assimilated via natural biodegradation and other weathering processes. Finally, to satisfy sustainability condition 3, the manufacture of synthetic chemicals (such as plastics), which do not readily degrade into harmless substances after their release into the environment, would have to be totally prohibited.

Transitioning to a circular flow economy, in which all three sustainability conditions are satisfied, would present a number of significant challenges.

Challenge #1: Serious Environmental Impacts of Large-Scale Renewable Energy Generation

Renewable energy is derived, directly or indirectly, from the continuous flow of energy from the sun. Renewable energy technologies include direct solar heating, solar thermal, photovoltaics, wind power, hydroelectric power and biomass conversion.[405] Given that long-term technical potentials for renewable energy (i.e., >4,200 EJ/yr*) exceed current worldwide energy consumption (i.e., 351 EJ/yr) by at least one order of magnitude, there is no doubt that renewable solar energy should in principle be able to supply present energy demands.[406] The primary reason that renewable energy sources have not been more widely used is their high cost relative to fossil fuels. However, renewable energy technologies could easily become commercially competitive in the near future with more aggressive investments in research and development, especially if they were subsidized as heavily as fossil fuels. The ratio of subsidies given for renewables versus non-renewables is on average 1:10, and it can be as high as 1:35 in some nations such as Germany.[407] Thus, the barriers to a rapid transition from fossil fuels to renewables are as much economic and political as technical in nature.

It is generally assumed that energy generation from renewables is more environmentally friendly than that from non-renewables energy sources such as fossil fuels or nuclear power.[408] While this assumption may be correct, it must be realized that the capture and conversion of solar energy can also have significant negative environmental impacts, especially if solar renewables are employed on a scale large enough to supply the growing demand for CO_2 emission-free power.[409] Consider, for example, that primary energy use is expected to increase about fivefold by 2100 relative to 1990 levels in most scenarios evaluated by the Intergovernmental Panel on Climate Change.[410]

* EJ = exa-Joule, or 10^{18} Joules.

Before discussing some of the potential negative impacts of various solar energy technologies, it is useful to review the implications of the second law of thermodynamics to show that environmental impacts of renewable energy generation are inherently unavoidable. Impacts are unavoidable because the flux of solar energy (neg-entropy) onto Earth is used to create highly ordered (low-entropy) "dissipative structures" in the environment.[411] Such structures include complex organisms, biodiversity in general, ecosystems, and carbon, nitrogen and other cycles, all of which are maintained by the constant inflow of solar energy.[412]

If the flow of solar energy were to cease, all these complex structures would decay and reach a final equilibrium state of maximum entropy. Similarly, if humans divert a fraction of solar energy away from the environment to create ordered structures for their own purposes (e.g., houses, appliances, transportation infrastructure, communication systems), less energy is available to maintain highly ordered structures in nature. The disturbance of these structures translates into the various environmental impacts that are associated with renewable energy generation.

Thus, the second law of thermodynamics guarantees that it is impossible to avoid negative environmental impacts (disorder) when diverting solar energy for human purposes. This should not be surprising given the important and numerous roles that solar-based energy flows play in the environment.[413] For example, direct solar energy radiation is responsible for the heating of land masses and oceans, the evaporation of water and therefore the functioning of the entire climatic system. Wind, which is the result of the differential heating of air masses by the sun, transports heat, water, dust, pollen and seeds. Rivers, which are part of the hydrogeological cycle driven by the sun, are responsible for oxygenation, nutrient and organism transport, erosion and sedimentation. The capture of solar energy via photosynthesis results in biomass that provides the primary energy source for all living matter and therefore plays a fundamental role in the maintenance of ecosystems.[414]

According to John Holdren, these potential environmental problems can be summarized as follows:

> Many of the potentially harnessable natural energy flows and stocks themselves play crucial roles in shaping environmental

conditions: sunlight, wind, ocean heat, and the hydrologic cycle are the central ingredients of climate; and biomass is not merely a potential fuel for civilization but the actual fuel of the entire biosphere. Clearly, large enough interventions in these natural energy flows and stocks can have immediate and adverse effects on environmental services essential to human well-being.[415]

In conclusion, the potential environmental impacts of large-scale biomass energy production, solar thermal plants, photovoltaic electricity generation as well as wind-energy capture and hydroelectric power, though presently not as obvious as those caused by fossil fuel use, may be significant and therefore must be carefully considered.

Biomass
Biomass currently provides about 3.6 quads (3.8 EJ) of energy in the US and about 55 EJ worldwide.[416] If biomass were to supply all primary energy currently used (i.e., 351 EJ in 1990[417]), biomass energy production would have to increase about sevenfold, and up to 40-fold by 2100 if growth in primary energy use continues as predicted by the IPCC.[418] Before discussing the feasibility of increasing biomass energy production in more detail, it must be kept in mind that humans already appropriate 30 to 40 percent of terrestrial primary productivity (photosynthetically fixed carbon) worldwide,[419] meaning that two fifths of the land's productive capacity is already tightly controlled and managed for supplying food, fiber and energy. Thus, it is unlikely that human appropriation of biomass for energy generation can be increased substantially without causing the collapse of complex ecosystems, depriving countless species of their photosynthetically fixed carbon energy food sources, thereby threatening or causing their extinction.[420]

The fact that biomass will not be able to supply a substantial fraction of the energy demand is particularly evident in the United States. Consider, for example, that the total amount of energy captured by vegetation in the US each year is about 58 quads (61.5 EJ), about half of which (28 quads or 29.7 EJ) is already harvested as agricultural crops and forest products and therefore not available for energy production.[421] Because most of the remaining non-harvested biomass is required to maintain ecosystem functions and biodiversity, very little will be available for additional energy generation. Given that the current US energy use is around 100 quads per year,[422] it follows that biomass could at best supply only a relatively small and insignificant fraction of the total US energy demand.[423] The limited potential of biomass energy becomes even more obvious if land requirements are considered. To supply the current worldwide energy demand of 351 EJ/yr solely with biomass would require more than 10 percent of the Earth's land surface, which is comparable to the area used for all of world agriculture, i.e., about 1,500 million hectares.[424] If ethanol from corn were to substitute for 100 percent of the gasoline consumption in the US, all of the available US cropland (both active and inactive, totaling about 190 million hectares) would have to be devoted to ethanol production, leaving no land for food production.[425] If all fossil fuels currently used in the United States were to be replaced by fuel from biomass grown on tree plantations, forests covering approximately the entire US land area would be required.[426] Clearly, there is simply not enough land for generating sufficient quantities of biomass energy to supply even the current energy demand, let alone the vastly increased (fivefold) energy needs that are predicted for the next 100 years.[427]

Any significant increase in biomass energy production would cause land use conflicts with agriculture, particularly given the increased demand for food expected due to future population growth.[428] In any conflict between food and energy, it is unlikely that energy will win in the long run, and if it were to, that should be cause for serious ethical concern. Consider, for example, that the corn used to produce the ethanol to fill a 25-gallon tank once could feed one person for a whole year.[429] Unfortunately, in free-market economies, agricultural commodities such as food and animal feed are produced for people who can afford to pay the highest prices for them (i.e., consumers in industrialized nations) and not for people who need them most (i.e., the world's poor and hungry). Thus, because of the rising demand for biofuels, food crops such as corn and soybeans might be increasingly converted into ethanol and biodiesel, respectively. As a result, prices of corn, soybeans and other energy food crops would most likely rise, creating hardships for the world's poor. Indeed, a recent study conducted by the International Food Policy Research Institute found that aggressive growth in biofuel production (i.e., 10 percent, 15 percent and 20 percent gasoline substitution by biofuels worldwide in 2010, 2015 and 2020, respectively) would lead to substantial price increases for various feed crops.[430] For example, prices for cassava, corn, oilseeds, sugar beets, sugar cane and wheat would increase by 62 percent on average. These significant increases in food crop prices are likely to lead to greater food insecurity worldwide. According to research conducted by Benjamin Senauer at the University of Minnesota,[431] a 1 percent price increase in global staple food prices would subject approximately 16 million people to hunger or starvation. Thus, even a small increase in food prices could have a major impact on the daily lives of the world's poorest. For example, a global increase in food prices of only 10 percent would translate into 160 million more hungry people, which is numerically equivalent to approximately half the population of the United States.

The World Energy Council estimated that an additional 1,700 million hectares of farmland and 690 to 1,350 million hectares of additional land will be required by 2100 in order for biomass to support a "high growth scenario."[432] To achieve high biomass productivities, it would be necessary to use large-scale, high-tech agricultural methods such as intensive irrigation as well as the application of synthetic fossil-fuel-derived fertilizers and toxic chemicals such as pesticides and herbicides. According to Giampietro and colleagues,[433] freshwater demand for the production of biomass fuels would exceed the current freshwater supply by a factor of 3 to 104 in the 148 countries included in their study, thus clearly indicating that in addition to land, water would become a significant limiting factor. As is generally the case in modern agriculture, the large-scale application of fertilizers and toxic pesticides and herbicides is likely to result in various negative impacts such as groundwater and surface water pollution, fish and bird kills, species extinction and increased health risks to biomass plantation workers.[434]

In conclusion, biomass could supply only a small fraction of total energy demand, and any significant expansion of biomass energy would be constrained by the availability of land and water, potential conflicts with food production, and environmental concerns about chemical pollution, habitat loss and species extinction. Consequently, other renewable energy sources would be needed to meet rising energy demands.

Other Renewable Energy Sources

In addition to biomass, other renewable energy sources such as passive solar, solar thermal, hydroelectric, photovoltaics and wind power can also provide carbon-free energy in the future. However, all of these alternative energy technologies would have significant environmental impacts, particularly if deployed on a large scale.

Approximately one quarter (i.e., 18.4 quads or 19.5 EJ) of all fossil energy consumed per year in the United States is used for space heating and cooling of buildings and for providing hot water.[435] Because only 0.3 quad of fossil energy is currently saved by using passive solar technologies, tremendous potential exists for capturing more solar energy by redesigning buildings, particularly when combined with improved insulation.[436] Environmental impacts are likely to be minimal and would be generally limited to visual glare from solar panels (in cases, such as on hillsides, where the panels could be seen by neighbors) and the removal of shade trees.

The direct capture of sunlight by solar thermal receivers consisting of computer-controlled sun-tracking parabolic mirrors that focus solar radiation to generate steam for electricity generation is another promising method for producing renewable energy, particularly in desert regions. Because solar thermal energy systems are capable of converting 22 percent of incoming sunlight into energy, compared to the approximate 0.1 percent efficiency of green plants, which convert light to biomass via photosynthesis, much less land is required.[437] For example, only about 1,100 hectares are needed to produce 1 billion KWh/yr (0.0036 EJ/yr).[438] Thus, if the entire annual US energy demand of about 100 quads, or 106 EJ,[439] were to be supplied by direct solar thermal energy, 32 million hectares, or an area the size of a 570 km × 570 km (354 miles × 354 miles) square, would be required. Potential environmental effects would be relatively minor for small solar thermal receivers, possibly incinerating birds and insects that accidentally fly into the high-temperature portion of the beam,[440] but impacts may become significant for very large installations. Fragile desert ecosystems may be destroyed and microclimates changed.[441]

In the United States, hydroelectric dams provide annually about 3 quads (3.2 EJ or 870 billion KWh/yr) or equivalently 3 percent of the total US energy demand.[442] Considering that the best candidate sites have already been exploited, and that many negative environmental impacts are associated with hydropower generation, it is unlikely that the hydroelectricity supply can be expanded significantly, i.e., probably at most 50-100 billion KWh/yr in the US.[443] In fact, because of severe environmental problems such as obstructing fish migration, it is likely that some hydroelectric dams will eventually be dismantled, thereby reducing the contribution of hydropower to the US electricity supply.

Photovoltaic (PV) cells could in theory satisfy all of the energy demand in the US, particularly if the generated electricity were to be converted into a more versatile energy carrier such as hydrogen, thereby avoiding problems of intermittency and storage. Assuming a 10 percent solar-to-electricity conversion efficiency, an area the size of a 161 km × 161 km (i.e., 100 miles × 100 miles) square would have to be covered in a desert region such as Nevada to produce all of the energy currently consumed in the US.[444] To cover such an area would be detrimental to the affected fragile desert ecosystems and might also cause changes in the microclimate. These environmental impacts could be reduced by collecting solar energy in widely dispersed small units rather than using PV installations covering very large land areas. The manufacture of extremely large numbers of photovoltaic cells, however, is likely to contribute to pollution problems.[445]

Finally, large windmills—if deployed by the millions—could also provide a fraction of renewable carbon-free power. According to a US wind energy potential study by Elliott and colleagues,[446] at least 80 million hectares (i.e., more than 40 percent of all available US farmland) would have to be covered with windmills (50 m hub height, 250-500 m apart) to generate about 100 quads (106 EJ) of electricity, which is equivalent to 100 percent of the current annual US energy demand.[447] It is highly questionable whether the public would tolerate huge wind farms that cover large land areas, given concerns about blade noise and aesthetics.[448] Consider, for example, that in Germany, where more than three times as much wind energy is currently generated than in the United States (i.e., 14,600 MW with about 15,000 large windmills), public resistance against wind farms has developed to an intensity comparable to that against nuclear power plants, making the further expansion of wind energy capture highly unlikely.[449] Thus, at best only a rather small fraction (i.e., ≤5%) of carbon-free energy is likely to be generated in the future by large land-based windmills. To avoid public resistance, there are now efforts to move windmills offshore, out of sight of concerned citizens, but large-scale offshore wind farms are likely to have negative impacts on marine life.

In summary, all renewable energy technologies are expected to have significant environmental impacts, particularly if deployed on a scale sufficient to supply most, if not all, of the energy needed for future industrial and economic activities. Thus, before assuming—as many environmentalists do—that renewable energy is the best solution to the problem of unsustainable fossil fuel use and global warming, it is necessary to carefully study the environmental impacts and the public attitudes toward large-scale energy installations, many of which may cover hundreds of square kilometers.[450] The finding that environmental impacts of renewable energy generation cannot be avoided is not surprising, a sobering reality expressed in Commoner's fourth law of ecology "There is No Such Thing as a Free Lunch."[451] Thus, while renewable energy will certainly be an important component in the design of more sustainable circular flow economies of the future, it has limited potential and cannot be expanded indefinitely because of land limitations, severe environmental impacts and public opposition. Given that the immense economic expansion during the past two centuries was driven primarily by the availability of cheap and abundant fossil energy resources, it is unlikely that economic growth can continue at its previous pace, if at all, by solely relying on renewable solar energy, whose potential is not only limited but which is also more expensive.[452] As Ted Trainer predicts in *The Conserver Society,*

> When we do live on renewable energy we will have to adopt very energy-frugal ways. It is unfortunate that in attempting to gain support for renewable energy sources their advocates usually give the impression that the wind and the sun can solve our energy problems without any need to change to simpler life styles or to scrap the growth economy.... The general conclusion, therefore, is that all people cannot expect to have affluent lifestyles on renewable energy.[453]

Challenge #2: Replacement of Non-Renewable Materials with Renewable Substitutes

Industrial societies are inherently unsustainable because they depend on non-renewable metals and minerals whose supplies are strictly limited.[454] Consider, for example, that currently identified reserves of bauxite, copper, iron, lead, nickel, silver, tin and zinc will last only another 15 to 80 years if growth in production continues at 2 percent per year.[455] Using even the most optimistic estimates about mineral deposits in the Earth's crust to a depth of 4,600 meters, the life expectancy of most metal reserves is less than a thousand years.[456] To achieve sustainability, industrial societies will either have to completely recycle these non-renewable metals or find renewable substitutes.

Although there is generally great optimism among economists that resource scarcity can be overcome by finding more abundant substitutes,[457] there are actually only very few cases where limited metals have been replaced by materials that are either renewable or unlimited in supply. Probably the best example is the replacement of copper wires used in communications by fiber optic cables that are made from extremely abundant silica (quartz sand).[458] In a few cases, such as in the construction of car, boat and airplane frames, metals have been replaced by high-strength plastic polymers made from non-renewable petrochemical feedstocks. Although it should be in principle possible to synthesize these polymers from renewable plant-based feedstocks instead of non-renewable petroleum, thereby providing renewable metal substitutes, it must be realized that, as in the case of renewable biomass energy generation, the large-scale production of plant-derived feedstocks for the chemical industry is likely to have significant negative environmental consequences.

To assure that the substitution of non-renewables by renewables is sustainable over the long run, all renewable resources must be harvested at rates not exceeding the maximum sustainable yield; that is, over-harvesting and over-exploitation of renewable natural resources is not permissible. Unfortunately, as discussed earlier, rapid economic growth and the resulting increase in affluence in many developed and developing nations is often linked to the over-exploitation of natural renewable resources such as water, agricultural soils and forests.[459] As Herman Daly observes,

At the heart of the current crisis in economic theory and practice is the fact that we are consuming the Earth's resource beyond its sustainable capacities of renewal, thus running down that capacity over time—that is, we are consuming capital while calling it income.[460]

Consequently, if there were to be large-scale efforts to replace non-renewable minerals, metals and energy sources with renewable substitutes, there is a very high risk that many theoretically renewable natural resources, "natural capital," will become depleted and exploited even faster, leading us even farther away from sustainability and toward environmental collapse.

Although substitution is supposed to be a solution to the problem of resource over-exploitation and depletion, substitution cannot continue indefinitely. As soon as one resource is threatened with depletion, a substitute resource must be identified. If the substitute is non-renewable, it will sooner or later be depleted as well, requiring the search for another substitute. If the substitute is renewable, there are inherent production limits dictated by the maximum sustainable yield. Furthermore, there are certain natural resources, called critical natural capital,[461] that can never be replaced through technological innovation. These are, as Paul and Anne Ehrlich point out, the various ecological systems that carry out many vitally important functions:

> These systems maintain the quality of the atmosphere, dispose of our wastes, recycle essential nutrients, control the vast majority of potential pests and disease vectors, supply us with food from the sea, generate soils, and run the hydrological cycle, among other vital services.... But civilization could not persist without those free life-support services rendered by ecological systems.... The ignorant claim that with application of more energy and technology, substitutes could be found to replace those ecological services.[462]

In conclusion, there are fundamental limits to the replacement of non-renewable materials and energy by renewable substitutes.

Challenge #3: Complete Recycling of Non-Renewable Materials and Wastes

Given that it would be a serious challenge, and in many cases impossible, to find renewable substitutes for the numerous non-renewable metals and minerals that are currently used in industrialized nations, it would be necessary—in order to meet the above mentioned "sustainable materials use condition"—to recycle all non-renewable materials after their use. Unfortunately, complete recycling of non-renewable materials and wastes ("closing the materials loop") is inherently impossible, and even partial recycling would be a significant challenge in terms of engineering, energy requirements, and economics. According to pioneering industrial ecologist Robert Ayres,

> There are only two possible long-run fates for waste materials: recycling and re-use or dissipative loss. The more materials are recycled, the less will be dissipated into the environment, and vice versa. Dissipative losses must be made up by replacement from virgin sources. A long-term sustainable state would be characterized by near total recycling of intrinsically toxic or hazardous materials, as well as a significant degree of recycling of plastics, paper and other materials whose disposal constitutes an environmental problem. Heavy metals are among the materials that would have to be almost totally recycled to satisfy the sustainability criteria.[463]

Non-renewable materials and wastes can be grouped into three classes based on their technical and economical potential for recycling.[464] For materials belonging to class I, recycling is both technically and economically feasible. For example, recycling rates for class I materials such as paper, aluminum, copper, lead, nickel, iron and zinc range from around 25 to 50 percent and therefore could still be improved significantly.[465] For materials belonging to class II, such as packaging materials, solvents, refrigerants, etc., recycling is possible from a technical standpoint, but there have been, with few exceptions (e.g., bottle deposits), little economic incentives in the United States to do so. However, considering the successful implementation of recycling regulations dealing with consumer packaging materials in Germany,[466] it should in principle be possible to recover most of them.

The recycling potential of waste materials belonging to class III is problematic both from a technical and economic perspective. Unfortunately, many materials belong to this category, such as paint coatings, pigments, pesticides, herbicides, preservatives, flocculants, antifreezes, explosives, propellants, fire-retardants, reagents, detergents, fertilizers, fuels, lubricants, etc.[467] The current use of these class III materials is inherently dissipative because they disperse widely in the environment and become so diluted that it is impractical to recover them for reuse. For example, it is hardly possible to recycle the potassium or phosphorus used in agricultural fertilizers, the copper dispersed in fungicides, the lead in widely applied paints or the zinc oxides that are present in the finely dispersed rubber powder abraded from vehicle tires.

It could conceivably be argued by enthusiastic techno-optimists that it is possible, at least in principle, to recycle these highly dispersed materials if new technologies are developed and if enough energy is applied to carry out the purification and recycling processes.[468] Complete (100%) recycling occurs in nature all the time, as evidenced by the presence of biogeochemical cycles for carbon, oxygen, nitrogen, phosphorus and many other elements. The cycling of these elements is carried out via a complex system of biological transformation pathways that are enzymatically catalyzed by microorganisms and plants and driven by solar energy. The problem with many class III materials is that no circular biological transformation pathways exist for them in nature, and it would be extremely difficult to design new solar-energy-driven biotechnologies to recycle these highly dissipated materials.[469] Even if some type of innovative recycling technologies for these class III compounds could be developed, which is highly questionable, it would take tremendous amounts of energy to carry out the recycling process.

Using the second law of thermodynamics, it is possible to estimate the minimum energy requirements for concentrating a highly dispersed non-renewable resource. As expected, the energy needed for purification increases drastically with decreasing material concentrations in the environment.[470] Consequently, it is clear that enormous amounts of energy would be required to collect, purify and recycle highly dissipative wastes such as those belonging to class III. Considering the potential negative environmental consequences associated with sustainable and renewable energy generation, it is extremely unlikely that there will ever be enough cheap energy available to recycle these highly dispersed waste materials.[471]

In summary, complete recycling of non-renewables is practically impossible, particularly for class III materials. The inherently dissipative use of non-renewable metals and minerals poses two major problems for achieving the long-term sustainability of industrialized nations. First, there is only a limited supply of these trace metals and minerals, and their costs will increase as high-grade ores become depleted with time.[472] Second, the dispersal of many of these materials is causing major environmental disruption. For example, anthropogenic worldwide atmospheric emissions of arsenic, copper, lead, manganese, mercury, nickel, vanadium and zinc all significantly exceed the natural (atmospheric) fluxes of these elements, indicating their likely potential for disturbing natural cycles and disrupting sensitive ecosystem functions.[473] It can therefore be concluded that long-term sustainability can be achieved only if the use of limited non-renewable metals and minerals is discontinued or severely curtailed. This would be a significant challenge given that innumerable industrial processes and products are dependent on the availability of these materials and that, as was discussed previously, it would be exceedingly difficult, if not impossible, to find renewable substitutes.

Sustainability or Collapse?

Based on the above analyses, it is clear that continued economic growth will result in resource depletion and negative environmental impacts, despite the application of mitigating technologies. Infinite economic growth on a finite planet is impossible.[474] The magnitude of human economic activities is inherently limited by the size of the supporting ecosystem. In the words of Herman Daly,

> The economy, in its physical dimensions, is an open subsystem of our finite and closed ecosystem, which is both the supplier of its low-entropy raw materials and the recipient of its high-entropy wastes. The growth of the economic system is limited by the fixed size of the host ecosystem, by its dependence on the ecosystem as a source of low-entropy inputs and as a sink for high-entropy wastes, and by the complex ecological connections that are more

easily disrupted as the scale of the economic subsystem grows relative to the total ecosystem.[475]

Thus, in terms of evaluating the sustainability of continued economic growth, the critical question becomes, How much further can the size of the human economy expand before impinging on the functioning of the supporting ecosystems? Or, alternatively, What is the size of the human ecological footprint compared to the total land area available on Planet Earth? Harvard University socio-biologist Edward O. Wilson provides an answer:

> The ecological footprint—the average amount of productive land and shallow sea appropriated by each person in bits and pieces from around the world for food, water, housing, energy, transportation, commerce, and waste absorption—is about one hectare (2.5 acres) in developing nations but about 9.6 hectares (24 acres) in the US. The footprint for the total human population is 2.1 hectares (5.2 acres). For every person in the world to reach present US levels of consumption with existing technology would require four more planet Earths. The five billion people of the developing countries may never wish to attain this level of profligacy. But in trying to achieve a decent standard of living, they have joined the industrial world in erasing the last of the natural environment.[476]

Similarly, Mathis Wackernagel and William Rees, originators of the ecological footprint analysis, caution,

> There is much evidence today that humanity's ecological footprint already exceeds global carrying capacity. Such overshoot is only possible temporarily and imposes high costs on future generations.... Ecological footprint analysis by no means implies that living at carrying capacity is a desirable target. Rather, the ecological footprint is intended to show how dangerously close we have come to nature's limit. Ecological resilience and social well-being are more likely to be assured if the total human load remains substantially below Earth's carrying capacity. Living at the ecological edge compromises ecosystem's adaptability, ro-

bustness, and regenerative capacity, thereby threatening other species, whole ecosystems, and ultimately humanity itself.[477]

In this context, it should be recognized that most societies, including those as complex as the Roman and Mayan, have collapsed and vanished after having enjoyed long periods of prosperity.[478] There is, therefore, no reason to believe that our extremely complex technological society could not collapse as well. Indeed, all three pre-conditions for environmental and societal collapse are present in current technological societies:[479] (i) rapid growth in resource use and pollution, (ii) limited resource availability and waste absorption capacity, and (iii) delayed responses by decision-makers when limits have already been exceeded or soon will be. According to Donella Meadows and colleagues,

> If the signal or response from the limit is delayed and if the environment is irreversibly eroded when overstressed, then the growing economy will overshoot its carrying capacity, degrade its resource base, and collapse. The result of overshoot and collapse is a permanently impoverished environment and a material standard of living much lower than would have been possible if the environment had never been overstressed.... On the local scale, overshoot and collapse can be seen in the processes of desertification, mineral or groundwater depletion, poisoning of agricultural soils or forest lands by long-lived toxic wastes, and extinction of species.... On the global scale, overshoot and collapse could mean the breakdown of the great supporting cycles of nature that regulate climate, purify air and water, regenerate biomass, preserve biodiversity, and turn wastes into nutrients.[480]

Similarly, Joseph Tainter warns in *The Collapse of Complex Societies*,

> Complex societies historically are vulnerable to collapse, and this fact alone is disturbing to many. Although collapse is an economic adjustment, it can nevertheless be devastating where much of the population does not have the opportunity or the ability to produce primary food resources. Many contemporary societies, particularly those that are highly industrialized, obviously fall into this class. Collapse for such societies would almost

certainly entail vast disruptions and overwhelming loss of life, not to mention a significantly lower standard of living for the survivors.[481]

As the analyses in this and the previous chapters have shown, the belief that the application of advanced science and technology, as currently practiced, will automatically result in sustainability and avert collapse is greatly mistaken.[482] Science and technology will certainly be necessary but alone will be unable to bring about sustainability. What is required, and is indeed absolutely essential, is a change in societal values and policies in order to consciously direct technological innovation toward the goal of decreasing total impact (i.e., resource depletion and environmental pollution). The importance of society's goals and values in terms of directing technological change was made clear by Donella Meadows and colleagues:

> One reason technology and markets are unlikely to prevent overshoot and collapse is that technology and markets are merely tools to serve the goals of society as a whole. If society's implicit goals are to exploit nature, enrich the elites, and ignore the long-term, then society will develop technologies and markets that destroy the environment, widen the gap between rich and poor, and optimize the short-term. In short, society develops technologies and markets that hasten a collapse instead of preventing it….Technology and markets typically serve the most powerful segments of society. If the primary goal is growth, they produce growth as long as they can. If the primary goals were equity and sustainability, they could also serve those goals.[483]

In short, without a significant change in society's values, the current direction of progress in science and technology will only implement the existing values of growth, exploitation and inequality, thereby accelerating us toward collapse. Consequently, the main challenge will be to change society's goals from growth to material sufficiency and appropriate stable population size; from exploitation to just treatment of labor, future generations and the environment; and from gross inequality to a more fair distribution of both income and wealth. These changes in societal values would then automatically translate into various policies that would redirect science and technology toward meeting these new goals.

Unfortunately, there is very little time left for the required value changes to occur. The longer we wait to change direction while economic growth continues at the present rate, the fewer choices for action we will have in the future. Although it is probably true that resistance to value change is present at all levels of society, the strongest opposition would be expected to come from those who benefit most from the current economic system. In earlier cultures now extinct, such individuals and groups also believed that they could isolate themselves from the consequences of their unsustainable actions.[484] Clearly, they were dead wrong. They and their entire societies descended into social and political chaos with a precipitous rise in death rates and the disintegration of infrastructure, traditions and accumulated knowledge.[485] Probably the most important step toward preventing the collapse of our technological society is to convince those who exert the greatest influence that it is in their own best interests to promote the necessary value changes.

PART II

THE UNCRITICAL ACCEPTANCE OF TECHNOLOGY

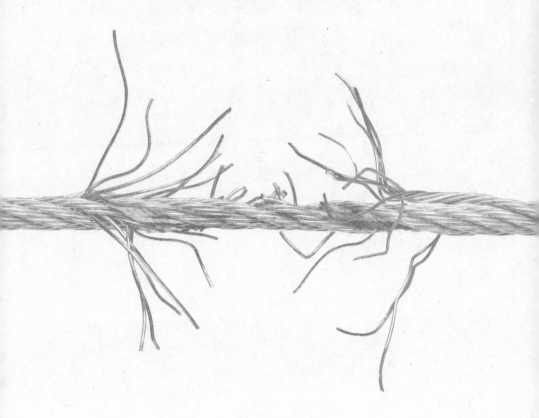

CHAPTER 7

Technological Optimism and Belief in Progress

A s DISCUSSED IN PART I, there is a remarkable confidence that science and technology will solve the major problems facing humanity, including those created in the first place by technologies. Environmental counter-technologies presumably will solve the problems created by polluting technologies. Medical and military technologies serve as social fixes that attempt to solve complicated social, cultural, economic and political problems. Improvements in technical efficiency are believed to be a panacea for almost any problem facing industrial societies, whether shortages of energy and mineral resources, environmental pollution, road congestion, rising health care costs or even long-term sustainability. Clearly, technological optimism and belief in continual progress permeate modern industrial societies. "Progress" is a doctrine, a dogma, accepted on faith and projected onto a hopeful but largely illusory future. Lewis Mumford describes this concept of progress in *Technics and Civilization*:

> According to philosophers and rationalists, man was climbing steadily out of the mire of superstition, ignorance, savagery, into a world that was to become ever more polished, human, and rational.... In the nature of progress, the world would go on forever and forever in the same direction, become more humane, more comfortable, more peaceful, more smooth to travel in, and above all, much more rich.[486]

Earnest Braun, in *Futile Progress*, defines technical progress as follows:

> Generally, technical progress can be seen as moving in the direction of higher performance, meaning greater speed; larger capacity; higher cost-effectiveness; higher efficiency; increasing reliability; reduced effort required from human labor; more autonomous acting machines, etc.[487]

Technological optimism and belief in progress has been expressed and re-affirmed by influential scientists, engineers, economists, politicians, and corporate propagandists. As Edward Teller, father of the hydrogen bomb, said in 1942, "The whole dynamic civilization of the West, for which America is the spearhead, is based upon scientific and technological advancements. We must trust our social processes to use these advancements in the right manner. We must not be deterred by arguments involving consequences or costs."[488] In the words of economist Frederick Benham, "There is one powerful force working constantly towards greater output per head and ever-rising standards of living. It is the march of science and invention; in technical progress lies the economic hope of mankind."[489] At the dawn of the nuclear age half a century ago, the chairman of the US Atomic Energy Commission envisioned a utopian future powered by thermonuclear power: "It is not too much to expect that our children will enjoy in their homes electrical energy too cheap to meter, will know of great periodic regional famines in the world only as matters of history, will travel effortlessly over the seas and under them and through the air with a minimum of dangers and at great speed, and will experience a life span longer than ours.... This is the forecast for an age of peace."[490] A major chemical company was promising "Better things for better living through chemistry."[491] President Kennedy announced in 1963 that "as we begin to master the potentialities of modern science, we move toward an era in which science can fulfill its creative promise and help bring into existence the happiest society the world has ever known."[492]

Economist Wilfred Beckerman enthusiastically observed, "After all, when we ponder on the fantastic technological progress that has been made in the last 20 or 50 years, the mind boggles at the progress that will be made over the next 100 million years."[493] Economist Julian Simon is even more optimistic: "The standard of living has risen along with the size of the world's population since the beginning of recorded time. There is no convincing economic reason why these trends toward a better life should not continue indefinitely.... The more [resources] we use, the better off we become—and there's no practical limit to improving our lot forever (or for at least seven billion years)."[494] Never mind that the present stability of the sun is predicted to continue for only four billion years,[495] that the probable persistence of any species, including our own, is only a small fraction of that time and that the survival of entire civilizations is almost ephemeral on that scale.[496]

These overly optimistic proclamations are not made in a cultural vacuum. After all, technological progress has delivered many impressive and tangible benefits in the industrialized nations, such as higher living standards and economic prosperity, decreased manual labor and increased leisure, greater mobility and faster communication, better medical treatment and increased life-expectancies. As Langdon Winner notes, "During the last two centuries, the factory system, railroads, the telephone, electricity, automobiles, airplanes, radio, television, and nuclear power have all figured prominently in the belief that a new and glorious age was about to begin."[497]

Similarly, Kenneth Stunkel and Saliba Sarsar observe,

> In the late 20th century, people in advanced industrial countries routinely expect science and technology to expand economic growth, relieve their medical difficulties, and provide an endless array of useful and diverting gadgets. Advances in telecommunications, computer and space technologies, genetic engineering, and a host of other fields have confirmed the ability of science and technology to come up with a seemingly inexhaustible array of surprises.[498]

John Platt observed more than 30 years ago that the speed of progress has been astounding, with many modern technological changes occurring on a scale wholly unknown to our ancestors:

> In the last century, we have increased our speeds of communication by a factor of 10 million; our speeds of travel by a hundred, our speeds of data handling by a million, our energy resources by a thousand, our power of weapons by a million, our ability to control diseases by something like a hundred, and our rate of

population growth to a thousand times what it was a few thousand years ago.[499]

Considering the rapidity and breadth of scientific discoveries and technological applications, it is not surprising that there is an enthusiastic anticipation of more to come, that the observed trends will continue into the indefinite future. While it is true that this belief is based on massive factual evidence, this evidence is of recent origin and may prove misleading.

Belief in Progress: A Brief History

The concept of progress did not exist in hunter-gatherer or traditional agricultural societies, most likely because the activities of these peoples were closely linked to the cycles of nature, the rhythms of day and night, of the moon and of the seasons, which served, year after year, as a reliable timetable for the sowing of seeds and the reaping of harvests. As a result of living in close connection with nature, where many events repeated in a predictable fashion, time was generally perceived as cyclical, and for countless generations, their own cycles of life and traditions remained relatively stable or changed extremely slowly. Even the early Greeks interpreted historical patterns as cyclical and did not believe in progress. In fact, the ambition of Greek philosophers was to rise above change and to find enduring truths. Similarly, the concept of progress was unknown in the Roman world.[500]

With the rise of Christianity, the traditional concept of cyclical time began to be slowly replaced by the view of time as linear. A possible explanation for this profound change is that, compared to most other world religions, Judaism and Christianity place great emphasis on historical narratives to strengthen the faith of their followers. Indeed, the Bible may be considered not only a religious scripture but also a historical document. Since the very recording of history suggests a linear view of time, the followers of the Judeo-Christian faiths began to see time as linear, not only in reference to the past but also the future. The Christian belief in eternal life, to be enjoyed or endured forever in either heaven or hell, is an example of extrapolating linear time into the indefinite future. This sharply contrasts with Eastern traditional beliefs in repetitive reincarnations, which reflect the earlier cyclical view of time.[501]

Although the new concept of linear time set the stage for the modern belief in progress, it was not until the Enlightenment of the 17th and 18th centuries that the idea of progress was applied to the material and technological. As was described in Chapter 3, Enlightenment philosophers René Descartes and Francis Bacon had developed the "New Science" (*Novum Organum*), which created great enthusiasm because it was believed that the newly gained scientific knowledge could be used, according to Bacon's dictum "knowledge is power," to exploit and control nature for the benefit of mankind.[502] The early successes in science and technology encouraged the belief that human reason was capable of generating not only scientific progress but also social and moral progress. In the words of Cambridge University historian J.B. Bury,

> [This] idea of human progress is a theory which involves a synthesis of the past and a prophecy of the future. It is based on an interpretation of history which regards men as slowly advancing... in a definite and desirable direction, and infers that this progress will continue indefinitely.... It implies that...a condition of general happiness will ultimately be enjoyed, which will justify the whole process of civilization.[503]

Because of the growing belief that reason could improve the human condition, a fundamental shift from a theocentric to an anthropocentric worldview occurred during the Enlightenment.[504] Faith in the Christian doctrine of salvation, which promised deliverance from suffering in an eternal life after death, was largely replaced by a popular belief in scientific, technical, social and material progress, which was soon to create a human-constructed paradise on Earth, to be enjoyed by all before death.[505] As Neil Postman summarizes in *Technopoly*,

> The decline of the great narrative of the Bible, which had provided answers to both fundamental and practical questions, was accompanied by the rise of the great narrative of Progress. The faith of those who believed in Progress was based on the assumption that one could discern a purpose to the human enterprise, even without the theological scaffolding that supported the Christian edifice of belief. Science and technology were the chief instruments of Progress, and in their accumulation of reliable

information about nature they would bring ignorance, superstition, and suffering to an end.[506]

Thus, the rational, secular approach, which had been so successful in advancing scientific knowledge, was expected to soon solve all of the problems facing society.[507] Progress in science and technology was the new creed, a statement of hope, the new socio-political gospel.[508] In the words of Leo Marx, "The new scientific knowledge and technological power was expected to make possible a comprehensive improvement in all the conditions of life—social, political, moral, and intellectual as well as material."[509]

To better appreciate the enthusiasm with which science was seen as a positive social and political force, consider the following excerpt from Denison Olmsted's article "On the Democratic Tendencies of Science," published in *Barnard's Journal of Education* in the mid-1850s:

> My object, in the present essay, is to prove that science, in its very nature, tends to promote political equality; to elevate the masses; to break down the spirit of aristocracy; and to abolish all those artificial distinctions in society which depend on differences of dress, equipage, style of living, and manners; to raise the industrial classes to a level with the professional; and to bring the country, in social rank and respectability, to a level with the city.[510]

Unfortunately, many of these and similar grand predictions of impending social, political and even moral progress did not come to pass. As a result, the earlier Enlightenment concept of universal human progress was quietly redefined and limited to material progress, which was to be brought about through innovative science and technology, and which could easily be measured in terms of economic growth.[511] Indeed, as Clive Hamilton writes in *Growth Fetish*, progress in today's world is judged almost exclusively in terms of economic growth and increasing material standards of living:

> The Enlightenment idea of progress is one of the ideological pillars of capitalism. The unspoken assumption motivating modern capitalism is that the world is evolving towards a better, more prosperous future and that the engine of this advance-

ment is economic growth—the expansion of the volume of goods and services at human's disposal.... It pervades our political discourse, the writing of our history, and the consciousness of ordinary people everywhere.... High and sustained rates of economic growth through each decade after the war rescued the idea of progress from historical oblivion. But not only was it rescued; it was reconceptualized. Applied science, evolutionary biology and ethics no longer powered the idea. The new engine was more mundane: material advancement would drive progress, and the measure of success became the standard of living.[512]

Nowhere is belief in material progress, economic growth and technological optimism more extreme than in the United States.[513] In recent public opinion surveys, 86 percent of Americans, compared to 71 percent of Europeans, agreed with the statement "Science and technology are making our lives healthier, easier, and more comfortable," and 72 percent of respondents in the United States, compared to only 50 percent in Europe, think that "the benefits of scientific research outweigh any harmful results."[514] There are a number of reasons why technological optimism and belief in material progress are so ingrained in the United States. First, much of the early history of the US was one of continuous territorial expansion into new frontiers that provided seemingly unlimited land and mineral resources that could be exploited for material gain.[515] Second, as a result of these apparently unlimited opportunities for development and the many successes that inevitably followed, a strong belief developed that the United States is exceptional, that anything can be accomplished here, particularly through the application of new technology.[516] Third, this belief in American exceptionalism was further strengthened by the "genius and audacity of American capitalists of the late 19th and early 20th centuries, men who were quicker and more focused than those of other nations in exploiting the economic possibilities of new technologies."[517] Finally, the success of 20th-century technology in providing Americans with undreamt-of material abundance has continued so steadily and has been so convincing that there appears little doubt that this type of progress will not continue forever.[518] Indeed, belief in scientific and technological progress as the savior of humanity has come to resemble religious faith in salvation.

Comparison of Belief in Progress to Religious Faith

Given that belief in progress has, since the Enlightenment, displaced, in a large segment of society, the earlier faith in salvation, it should be no surprise that our present belief in science and technology exhibits many features normally associated with religion. Here we will briefly discuss four similarities between religious faith and faith in scientific and technological progress: the promise of salvation, the means of controlling and maintaining systems of mass acceptance, the reliance on the wisdom and authority of "experts" and the ignorance of believers.

To address the ubiquitous fear of death, most traditional religions offer promises of some kind of life after death. Christianity specifically promises the salvation of one's soul from the consequences of sins committed on Earth and a pleasant eternal life in heaven. Science and technology also offer many promises, but these are much more tangible and immediate: freedom from toil, boundless leisure, faster travel, instant communication, absence of sickness and debility, delay of death and, of course, unlimited material affluence. However, in accordance with the utilitarian philosophy with which the belief in progress is associated, the key promise of this human-built paradise, this "industrial version of heaven,"[519] is happiness to be attained here and now. Religious beliefs in eternal life have the distinct advantage that they are intrinsically untestable because the deceased cannot report to us whether there is indeed life after death. This untestable nature of religious beliefs has made it possible for them to persist in various forms for thousands of years. By contrast, many earlier prophecies of scientific and technological progress have been demonstrably fulfilled, and the benefits of science and technology can be objectively quantified. As a result of past successes, belief in future progress has been strengthened. However, as will be discussed in Chapter 9, one of the important promises of material progress, happiness, has not resulted, and this important reality, along with the deterioration of the environment and decreasing affluence, is likely to weaken contemporary faith in scientific and technical progress and expose it to increasing doubt.

To persist and maintain influence, all systems of mass faith require protection from dissenting views and contradictory facts. Organized religions such as Christianity have for centuries controlled the flow of information to the masses. For example, until relatively recently, the

Catholic Church disapproved of the reading of the scriptures of competing religions, forbade marriage or extended contact with adherents of other faiths, used the power of repetitive prayers and rituals to indoctrinate dogmas and disseminated doctrine along tightly controlled channels in a centralized hierarchy. Even today, in order to maintain faith, many fundamentalist Christians in the United States home-school their children in an attempt to insulate them from scientific facts that contradict scripture, such as the theory of biological evolution. Similarly, belief in technological progress is also maintained and continually promoted by controlling the flow of information to the public. A strong belief in progress and technological optimism continues unabated because most reports on science and technology in the mass media are biased toward positive results and wishful thinking, while discussions regarding negative long-term consequences are infrequent or absent.

Systems of mass faith are also strengthened by invoking the authority of their leaders. Again, the Catholic Church provides a good example. The authority of the pope is derived directly from St. Peter, and this authority is passed down through the church hierarchy to cardinals, bishops and priests. In addition, for hundreds of years, the clergy conducted weekly masses in Latin, a language that was generally not understood by the public. This "language barrier" served as a tool for establishing authority and for keeping believers mystified, thereby limiting comprehension and avoiding critical analysis. Similarly, scientists often appear on the media to authoritatively communicate their findings and promises, largely in a language not understandable to the scientifically illiterate which, unfortunately, is most of the public. Awe can also be inspired by majestic buildings. As the great gothic cathedrals were built in Europe during the Middle Ages to strengthen belief in Church and God, our present high-technology research laboratories and hospitals serve as awe-inspiring edifices that nurture our faith in progress. As Melvin Konner writes in *The Trouble with Medicine*,

> As countless people have crossed the threshold of cathedrals or
> mosques or synagogues, desperately seeking a remedy for illness,
> so we in the modern world cross the threshold of a great and powerful hospital. It is our modern cathedral, embodying all the awe
> and mystery of modern science, all its force, real and imagined.

And best of all, it often does what it is supposed to do: it works.... And we may well feel awe, because this is one of the places in our modern world where we experience faith; it is, in effect, our temple of science.[520]

Finally, systems of mass faith can persist only if their followers have a strong need to believe in the authoritative promises and remain ignorant of contradictory facts. As Robert Greene explains in *The 48 Laws of Power*, "People have an overwhelming desire to believe in something,"[521] and this is often exploited by those in power. The average person today is probably as credulous as was the average person in the Middle Ages. As Neil Postman writes, referring to comments made earlier by George Bernhard Shaw, "In the Middle Ages, people believed in the authority of their religion, no matter what. Today, we believe in the authority of our science, no matter what."[522] Such a high degree of gullibility is generally associated with an unwillingness or inability to be informed about all the relevant facts in support of or in opposition to the particular belief system. Throughout the ages, most people were illiterate and depended on the clergy for the transmission and interpretation of scripture. Even today, most Christians are not aware that many historical facts do not support their religious faith and that many beliefs and doctrines that are taken for granted today emerged as a result of fierce political battles and compromises during the early years of the church.[523] Similarly, most people today are insufficiently educated to fully comprehend and evaluate the implications of modern science and technology; they depend instead on the promises of scientists, technical authorities and the media.

Ignorance: The Basis of Most Techno-Optimism

Ignorance of relevant scientific and technical detail underlies most expressions of exuberant technological optimism, not only among laypeople but even among highly educated scientists and engineers. Over the years, it has been our observation when attending scientific conferences, technical meetings and briefings with entrepreneurs, that participants express the most hope about a particular technology when they know the least about it.[524] We gave this observation the amusing title: "Huesemann's Law of Techno-Optimism" which states that optimism is inversely proportional to knowledge. Experts who have extensive knowledge of a specific technology express much more guarded optimism, if not pessimism, and generally provide a more balanced view. Given the extreme complexity of industrialized societies, nearly everyone, including highly educated scientists and engineers, who are essentially laypeople in any field other than that in which they specialize, is ignorant of the intricacies and possible side effects of most advanced technologies outside their narrow fields of expertise. In the absence of knowledge, it is easy to succumb to technological optimism.

Because optimism appears to be greatest when understanding of relevant facts is least, we often find expressions of the most rampant technological optimism among people who are scientifically illiterate. According to a recent report by the National Science Foundation, more than 80 percent of Americans do not meet minimal standards of scientific literacy, defined as knowing the basic facts and concepts of science as well as understanding how the scientific process works.[525] This high degree of scientific ignorance has led to the result that half of all US citizens do not believe in the reality of biological evolution,[526] and to the widespread belief in pseudo-science and paranormal phenomena, with, as recently as 2001, large percentages of Americans believing in astrology (28%), witches (25%), extra-terrestrial beings visiting the Earth (33%) and people being possessed by the devil (40%).[527]

It would be arrogant to suppose that the problem lies only with the scientifically illiterate. Even those educated in science and engineering find themselves quite ignorant when required to critically evaluate new technologies outside their immediate areas of expertise. It has been our own experience, that there are often only a very few experts worldwide, generally less than 100 and sometimes less than 10, who are truly qualified to evaluate, in the necessary technical detail, the positive and negative aspects of a particular new technology. Virtually no one is sufficiently educated, nor can one be, to fully understand and critically judge the enormous number of new scientific and technological developments, a fact that has led to excessive technological optimism.

Ignorance of history may also promote technological optimism. We frequently encounter a mistaken overestimation of the contributions of science and technology to human progress, as for example, the common but false belief that advances in medicine have been the main reason for declining death rates and increasing life expectancies in industrialized societies during the past century. Ignorance of anthropological data allows the common myth to be perpetuated that life in primitive cultures was extremely miserable,[528] a convenient myth that supports the ethnocentric belief that life in today's technological society is a great improvement over earlier times. Some "primitive" peoples even today, not heavily propagandized and given the choice, do not want to be "developed" and integrated into a high-technology society.[529]

As we saw in Part I, there is also limited understanding of (a) the principles governing real-world applications of science and technology (Chapter 1), (b) the problem of unintended consequences (Chapter 2), (c) the degree of technological exploitation and control (Chapter 3), (d) the dubious "solutions" brought about by counter-technologies and social fixes (Chapter 4), (e) the inherent thermodynamic and practical limitations of efficiency improvements (Chapter 5) and (f) the fact that technological progress alone will be insufficient to bring about sustainability (Chapter 6).

Finally, because of an incomplete understanding of the inherent limitations of science and technology, the mistake is often made of extrapolating past trends into the indefinite future. Consider these silly extrapolations by overly techno-optimistic economists: "The costs of raw materials have fallen sharply over the period of recorded history, no matter which reasonable measure of cost one chooses to use. These historical trends are the best basis for predicting the trends of future costs,"[530] and "The most reliable method of forecasting the future cost and scarcity of energy is to extrapolate the historical trends of energy costs."[531]

In general, extrapolations from historical trends are naïve because they do not take into account the fact that the underlying causes of positive developments are not necessarily permanent and may change unexpectedly, thereby resulting in the stagnation or reversal of earlier achievements. For example, population growth has been possible because sufficient food could be produced by converting more land into farmland by clearing forests and irrigating deserts, and by applying petroleum-based fertilizers, pesticides and herbicides, but this cannot continue forever given inherent oil, land and freshwater resource limitations. Similarly, the rise in material affluence during the past century has been driven to a large extent by improvements in technological efficiencies, but, as was shown in Chapter 5, there are inherent thermodynamic and practical limits to efficiency gains. Likewise, as was shown in Chapter 6, economic growth cannot continue indefinitely on a finite planet because of ecological constraints. Thus, simple-minded extrapolations of past trends into the future will most certainly be wrong because the underlying causative mechanisms were not taken into account. As Paul Ehrlich points out, "Proof of past success is no assurance of future well-being, and the mechanical projection of economic curves is hardly a reliable guide to the future."[532] Thus, to avoid the error of simplifying and projecting trends that have complex causes, we should heed the advice given to financial investors: "Past performance does not guarantee future success."[533] The same holds true for technological achievements, given environmental constraints.

Finally, when evaluating progress, one common approach is to compare our modern way of life to that of people in pre-industrial, traditional or so-called "primitive" societies. Since it is often implicitly assumed, as English Enlightenment philosopher Thomas Hobbes proclaimed four centuries ago, that the life of primitive man is "solitary, poor, nasty, brutish, and short,"[534] it seems obvious that hundreds of years of social and technological progress have delivered us from the unbearable living conditions of earlier times. Though it is convenient to belittle the living conditions of primitive and so-called "underdeveloped" societies, thereby automatically exaggerating the comparative benefits of technological progress, it is highly unlikely that life in pre-industrial cultures was as dismal as is claimed by the true believers and promoters of "progress." As E. F. Schumacher writes,

> It is a grave error to imagine that there had never been anything but dire misery for the great mass of humanity in traditional societies. This is a part of the modernistic prejudice which completely falsifies history.[535]

And indeed, the testimony of archeologists and anthropologists who have studied traditional societies around the world substantially falsify claims that life in pre-industrial cultures was full of misery.[536] While ownership of material goods was often minimal, and high child mortality assured natural selection for good health, people were able to meet their basic needs and generally enjoyed more than enough time to maintain a rich and satisfying social and cultural life. Indeed, life for many indigenous peoples worsened precipitously after being invaded by colonial nations, who employed their superior technologies to exploit in the name of "development" (see Chapter 3).

The pervasive myth that life is miserable in "primitive" cultures not only leads us to overestimate the benefits of technological progress, it may also be used to legitimize the "development" of "underdeveloped" people against their will. As Jerry Mander writes,

> I am still astonished when intelligent people describe life in pre-industrial times as dirty, miserable, poor, and subject to the awful expressions of nature. Surely they must be aware that indigenous people of the temperate zones of the planet—long before the harshness of 16th and 17th century Europe—lived very pleasant and relatively easy lives.... Our mythology has been that native people live with the awful oppression of "subsistence economics"—a term that by its mere utterance invokes feelings of pity and images of squalor.... Pre-technological peoples, living hand to mouth in a never-ending search for food and protection from the elements, need and want what Western society brings. So goes the story. Given this logic, most Westerners are shocked to find that the majority of indigenous peoples on the Earth do not wish to climb onto the Western economic machine. They say their traditional ways have served them well for thousands of years and that our ways are doomed to fail.... The familiar assumption that everything before industrialization was pain, poverty, slavery, and victimization by nature is the assumption that works best for the technological-capitalist agenda and its massive invasion of these "afflicted" societies. It makes it seem as if capitalism and industrialization were altruistically motivated, do-gooder activities.[537]

To emphasize the achievements of industrialization, the substantial reduction in work time and the concomitant increase in leisure time during the past 150 years are often cited as proof of technological progress.[538] For example, during the period from 1870 to 1938, annual hours worked per capita fell in France by nearly 1,100, from 2,945 to 1,848. Similarly, from a common base of about 2,950, annual hours worked per person declined by 625, 296, 720, 717 and 902 in Germany, Japan, the

Netherlands, the United Kingdom and the United States, respectively.[539] These trends in work-time reduction have continued at a slower pace in most industrialized nations, with the notable exception of the United States, where annual hours worked per person increased from 1,716 to 1,878 between the years 1967 and 2000.[540]

The problem with these optimistic work-time reduction statistics is that their reference point is the middle of the 19th century, when annual labor hours had reached a peak as a result of large-scale industrialization. It should be noted that the peasants and artisans of the Middle Ages worked considerably fewer hours per year than the factory workers at the height of the Industrial Revolution.[541] As Marshall Sahlins discusses in *Stone Age Economics*, even today's hunter-gatherer tribes spend only three to four hours per day on all economic activities such as plant collection, hunting, food preparation and weapon repair. The Dobe Bushmen of southern Africa worked only 15 hours per week, with 65 percent of the population not working at all.[542] These so-called "primitive" people worked just enough to meet their basic needs, which allowed them large amounts of time for enjoyable leisure activities such as dancing, games, rest and ritual. Even with this minimal effort, their food consumption was varied and adequate, with hunger being less prevalent than in the world at large, where at least 800 million poor people are undernourished.[543] Therefore, when we use as a reference point the life of indigenous peoples instead of the exploited factory workers of England and the United States at the height of the Industrial Revolution, no progress has been made since pre-industrial times in terms of the total number of hours worked per day, the amount of leisure time available for activities that enhance psychological well-being or even food security.

Medical Techno-Optimism

Nowhere is faith in medical technology stronger than in the United States, where approximately two thirds of the public is interested in news about medical discoveries and about one third believe that modern medicine "can cure almost any illness for people who have access to the most advanced technology and treatment."[544] This type of optimism is misplaced and most likely caused by ignorance of the fact that medical advances have had very little to do with the large decline in mortality and concomitant rise in life expectancy during the past 150 years, that the effectiveness of most medical treatments has never been rigorously tested, that a significant fraction of cures are brought about by the placebo effect and that there are inherent limits to increasing the human lifespan.

There has, indeed, been an unprecedented increase in life-expectancy in industrialized nations during the past 150 years. The average life expectancy at birth has increased from about 41 years in 1840 to about 75-80 years (for males and females, respectively) today.[545] The main reason for this significant increase in life-expectancy at birth is the large decline in childhood mortality.[546] For example, death rates for children in the United States fell from 900 per 100,000 in 1900 to 40 per 100,000 in 1985.[547] The critical question, in the context of this discussion, is whether medical advances can be credited with this large reduction in childhood mortality.

Despite the general belief that medical science is responsible for the drastic reduction in death rates and improvements in health in industrialized nations, medical scholars have repeatedly pointed out that modern medicine should not be credited with these achievements. As Richard Taylor points out,

> The majority of the decline in mortality in Western nations over the last 150 years occurred prior to the advent of modern medicine and in association with substantial improvements in nutrition, hygiene, sanitation, water supply, housing, and general social conditions. Unfortunately these factors have not received the attention they deserve. Rather, medical science has been given the credit for spectacular improvements in health which it could not possibly have influenced.... The corollary is that if it were not for doctors, medicines, and vaccines, we would die off like flies.[548]

Although the general improvement in public health in industrialized nations can be attributed to positive changes in economic, social and environmental conditions, the large increase in life expectancy was primarily due to improved nutrition and better hygiene, both of which decreased mortality caused by infectious diseases.[549] As Thomas McKeown summarizes in *The Origins of Human Disease*,

The transformation of health and rapid rise of population in the Western world during the last three centuries have a common explanation: they resulted from a decline in mortality from infectious diseases. The infections declined mainly for two reasons: increased resistance to the diseases due to improved nutrition, and reduced exposure to infection which followed the hygienic measures introduced progressively from the late 19th century. The contribution of medical treatment and immunization to the decline of mortality was delayed until the 20th century, and was small in relation to that of other influences.[550]

It is interesting to note that most of the decline in mortality from infectious disease occurred *prior* to the introduction of antibiotics in the 1940s. Thus, these "miracle" drugs should not be credited with the decrease in childhood death rates and increase in life expectancy. For example, death rates from respiratory tuberculosis in England declined by more than 80 percent between 1855 and 1965 without antibiotics.[551] Similarly, death rates from tuberculosis in the United States decreased about eightfold from 1900 to 1950, indicating that about 90 percent of the improvement had occurred prior to the introduction of drugs that could kill the tuberculosis bacterium.[552] In addition, the combined death rates from scarlet fever, diphtheria, whooping cough and measles from 1860 to 1965 for children up to 15 years of age show that nearly 90 percent of the observed decline in mortality from these conditions occurred prior to the advent of antibiotics or widespread immunization.[553] Similarly, incidence rates for cholera, dysentery and typhoid peaked and declined precipitously prior to the introduction of antibiotics.[554] In conclusion, antibiotics and immunizations have contributed very little, possibly less than 5 percent, to the total decline in death rates from infectious diseases during the past 100 years.[555]

Another indication that modern medical practice has not, to any significant extent, improved life expectancies is the absence of any correlation between health care expenditures and mortality rates.[556] For example, in a survey of nations, infant mortality rates show an inverse relationship to the number of doctors per capita, indicating that social, lifestyle and other factors are more important for infant survival than the delivery of medical care.[557] Similarly, while annual death rates in the United States declined by almost 50 percent (from 17.2 to 9.5 per 1000) between 1900 and 1960,[558] the subsequent introduction of high-tech medicine and the associated tripling of medical care expenditures from 1960 to 2003 (i.e., from 5.2% to 15.8% GDP[559]) has brought about only the rather insignificant decline in death rate from 9.5 to 8.4 per 1000. In summary, the belief that modern medicine is responsible for the dramatic decline in death rate and the concomitant increase in life expectancy that were observed during the past 150 years is unfortunately, like most techno-optimism, based on ignorance.

It is highly questionable whether modern medical practice should be credited with healing common minor ailments or even curing many serious degenerative diseases because most medical treatments and procedures have never been rigorously tested for their effectiveness (see Chapter 8). Consequently, it is possible that a disease condition may improve independently of, or in spite of, high-technology medical treatment. For example, many diseases are cyclical and improve on their own through natural healing mechanisms, which are often assisted by the placebo effect. Indeed, it is highly likely that the placebo effect is responsible for a large percentage of cures that are erroneously credited to high-technology medicine.

According to Steward Wolf, the placebo effect is "any effect attributable to a pill, potion, or procedure, but not to its pharmacodynamic or specific properties."[560] Placebos provide beneficial as well as negative side effects.[561] As Wolf notes, the history of medicine can be viewed as the history of placebos:[562]

> Placebos have been used to alleviate human suffering since the beginnings of medicine, but not usually knowingly. Each medical era has brought forward chemical agents, efficacious at the time, but later found to lack the pertinent pharmacodynamic property. Countless herbs and potions fill the pages of textbooks as prevailing fashions have changed with each generation, their placebo action deriving in part from the faith and enthusiasm of earnest physicians. Many such agents have died out with difficulty. When, from time to time, various ones of them have been exposed as chemically useless, new, often more expensive but equally intrin-

sically inert nostrums have taken their places, each enjoying its day of clinical effectiveness.[563]

As is the case for primitive magical and faith healing, anyone may respond to a placebo under conditions that galvanize that individual's belief. The placebo response is strongest when both the patient and doctor believe in the treatment, and the patient and doctor believe in each other.[564] The doctor's belief in the treatment is probably the most important factor, because the patient's faith and unconscious placebo response is unlikely to be activated without it.[565] Indeed, the American Surgical Association showed more than 50 years ago that skeptical surgeons obtained only half as many five-year cures and 20 times the incidence of recurrent ulcers than did their more enthusiastic colleagues.[566]

As Andrew Weil points out, belief alone can bring about medical cures, and differences in beliefs may be responsible for the variability in treatment success:

> Cures of organic illness following visits to miracle shrines, faith healers, and Christian Science practitioners demonstrate that belief alone, without any physical intervention, is enough to bring about therapeutic success.... Belief in a system of treatment varies from practitioner to practitioner and patient to patient. Such variation can explain why any system works some times and not others. Since belief alone can elicit healing, the occasional success of treatments based on absurd theories is not mysterious. Since belief in a new system is greatest in its inventor and those in direct contact with that person, the declining efficacy of a system over time is also understandable.[567]

Furthermore, the placebo response can cause a wide range of both desirable and undesirable effects. Dr. Weil continues,

> As far as the strength of the placebo response goes, it can be of any magnitude. There are placebo deaths (i.e., voodoo deaths) and placebo cures of cancer.... There is no direct physical response of the human body to any therapeutic procedure that cannot occur with equal form and magnitude in response to an inert placebo. Placebos can relieve severe postoperative pain, induce

sleep or mental alertness, bring about dramatic remissions in both symptoms and objective signs of chronic disease, initiate the rejection of warts and other abnormal growth, and so forth. They can equally elicit all the undesired consequences of treatment with real drugs, including nausea, headaches, skin rashes, hives, and more serious allergic reactions, damage to organs, and addiction.[568]

The placebo response is even stronger in the case of stressing and painful treatments.[569] Consequently, many modern medical procedures are likely to elicit a strong placebo effects. As Dr. Weil notes,

> In our society, widespread faith in technology and the importance of physical reality makes relatively traumatic procedures more effective at mobilizing mind-mediated healing than relatively gentle ones....A drug that produces nausea or dizziness, or an injection that leaves the arm sore, is a convincing treatment. So is any form of surgery. If a mere injection can elicit a placebo response, the mind-focusing power of a full-scale operation, with all of its drama and violation of the body, must be formidable. Any surgery, in addition to its direct effect, is likely to generate a placebo response.[570]

In the few cases where systematic studies have been conducted, there is evidence that the effectiveness of certain surgeries is entirely due to the placebo effect. In the 1950s, a team of cardiologists at the University of Washington, led by Dr. Leonard Cobb, carried out a carefully designed study to test the effectiveness of internal mammary ligation surgery on the relief of angina pectoris. Remarkably, the control group that received a sham operation (i.e., just a skin incision without ligation surgery) had the same degree of improvement (i.e., reduced chest pain, less drug use, and positive changes in the electrocardiogram) as the patients who received the full operation.[571] Similarly, more recently a study was conducted to evaluate the effectiveness of arthroscopic surgery for arthritis of the knee. Again, patients undergoing simulated surgery involving just a skin incision reported the same improvement as those undergoing the real surgery, which entailed removing rough joint cartilage. As one patient who unknowingly received the sham surgery told a reporter: "The surgery was two years ago and the knee never has bothered me since."[572]

The prevalence of the placebo response is significant. In 15 studies involving more than 1,000 patients, the placebo effect accounted for significant improvements in conditions ranging from serious post-operative pain and angina pectoris to nausea and even the common cold in about 35 percent of all cases.[573] Similarly, surgery was found to bring about a placebo response in approximately 37 percent of all patients.[574] Thus, over one third of the observed improvements following medical treatments are likely attributable to the placebo effect. Therefore, until most medical treatments are rigorously tested in double-blind placebo-controlled trials, there is no reason to give modern high-tech medicine full credit for eliciting cures. As in the earlier history of medicine, they may very well be due in large degree to the placebo effect.

Because life-expectancies have increased continually during the past 150 years, many believe that these trends will continue indefinitely and that modern medicine will, sooner or later, conquer death altogether. The age-old quest for immortality, this time with the help of science and technology instead of religion, is misguided for several reasons. The mistaken belief that humans will continue to live increasingly longer lives is the result of confusing two related terms: life expectancy and life span. As James Fries and Lawrence Crapo explain,

> The life span is the biological limit to length of life. Each species has a characteristic average life span. This life span appears to be a fixed, biological constant for the species. For human beings, the life span has not changed for millennia. Life expectancy refers to the number of years of life expected from birth for an individual or group. Life expectancy from birth cannot exceed the life span, but it can closely approximate the life span if there is little death at early ages.[575]

Although life expectancies have increased during the past 150 years, the life span has not changed for at least 100,000 years and is about 85 years.[576] Although diets, lifestyles, vitamins, drugs and medical treatments may increase life expectancies, they have not been effective in extending the human life span.[577] Therefore, the quest for immortality or for significantly extending the human life span, will remain what it has always been: a futile search motivated by hope, without an appreciation of biological facts.

Death is also unavoidable because physiological functions and organ reserves decline with age. Death occurs when the body can no longer maintain the necessary internal balance, or homeostasis, in the presence of external perturbations. As James Fries and Lawrence Crapo explain,

> Living organisms under threat from an extraordinary array of destructive sources maintain their internal milieu despite the perturbations.... The ability of the body to maintain homeostasis declines inevitably with decreasing organ reserve.... The important point, however, is that with age there is a decline in the ability to respond to perturbations. With the decline in organ reserve, the protective envelope within which a disturbance may be restored becomes smaller.... If homeostasis cannot be maintained, life is over.... Death must inevitably result when organ function declines below the levels necessary to sustain life.[578]

Given that medical science and technology has not made progress in restoring physiological functions and organ reserves to youthful levels, death will remain inevitable in the foreseeable future.

Finally, even if the human life span could be extended indefinitely, fatal accidents would still happen, with the result that the average life expectancy would be "only" around 1,200 years.[579] It must be recognized, however, that such a large increase in human longevity would have many unexpected, somewhat depressing consequences. For example, to avoid human overpopulation, the birth rate would have to be as low as the death rate, resulting in a situation in which most couples could not have children and most people would be very old.

In summary, beliefs that progress in medical science and technology is the primary cause for the large increase in life-expectancy during the past 150 years, that high-tech medical treatments are solely responsible for cures and that continued medical research may ultimately conquer death are all based on ignorance. Technological optimism and belief in progress in general are, to a large degree, based on ignorance. Unfortunately, this ignorance is created and maintained by the mass media, which are positively biased because they are controlled by those who profit most from the introduction of new technologies.

Techno-Optimism and the Mass Media

There are few who are not impressed by the seemingly endless stream of technological innovations: computers, cell-phones, video cameras, the Internet and countless electronic devices. New technologies often have an immediate appeal, with attention being focused on short-term benefits while long-term negative consequences are not anticipated.[580] As Jerry Mander observes,

> We are hypnotized by the newness of the machine, dazzled by its flash and impressed with its promise.... Compounding this problem is the fact that every technology presents itself in the best possible light. Each technology is invented for a purpose and it announces itself, as it were, in these terms. It arrives on the scene as a "friend", promising to solve a problem.... What's more, the new machines actually do what they promise to do, which leaves us feeling pleased and impressed. It is not until much later, after a technology has been around for a while—bringing with it other compatible technologies, altering economic arrangements and family and community life, affecting culture, and having unpredictable impact on the land—that societies both familiar and unfamiliar with the machine begin to realize that a Faustian bargain has been made.[581]

As was pointed out earlier, it is inherently impossible to design new technologies that are guaranteed to never have negative effects of any kind (Chapter 1). Certain segments of the population will profit and other segments will not. Those who enjoy the benefits are likely to be techno-optimists. Unfortunately, the mass media exploit the public's ignorance and the inherent fascination with novelty to promote the false view that there will be only positive effects both now and in the future. This is done by presenting the most positive aspects of a new technology while omitting to mention potential negative side effects. As Jerry Mander writes,

> That our society would tend to view new technology favorably is understandable. The first waves of news concerning any technical innovation are invariably positive and optimistic. That's because, in our society, the information is purveyed by those who stand to gain from our acceptance of it: corporations and their

retainers in the government and scientific communities. None is motivated to report the negative sides of new technologies, so the public gets its first insights and expectations from sources that are clearly biased.[582]

Why do the mass media exhibit such a positive bias toward new technologies? Before attempting to answer this question, we need to understand the function of mass media in industrialized societies. According to the model of propaganda developed by Edward Herman and Noam Chomsky, "The 'societal purpose' of the media is to inculcate and defend the economic, social, and political agenda of privileged groups that dominate the domestic society and the state."[583] The mass media are controlled by those who are able to pay for access to it. This situation is described by Edward Herman's "power law of access" and the "inverse power law of truthfulness," which can be summarized as the greater the economic and political power, the easier it is to gain access to the media and the more freedom is granted to tell complete lies.[584] Information distributed by the mass media is often greatly biased because it has first to pass through a number of filters, such as inoffensiveness to influential owners of media conglomerates or to corporations that pay for the advertisements. Anything that even indirectly places media profits at risk has a very small chance of being broadcast.[585] The result is that through careful selection of topics, framing of issues, filtering of information, special emphasis and tone, the public debate about controversial topics such as the introduction of new technologies is kept within the bounds of the prevailing paradigm: any technological innovation is "good" and its value is not open to question. As Herman and Chomsky conclude,

> In sum, the mass media of the United States are effective and powerful ideological institutions that carry out a system-supportive propaganda function by reliance on market forces, internalized assumptions, and self-censorship, and without significant overt coercion.[586]

Corporations develop and market new technologies for a very simple reason: to increase their sales volumes and profit margins. Technological innovation is a proven way to increase a company's market share, and competitiveness, but only if the new products can be advertised heavily through the mass media. Since many corporations have large marketing budgets, they have easy access to the mass media by purchasing advertising space and time. Since the goal of all corporations is to increase profits, it is necessary to present new technologies only in the most favorable light. The negative side effects of technological innovations, even when known, are omitted or minimized. The mass media, therefore, convey an unrealistically positive view of new technologies because they are, to a large degree, controlled by the very corporations that stand to profit from them.

It would be a mistake, however, to hold corporations and the mass media entirely responsible for the rampant technological optimism in the United States. As Neil Postman observes in *Technopoly*, the belief that a new technology creates only positive effects is much more than a conspiracy of entrepreneurs and corporations against the public. It is, in fact, a deeply embedded cultural phenomenon:

> In cultures that have a democratic ethos, relatively weak traditions, and a high receptivity to new technologies, everyone is inclined to be enthusiastic about technological change, believing that its benefits will eventually spread evenly among the entire population. Especially in the United States, where the lust for what is new has no bounds, do we find this childlike conviction most widely held.... This naive optimism is exploited by entrepreneurs, who work hard to infuse the population with a unity of improbable hope, for they know it is economically unwise to reveal the price to be paid for technological change. One might say, then, that if there is a conspiracy of any kind, it is that of a culture conspiring against itself.[587]

The Decline of Techno-Optimism

In order to suppress anything that might threaten social cohesion or challenge the power structure, every society has taboos related to certain kinds of discourse and action.[588] In modern industrialized societies, there is a strong taboo against challenging the faith in science and technology and their supposed contribution to "progress." Any questioning of that faith is seen as heresy. Those who criticize new technologies are labeled "anti-progress" or, in more derogatory terms, "Luddites," after the machine-smashers who opposed the mechanization of labor during the Industrial Revolution of 19th-century England. Indeed, the idea of "progress" is used to suppress criticism, to enforce passivity and to avoid debate about the introduction of new technologies. Criticism of technological and industrial development is often stifled by invoking the illusion of inevitability, the "you can't turn back the clock" argument (see Chapter 10). As Jerry Mander observes,

> The operating homilies remain the same: "You can't stop progress", "Once the genie is out of the bottle you cannot put it back", "Technology is here to stay, so we have to find ways to use it better". In reality, these are all rationalizations to cover up a culture-wide passivity; a failure to take a hard look at technology in all of its dimensions, or to draw the obvious conclusions from the evidence at hand.[589]

In addition to "there is no turning back" arguments, technological developments are often promoted as value neutral. As will be discussed in Chapter 10, the myth of value-neutrality is employed to convince people that technological change occurs in an objective and rational way, thereby presenting it as inevitable and legitimate, and deflecting potential criticism.[590] Finally, even if critical attitudes toward technological development are expressed, most dissenters have insufficient financial means to access the media.[591] Individuals in our society generally lack the freedom of *public* speech, which is effectively reserved, as mentioned above, to those who can pay for it. The general public is, therefore, unlikely to hear about the potential negative aspects of new technologies, thereby promoting the impression that all technological change is good.

Despite the many ways in which criticism of technology is discouraged and even suppressed, there has nevertheless been a decline of faith in science and technology during the past century. The idea of progress being brought about by science and technology was dealt a number of severe blows, first by the technologically facilitated violence and destruction witnessed during World War I and World War II and later by the many serious environmental problems precipitated by synthetic organic chemicals, large-scale industrialization and the adoption of materialistic, resource-intensive lifestyles. As Thomas Parke Hughes describes in *Changing Attitudes Toward American Technology*,

> The carnage of World War I and the well-publicized exploitation of technology brought a new ambiguity, the legacy of a century of rarely qualified praise confronted by the obvious horrors of total war. It was further reinforced by the growing belief that technology was not directed by benign providence or inexorable laws of progress towards social objectives believed generally—and vaguely—desirable. The growing conviction that technology and progress were not synonyms and that man needed to assume control of, and make decisions about, the use of technology brought less euphoric prophesies for the future.[592]

The beginning of the environmental movement about half a century ago followed by the atrocities of another high-tech war in Vietnam engendered a highly critical attitude toward technology and the industrialized culture it reflected. As Kenneth Stunkel and Saliba Sarsar wrote in 1994,

> The insight that all technological development is not a blessing has emerged in the past 30 years. Suddenly many technologies do not seem benign.... If one means by "progress" improving and dignifying the human condition, its identification with technological development has become problematic and ambiguous.[593]

Similarly, Jerry Mander observes,

> Technological society, during the past half-century, has demonstrably not achieved the benefits it advertised for itself. Peace, security, public and planetary health, sanity, happiness, fulfillment are arguably less close at hand than they ever were in the past.[594]

Compared to religions that make grandiose promises that are inherently untestable, thereby protecting themselves against criticism and maintaining their authority and belief systems for thousands of years, the promises made by science and technology can be readily and objectively evaluated. In the early stages of scientific and technological development, when many promises were fulfilled or even exceeded, the countless negative consequences that later appeared have also been objectively quantified and to some extent publicized, thereby eroding somewhat the earlier optimistic belief in scientific and technological progress. As Jeremy Ravetz observes, "'Losing faith' is supposed to be a problem that afflicts the religious; but on reflection we see that belief in progress and science is also vulnerable in its own way."[595] Indeed, science and technology has come under increasing attack from various quarters: the environmental and counter-culture movements, anti-establishment intellectuals and religious fundamentalists.[596]

These critics point to overwhelming evidence that advances in science and technology have caused not only many unintended negative and often irreversible consequences but also have not fulfilled the earlier promise of greater human happiness and well-being. As discussed in Chapter 2, the number of unintended consequences is large: environmental pollution, ecosystem destruction, species extinction, global climate change, human overpopulation, increased violence and destruction, as well as social and moral dysfunction, which result from the adoption of an overly materialistic lifestyle. As will be discussed in Chapter 9, despite rising material affluence in industrialized nations, people are not happier than before, indicating that much of the "progress" brought about by science and technology has been an utter failure in terms of human psychological well-being.

Georg Friedrich Juenger observed more than 50 years ago in *The Failure of Technology* that "progress is an optical illusion."[597] It is an illusion embraced with the fervor of a religious faith, and it is promoted and exploited by those who profit most from the development and distribution of technological innovations.

Clearly, a redefinition of progress is needed. Hopefully, our improved understanding of the limitations of science and technology will result in a new paradigm of progress: progress will no longer be seen as the technological control and exploitation of nature and people for the benefit of the few with negative consequences for the many. Instead, progress will consist of increasing our awareness and understanding of how to adapt to our natural environment and live within its limits, and how to improve our well-being and happiness in non-materialistic ways. This paradigm shift and its implications for our lifestyle choices as well as for the practice of science and technology will be discussed in more detail in Part III.

CHAPTER 8

The Positive Biases
of Technology Assessments
and Cost-Benefit Analyses

NATURE MAY BE SEEN as a complex interconnected network of cause-and-effect relationships that is continuously adjusted by the process of evolution. It follows that any attempts to modify the natural order through science and technology in order to bring about perceived benefits will likely produce negative effects as well. As Jeremy Ravetz observes, "It is impossible to segregate…the 'negative' aspects of the phenomenon from the 'positive.'"[598]

This aspect of technology has been described by various authors in different ways: "Every technical advance is matched by a negative reverse side,"[599] "Every technology is both a burden and a blessing; not either-or, but this and that,"[600] "Technology, in sum, is both friend and enemy,"[601] "There is no such thing as a free lunch" (Commoner's fourth law of ecology),[602] "The central fact about modern technology is that its powers for both good and evil increase as it evolves,"[603] "If history has taught us anything, it is that every new technological revolution brings with it both benefits and costs,"[604] and finally, "Can any reasonable person believe for a moment that such unprecedented [technological] power is without substantial risks?… The point is, power is never neutral. There are always winners and losers whenever power is applied."[605]

Technological manipulations amount, at best, to zero sum games in which costs balance benefits. Of course, if mega-technology is developed in the presence of enormous ignorance of the myriad complex processes of nature, it is also possible that technological innovations will result in negative sum outcomes where costs exceed benefits by a large margin, as in the case of the irreversible negative consequences of species extinction or the potentially irreversible consequences of global climate change (see also Chapter 2). As was shown in Chapter 7, it is ignorance that fuels the belief in perpetual technological progress, a belief that technological developments result in overall net improvements, that benefits always exceed costs.

The illusion of progress is maintained and further advanced by technology assessments, risk assessments, cost-benefit analyses and life cycle analyses, most of which are biased in favor of technological development. In reviewing the standard procedures involved in cost-benefit analyses, we find that each step in technology assessment has its inherent problems and ambiguities, some of which are specifically exploited, knowingly or unknowingly, to produce positive recommendations for new technological developments and deployments. We conclude with a number of specific examples in which either positively biased or poorly performed cost-benefit analyses were responsible for the development and diffusion of technologies of marginal or no benefit.

An Overview of Cost-Benefit Analysis

Before discussing the inherent limitations of cost-benefit analysis, often referred to as technology assessment or risk assessment, we need to understand the basic procedure. It is important to recognize that the entire concept of cost-benefit analysis is implicitly based on the underlying moral philosophy of utilitarianism, which claims that ethically preferred choices and actions are those which maximize overall benefits, or to quote Jeremy Bentham's maxim, to generate "the greatest good for the greatest number."[606] Thus, the overall goal of cost-benefit analysis is to select those policy options with respect to technology development and deployment that maximize overall benefits and minimize costs and potential risks.

A comprehensive and systematic cost-benefit analysis involves the following steps:[607]

1. Decide whose benefits and costs are to be taken into consideration (i.e., who has "standing"). In most cases, these are the persons who are most directly impacted, either positively or negatively, by the

technology under consideration. For example, in the case of siting a planned nuclear power plant, people who live in close proximity to the proposed plant and waste-handling facilities should have "standing" because they bear most of the risks related to radioactivity. In addition, people who will benefit from the electricity provided should also be considered in the cost-benefit analysis. In principle, future generations, as well as animal and plant species, or even the environment in general, should be considered to have standing if any of these are impacted by specific technological developments.[608]

2. Construct a portfolio of possible alternatives. These could consist of alternatives involving different technologies or different deployments of the same technology. For example, the costs and benefits of nuclear power generation could be compared to power generation from fossil fuels, hydroelectric dams or windmills. The costs and benefits of siting the same nuclear power plant in different locations could also be evaluated and compared.

3. Catalog all potential impacts and select measurement indicators. For example, the presence of a nuclear power plant might depress nearby property values and could result in various radiation-induced diseases, including premature death, in the exposed population, particularly in case of plant malfunctioning or reactor core meltdown. The creation of new jobs and the generation of electricity could be considered positive impacts. Measurement indicators for these examples could be percent reduction in real estate values near the power plant, health care costs resulting from radiation-induced diseases, lives lost prematurely, number of jobs created and money saved per kWhr electricity, respectively.

4. Predict quantitative impacts over the life of the project. For example, to estimate the additional health care costs of radiation-induced diseases, it would be necessary to have a sound quantitative estimate of the risk of nuclear plant accidents. In addition, negative impacts related to radiation may continue beyond the useful life of the power plant because the radioactive wastes could cause harm, if not carefully contained and monitored, for thousands of years.

5. Monetize (attach dollar values to) all impacts, for example, the positive financial impact of job creation or the cost of lives lost because of radiation-induced diseases.

6. Discount for time to find present values. The need to discount arises because of the preference of people to consume now rather than later. A more subtle reason for discounting is that the present is generally considered, with all its benefits and costs, to be more important than an unknown future. Thus, for any project that has either costs or benefits arising over many years, it is necessary to have a method for aggregating the benefits and costs that occur at different times. Generally, future benefits and costs are discounted relative to present benefits and costs. For example, while people who live presently in the vicinity of a nuclear power plant may have the same risk of radiation-induced diseases as those who will live there 30 years from now, the risk to people in the future is discounted significantly in most cost-benefit analyses, often to negligible levels or zero.

7. Sum up all monetized benefits and costs.

8. Perform a sensitivity analysis. For example, it may be possible to evaluate the effect of various nuclear reactor safety designs on the risk of radiation accidents, which in turn will affect the outcome of the cost-benefit analysis.

9. Recommend the alternatives with the largest net monetary, social and environmental benefits.

Although this step-wise procedure appears to be straightforward, many difficult problems arise as soon as one actually tries to carry out a real cost-benefit analysis.

Several major inherent problems and ambiguities in cost-benefit analyses erroneously bias risk assessments in favor of new technology.

Problem #1: Boundary Selection and Externalization of Costs

The selection of spatial and temporal boundaries for a given cost-benefit analysis has a significant impact on the final outcome. It is extremely difficult to determine whose benefits and costs should be counted, (i.e., who has "standing" (Step 1)). Should it be a selected group of people, all humans presently living, future generations, certain animal species, all animals, all plants, the entire environment? Should the boundaries of the analysis be at the local, state, national or international level? The answers to these questions cannot be found by objective analyses but depend very heavily on subjective judgments,[609] thereby greatly increasing the likelihood of value conflicts and power struggles.

Because the use of any technology results in a spatial and temporal separation of costs and benefits,[610] the cost-benefit ratio can be manipulated by adjusting the spatial and temporal boundaries of the analysis.[611] Given that many technological benefits are immediate and obvious, while potential negative consequences may be delayed and difficult to predict or detect (Chapter 2), the cost-benefit ratio generally becomes less favorable as spatial and temporal boundaries of the analysis are widened in space or time to include more distantly affected parties. For example, as long as we evaluate the benefits and drawbacks of specific technologies, such as the automobile or the television, in very narrow personal terms by asking "How are they useful to me right now?" we will see these technologies in too positive a light and thus accept them readily.[612] If we were to consider all potential negative short- and long-term social and environmental consequences of these technologies (Chapter 2), the positive aspects may no longer outweigh the negative ones, and we may be much more hesitant to adopt them. Similarly, at the corporate level, technological innovations are judged primarily in terms of whether they will result in an overall net benefit to the company; that is, they will be adopted only if they are able to generate profits. If all environmental and social costs, now and in the future, were taken into account, the cost-benefit ratio would not necessarily be favorable, and many innovative technologies would never reach the market.[613]

Although in recent years companies have been forced by environmental laws and regulations to consider the various potential environmental costs of their business decisions, the costs and benefits that may accrue to future generations are rarely considered. In fact, our present political and economic institutions discount the future so heavily that consequences more distant than a few decades are essentially ignored.[614] This discounting of the future to obtain the present value (Step 6 above) is ethically questionable because potential negative consequences to future generations may be greatly underestimated. In this context it should be recognized that many unsustainable and environmentally polluting technologies generate benefits to presently living people at the expense of future generations. Thus, it is highly likely that our affluent lifestyle of today will deprive people of the future a similar lifestyle.[615]

Future generations will have to deal with a number of serious negative consequences that have been created by our present technologies, for example, the depletion of non-renewable energy and mineral resources, soil erosion from unsustainable industrial farming practices, global climate change caused by the burning of non-renewable fossil fuels, human overpopulation, species extinction, pollution by resistant synthetic organic chemicals, radiation from nuclear wastes and so on (see also Chapter 2). The problem of nuclear waste storage is a good example of how hundreds of future generations will have to bear the risk of radiation-induced illness, permanent damage to the gene pool and perhaps premature death without receiving any benefits, all of which accrued instead to a few generations that used the relatively inexpensive electricity to support their materialistic lifestyle.[616] Similarly, while hundreds of millions of people may have benefited during the past 100 to 150 years from the burning of fossil fuels, many billions of future people will for hundreds, if not thousands of years, have to bear the environmental and social costs brought about by global climate change.

The discounting of the future to obtain present value (Step 6 above) and the resulting transfer of significant costs and risks to future generations has been justified by claiming that in practice the more remote effects of our technological choices and actions are usually too small or too uncertain to be taken into account.[617] While this may be true in some cases, discounting of the future clearly violates the ethical principle of intergenerational justice[618] because it basically amounts to the exploitation of future generations by the present generation (Chapter 3). In addition, it has been claimed that because the needs of future generations cannot be known with absolute certainty, current people have no obligation to consider them. But clearly, people in the future will most likely value their health and life as much as people do at present.[619] As R. and V. Routley conclude,[620]

> The question we must ask is what features of future people could disqualify them from moral consideration or reduce their claims to it to below those of present people? The answer is: in principle none. Prima facie moral principles are universalizable, and

lawlike, in that they apply independently of position in space or in time.... As a result of this universalizability, there is the same obligation to future people as to the present.[621]

The transfer of risks to future generations is an excellent example of cost externalization. Negative externalities occur when one party is adversely affected by another's action but receives no payment or compensation in return.[622] Externalized costs are not captured within the economic system through prices. As a result, consumer prices are generally lower than they would be if all costs were internalized, thus, externalization stimulates commodity consumption. For example, if all the environmental and social costs of gasoline manufacture and use were to be internalized, the cost of gasoline would be much higher than it is today, which, in turn, would discourage the wanton overconsumption of this non-renewable fuel.[623]

Given that profits are generally defined as the difference between revenues from product sales and the overall cost of operating a business,[624] the externalization of costs increases profits in two ways. First, as just indicated, by artificially depressing the price below the true market value, cost externalization spurs consumption and thus increases sales revenues and, by implication, profits. Second, and more importantly, any effort to reduce business costs by externalizing some of them onto unsuspecting victims will immediately and automatically increase profits. Indeed, from a business perspective, the best way to maximize profits is by having someone else shoulder the environmental and social costs of an enterprise.[625] Thus, there is generally strong opposition by business interests to any laws and regulations that attempt to mandate the internalization of previously externalized costs. As Massachusetts Institute of Technology (MIT) professor Noam Chomsky notes, "Costs and risks are socialized, and the profit is privatized. That's called capitalism."[626]

The extent of cost externalization by US corporations is enormous. As David Korten writes in *Agenda for a New Economy,*

In 1996, Ralph Estes, a CPA and a professor of accounting, published an inventory of the results of formal studies documenting uncompensated costs that corporations pass on to the public each year—not including direct subsidies and tax breaks.... The total for the United States came to $2.6 trillion a year in 1994 dollars. This was roughly five times the reported corporate profits in 1994 ($530 billion), and 34 percent of 1994 US GDP of $6.9 trillion. Most of the documented externalized costs reflected real-wealth losses, such as worker injuries; the toxic contamination of land, air, and water; and unrecoverable resource depletion. These results reveal a pattern of massive market distortion that put a lie to corporate claims of efficiency and public benefit.[627]

It is clear from these statistics that the reported corporate profits could be realized only in the presence of massive cost externalization. If corporations were forced to pay these externalized costs, they would no longer be profitable and thus would cease to exist.

The ability to externalize costs is also related to power.[628] In general, the powerful exploit the weak and those who cannot defend themselves by externalizing costs onto them. The exploited include uninformed people, the poor, people not yet born and the environment. The more powerful actors in society (i.e., large corporations and the government agencies serving them) control the technology assessment process with the goal of biasing it in favor of new technology development and deployment by defining the boundaries of the cost-benefit analysis so as to allow for the externalization of costs. (See Problem #3 on institutional biases below.)

If the boundaries of the cost-benefit analyses were expanded to include all impacted parties, the cost-benefit ratio of many innovative technologies would be much less favorable. As a result, many technological innovations would not be developed further and, as mentioned above, would never reach the market.

Problem #2: Prediction of Potential Impacts and Selection of Appropriate Indicators

After the boundaries of the cost-benefit analysis have been selected, it is then necessary to predict all potential positive and negative impacts over the lifetime of the proposed project or technology deployment and to select appropriate indicators for them (Steps 3 and 4 above). As Charles Perrow points out in his classic *Normal Accidents: Living with High-Risk Technologies*, serious accidents are inherently unavoidable in complex technological systems such as petrochemical plants, refineries, nuclear reactors, and so forth.[629] The question is not *if* complex technological systems will fail but rather *when* they will fail. For example, despite repeated safety assurances by the nuclear-industrial complex, there have been at least three serious nuclear power plant accidents resulting in the release of dangerous radiation affecting thousands, and in some cases, potentially millions of citizens: Three Mile Island (1979), Chernobyl (1986), and Fukushima (2011).

As was discussed in Chapter 1, the prediction of all possible consequences is impossible, particularly given the inherent limitations of current mechanistic, reductionist science. Because there will never be sufficient research funding to elucidate all possible interactions and cause-effect relationships within complex environmental and social systems, there will always be a high degree of ignorance and uncertainty with respect to the prediction of unintended side effects.[630] As Jeremy Ravetz observes,

> The pervasiveness of ignorance concerning the interactions of our technology with its environment, natural and social, is a very new theme. "Scientific ignorance" is paradoxical in itself and directly contradictory to the image and sensibility of our inherited style of science and its associated technology. Coping with ignorance in the formation of policy for science, technology, and environment is an art that we have barely begun to recognize, let alone master. Yet ignorance dominates the sciences of the biosphere.[631]

As a result of this insurmountable ignorance, many less obvious but potentially important environmental, social, cultural, political, psychological and spiritual consequences of new technologies and their interactions are easily overlooked.[632] The long list of unintended consequences society is experiencing today is a sobering reminder of our collective ignorance of the long-term effects and second-order consequences of technological innovations.

Even if all potential positive and negative impacts could be predicted, we would still have the substantial challenge of finding specific measurement indicators that would properly quantify each impact. (See Problem #4 on monetization below.) The selection of measurement indicators depends on our current state of knowledge, which, because of its perpetual incompleteness, could result in a situation in which the most important impacts would not be measurable.[633] For example, how is one to quantify a measurement indicator for species extinction if one does not even know whether a specific species exists or whether it will be negatively affected by human activities? In addition, because the choice of a specific indicator will usually depend on what is easy to measure rather than what is an appropriate estimate of the impact, it is likely that important consequences will not be quantified. For example, in the past, the effects of chemical pollution on rivers and lakes was simply quantified as "fish kill," because this was all that could be measured. Today, because of the development of much more sophisticated instrumentation and techniques, more subtle effects of chemical pollution can be detected, such as the bioaccumulation of organic chemicals in the fatty tissues of aquatic species or the disturbance of sexual development in fish by endocrine disruptors.[634]

Because technological benefits are often immediate and obvious, while the negative consequences are delayed and less obvious, it follows that our collective ignorance causes most cost-benefit analyses to be biased in favor of technology development and deployment. As Harold Green points out,

> In the early states of development and introduction of a new technology, potential adverse effects are uncertain and speculative. This means that in any objective balancing of benefits against adverse consequences at this stage, obvious and immediate benefits will almost always outweigh adverse consequences. Moreover, there are always well-financed vested interests in the technology to articulate and press the benefits, while minimizing the risks, but there rarely are any knowledgeable persons with sufficient interest and resources to articulate the risks.... Controls will be deferred until adverse consequences have been demonstrated through experience. By that time, however, the vested interests in use of the technology have become more powerful, and the public has become accustomed to its benefits, making the political effort to impose controls much more difficult.[635]

Problem #3: Institutional Biases and the Perception of Costs and Benefits

In the preceding discussion of cost-benefit analysis, it has been implicitly assumed that it is obvious whether a specific consequence should be defined as a cost or a benefit. Unfortunately, this is often not the case because diverse stakeholders may have vastly different views as to what constitute costs and benefits, and their relative importance. Consider this example from a standard textbook on cost-benefit analysis:

> Some may view the impact as a benefit while others may view it as a cost. Consider, for example, "flooded land" as a potential impact category. Residents of the flood plain will generally view floods as a cost, while duck hunters will regard them as a benefit.... Sometimes such differences in valuations flag issues relevant to standing or the issue of weighting costs or benefits that accrue to particular individuals or groups.[636]

In most cases, cost-benefit analyses are carried out by stakeholders consisting of major institutions such as government agencies, corporations or environmental organizations. Because each institution has a specific stake in the final outcome of the analysis, it is extremely difficult to control for bias. Indeed, each institutional stakeholder attempts to define the problem in such a way as to maximize its own specific objectives such as maintenance of power, profit maximization or promotion of certain ideologies. And because most employees are financially dependent on the institution for which they work, they may have great difficulty carrying out objective analyses and may instead choose to promote and defend the highly biased positions of their respective institutions.[637] Consider, for example, the conflicting perspectives on costs and benefits among different government agencies, some of which have been classified in their bureaucratic role as "fiscal guardians" and others as "spenders":

> Guardians tend to be found in central budgetary agencies.... Their natural tendency is to equate benefits with revenue inflows to their agency or other government coffers and costs with revenue outflows from their spending agency or government coffers. Thus, they engage in revenue-expenditure analysis...

and they ignore non-financial social benefits.... Spenders tend
to come from service or line departments.... Most importantly,
spenders have a natural tendency to regard expenditures on con-
stituents as benefits rather than as costs. Thus, for example, they
typically see expenditures on labor as benefits rather than costs.
Spenders regard themselves as builders or professional deliver-
ers of government-mandated services.... Guardians and spenders
almost always oppose one another.[638]

In addition, representatives from different organizations focus on different kinds of risks according to what they perceive to be most threatening to their respective institution's objectives and mission. For example,
corporations stress technical risks because these are prone to reduce
profits; government agencies focus on societal risks; and environmental
organizations on environmental risks.[639] Furthermore, different organizations have opposing views of nature, which, in turn affect their attitudes toward environmental risks. For example, government agencies
consider nature to be resilient within reasonable limits, requiring them
to impose certain rules and regulations to contain risk within these limits. Egalitarian environmental groups see nature as fragile, requiring
extensive protection from human interference. Business interests often
view nature as an unlimited source of wealth if all "artificial" constraints
imposed by government agencies and environmental organizations
were removed.[640]

In many cases, the underlying value conflicts among different stakeholder organizations are impossible to resolve. As a result, the final outcomes of cost-benefit analyses and technological risk assessments are
often determined by power and politics rather than by rational and objective examination of all relevant issues. The resulting struggle is often
won by the organizations with the greatest financial resources and, thus,
the greatest ability to define the problem in such a way as to sway the outcome of the analysis in their favor.[641] For example, the way the boundaries of the analysis are selected, who is considered to have standing, which
potential alternatives are chosen and whether and how to attach prices
to non-market values such as health and environmental quality all have
significant impacts on the outcome of technology assessments. Given
that, in most cases, the large corporations that develop new technologies

have much greater financial resources and often superior scientific and technical expertise with which to define and shape the issues than either regulatory agencies or environmental organizations, it is not surprising that the outcomes of many cost-benefit analyses are biased in favor of technology development and deployment.

In conclusion, the entire cost-benefit analysis procedure is neither an objective nor a value-neutral exercise (see Chapter 10) but, instead, is greatly influenced by the values of the most powerful stakeholder. The underlying value conflicts and institutional biases are often not readily apparent because they are obscured by the seemingly objective and rational cost-benefit analysis procedure. (See Steps 1–9 above.) As Robert Proctor observes, "Risk assessment provides an ability to cloak political or value decisions in the false objectivity of science."[642] The entity with the greatest financial resources and influence, most often the very corporation that has developed the new technology in question, defines the key issues and controls the procedure, thereby biasing the outcome of the cost-benefit analysis. In summary, power rather than objective analysis determines which innovative technologies will ultimately be adopted by society.

Problem #4: Monetization of Non-Market Values

Step 5 of the cost-benefit analysis procedure (above) involves the monetization of all projected positive and negative impacts. The attachment of a dollar value to widely divergent impacts provides a measure that is applicable to all of them. This allows the summing up of all positive and negative effects (Step 7). Generally, the policy option that has the greatest net positive monetary benefit (the sum of all monetary benefits minus the sum of all monetary costs) will be preferred. Monetization enables the comparison of various benefits and costs, which, because of their vastly different characteristics, would otherwise not be comparable. For example, through monetization, it is supposedly possible to compare the relative costs of different negative technology-induced impacts such as habitat loss, species extinction, increases in various human illnesses and the increases in human mortality.[643]

Economists use a number of techniques for assessing the market value of various benefits and costs, such as determining the actual market price, identifying the price of a market substitute or interviewing people in order to estimate their "willingness to pay."[644] The main problem with these and other monetization procedures is the underlying assumption that economic markets are appropriate measures of what is valuable.[645] Indeed, it is exceedingly difficult, if not impossible, to assign a price to non-market values.[646] As Kenneth Stunkel and Saliba Sarsar summarize,

> The inclination to translate harms and goods into monetary terms ignores a large field of intangible goods, such as the beauty of natural areas, the closeness of communities, attachment to historical sites that link people to the past, and so on. Everything that people value, and for which they would sacrifice a great deal, cannot be converted to market values.... The most difficult intangible to cope with is human life. How much disease and loss of life is a policy or technology worth? How does one determine "worth"?[647]

And, as John Peet points out, the very idea of attaching prices to highly valued items that should never be considered "for sale," such as health or life, is seen by many people as completely inappropriate, even distasteful:

> To most of us, some things or people in our lives or in nature are priceless. That does not mean that they have no value, nor does it mean that their value is infinite. It just means what is says—that they have no price.... No numerical value can possibly take account of people's spiritual relations with their environment or with one another. To attempt to reduce emotions and spirituality to numbers leads to the debasement of everything.[648]

To provide an example of the difficulty in determining the market value of risks, consider that most people, for a given benefit, are willing to accept much greater risks if the risks are voluntary rather than involuntary. For example, it has been shown that the public is willing to accept voluntary risks (e.g., from sport activities, hunting and smoking) that are approximately 1,000 times greater than involuntary risks (e.g., from environmental pollution, proximity of industrial plants).[649] As Barry Commoner observes, "We are loath to let others do unto us what we happily do unto ourselves,"[650] indicating that personal autonomy with respect to choosing risks is greatly valued. The monetization of indeterminate values such as autonomy may be impossible. A National Academy of Sciences report concludes, "There is no satisfactory way to summarize all the costs or benefits of regulatory options in dollars or other terms which can be mathematically added, subtracted, or compared."[651]

Despite this condemnation by one of the most prestigious institutions of science, cost-benefit analyses continue to be carried out, using various monetization procedures that attach prices to intangible values. Because many technological benefits are easily quantifiable in economic terms, the monetization step within cost-benefit analyses is inherently biased in favor of the development of new technologies, even if they are destructive to human and environmental values.

Problem #5: The Ethics of Cost-Benefit Analyses

Finally, the entire process of cost-benefit analysis, which is based on utilitarian philosophy, is an attempt to maximize overall benefits to society (Step 7) while at the same time ignoring issues of equity and distributional justice; that is, cost-benefit analysis is insensitive to the fact that benefits may accrue to some individuals or groups at the expense of others.[652] As Corlann Gee Bush observes,

> Technology is, therefore, an equity issue. Technology has everything to do with who benefits and who suffers, whose opportunities increase and whose decrease, who creates and who accommodates.... Equity has not been a major concern of either technophobes or technophiles.[653]

There are a number of reasons why issues of equity and justice are insufficiently addressed in most technology assessments. First, are the obvious problems of predicting all impacts (Problem #2) and monetizing them correctly (Problem #4). Second, even if that were possible, there are often no mechanisms for compensating potential victims for the various costs they will have to bear.[654] For example, it is impossible to compensate future generations or endangered species for the risks we impose on them. Finally, even if negatively affected parties, such as the poor who live next to polluting chemical plants, could be identified and thus in principle be compensated financially, there would be strong resistance.[655] As discussed above (Problem #1), corporations that develop and market new technologies are strongly motivated to externalize as many costs as possible onto unsuspecting and powerless victims, with the goal of optimizing corporate profits. Thus, any attempt to address distributional injustices by internalizing previously externalized costs via financially compensating affected parties will be strongly opposed because of significant profit reduction. The entire cost-benefit analysis procedure will remain ethically flawed because it violates principles of equity and justice.

In summary, we have found that the entire technology assessment procedure is inherently biased in favor of technology development and deployment. First, because profit maximization is the primary motive behind the development of new technologies, there is a strong incentive to ignore as many environmental and social costs as possible by externalizing them onto unsuspecting people, future generations and the environment. Second, there is and always will be considerable ignorance about the long-term negative consequences of new technologies while short-term benefits are generally clear and obvious. Third, it is difficult, if not impossible to assign market prices to indeterminate values that may well be harmed by new technologies, while at the same time many technological benefits are easily quantifiable in economic terms. Finally, the most powerful stakeholders, generally the corporations that develop new technologies, define the key issues and control the technology assessment procedure, thereby biasing the entire analysis.

The principal problem with cost-benefit analysis is that it gives the impression of being objective and unbiased, thereby legitimizing decisions that are made primarily for narrow financial gain.[656] As Edith Stokey and Richard Zeckhauser observe in A Primer for Policy Analysis,

> Benefit-cost analysis is especially vulnerable to misapplication through carelessness, naivete, or outright deception. The techniques are potentially dangerous to the extent that they convey an aura of precision and objectivity. Logically they can be no more precise than the assumptions and valuations that they employ.[657]

Ian Barbour concludes that technology assessment is "a one-sided apology for contemporary technology by people with a stake in its continuation. The real expertise needed for critical assessment is social and moral, not technical."[658] We will discuss the need for the critical social and moral assessment of technology further in Part III of this book. Before doing so, let us examine how the externalization of long-term environmental and social costs led to the enthusiastic acceptance of a technology that probably never should have been adopted: The automobile.

The Uncritical Adoption of the Automobile

When the first automobiles became available more than a century ago, they were enthusiastically accepted by the public because they promised to increase the speed of travel considerably. Many new technologies are judged only in terms of their personal and immediate benefits, while long-term social and environmental consequences are either unknown or ignored. Thus, the cost-benefit ratio of the newly available automobiles appeared highly favorable, leading to their rapid acceptance as the primary means of transportation. The "car culture" has spread across the entire planet, with more than 700 million cars on the road today.[659]

Only after many unforeseen negative environmental and social consequences had become apparent (see Chapter 2 for examples) was the true cost of road transportation estimated. The costs of transportation by automobile may be categorized as direct, internal costs and indirect, external ones. Internal costs are those paid directly by the owner and operator of the automobile and consist of expenses related to car ownership, maintenance, repair, insurance and fuel consumption. The hidden or indirect environmental and social costs are externalized onto society at large and consist of expenses related to road construction and maintenance, road services, parking, congestion, medical and other costs related to accidents, soil, air, and water pollution, as well as military defense of oil supplies.[660]

A number of recent economic analyses have shown that the indirect costs of road transportation are about equal to the direct cost borne by individual car drivers.[661] Thus, if all indirect externalized costs were to be internalized and captured by fuel taxes, the cost of driving a car would be approximately twice as high as it is today. For example, eight independent analyses conducted during the 1990s concluded that the

total subsidies and other externalized costs of motor vehicle use in the
United States ranged from 378 billion to 1.69 trillion dollars per year
which, when added to the price of gasoline, would result in an average
fuel price of $5.65 per gallon.[662] Given that during the 1990s, the aver-
age price of gasoline in the US was around $1.20 per gallon,[663] the price
of gasoline should have been at least four times higher to capture all
indirect costs associated with driving.

Without efforts to internalize social and environmental costs by
imposing higher gasoline taxes, the price of gasoline will continue to remain artificially low[664] and thus promote not only excessive driving[665]
but also curtail public transportation initiatives.[666] As David Maddison
observes,

> Higher taxes would close the gap between private costs and social
> costs and curtail these socially wasteful journeys. Their absence
> distorts price signals which are used as the basis for investment
> in infrastructure and public transport systems. If motorists had
> always paid the full costs of their journeys, urban geography and
> commuting patterns might be very different to those observed
> today.[667]

Similarly, Andrew Kimbrell writes,

> When the price of gasoline is so drastically underestimated in
> the minds of the drivers, it becomes difficult if not impossible to
> convince them to change their driving habits, accept alternative-
> fuel vehicles, support mass transit, or consider progressive resi-
> dential and urban development strategies.[668]

One must wonder whether the automobile would have been adopted
so enthusiastically if the true cost of highway transportation had been
known from the outset and drivers had been required to pay that cost.
Indeed, the benefits of automobile transport look much less appealing,
particularly in terms of travel speed, if one were to take into account
the many hours of paid work needed to finance the car. According to a
simple analysis carried out by Ivan Illich more than 30 years ago, the typical American male devotes 1,600 hours per year to his car, which is the
total time spent driving and working to pay for all associated expenses.
In return for this substantial time investment, he enjoys the benefit of

driving about 7,500 miles per year, resulting in an average speed of less than five miles per hour.[669]

Given that Illich's projected average travel speed of five miles per hour appears unrealistically low, we performed our own analysis, using relevant statistics from the mid-1990s: The annual fixed cost to operate a car in 1995 was $3,891[670] and the average vehicle miles traveled per driver per day was 32 miles, which is equivalent to 11,700 miles per year.[671] Annual expenses for maintenance and tires amounted to 4 cents per mile, or $470 per year.[672] Assuming an average gas mileage of 21.1 miles per gallon (1995), approximately 550 gallons of gasoline needed to be purchased per year.[673] If the true price of gasoline was around $5.65 per gal (see average of eight studies above), then the annual gasoline expense per vehicle amounted to $3,140. In summary, the total fixed and variable costs of owning and operating a car amounted to about $7,500 per year. Given that the median net income per worker was about $22,000 in 1995,[674] approximately 35 percent of this net income was directly and indirectly spent on car transportation expenses. This is equivalent to 760 hours of work per year. In addition, assuming the average driver spent 1 to 1.5 hours per day commuting to work, shopping and recreation, another 365 to 547 hours were spent in the car. In summary, a car owner of median income spent between 1,125 and 1,307 hours per year (1995) related to owning and driving a car. Given that this car owner drove about 11,700 miles per year, the average speed of travel was about 9.0 to 10.4 miles per hour (or an average of 9.7 mph).

Although there are a number of uncertainties and assumptions in this analysis, the overall results confirm, at least in principle, the earlier assertion made by Illich that the average speed of automobile transport is considerably less than normally thought: faster than a pedestrian but slower than a bicycle.[675] As will be discussed further in Part III, it would be much better to redesign cities to have work locations and living quarters in close proximity in order to allow personal transport to be carried out either by foot or by bicycle. This would also free up large amounts of working time (ca. 760 hours per year, see above) that is currently devoted to financing automobiles and fuel. Long-distance transport would still have to be carried out by cars, trucks, trains and ships, which, however, would no longer run on fossil fuels but on very limited amounts of biofuels, whose hidden costs will be discussed next.

The Hidden Costs of Biofuels

As was discussed in Chapter 6, the large-scale generation of renewable energy, including biomass for biofuel production, is likely to have not only significant resource constraints but also numerous negative environmental impacts.[676] The manufacture of biofuels, such as ethanol from corn or biodiesel from soybeans, is currently at an early stage of development and, like most new technologies, is viewed much too positively because the boundaries of the cost-benefit analyses are drawn too narrowly, which, in turn, results in the omission of many costs that appear distant, both in space and time. For a biofuel to become a widespread substitute for fossil fuels and thereby reduce greenhouse gas emissions, it must have three demonstrable benefits: (a) it must be economically competitive with fossil fuels, (b) it must provide net renewable energy and (c) it must result in a net reduction of greenhouse gases. All of these benefits have been overestimated because various hidden costs were ignored, which, in turn, has resulted in favorable cost-benefit ratios and thus has led to the rapid, but premature, growth of biofuel production in recent years.

Regarding economic competitiveness, biofuels currently face the problem of relatively high production costs due to significant expenses for manufacturing capital and labor as well as increasing prices for agricultural feedstocks such as corn and soybeans, which are normally used as animal feed. If crude oil prices continue to increase, biofuels should become cost-competitive with fossil fuels in the near future. The cost of biofuels may be already competitive if compared to the true cost of fossil fuel (see above). However, it must be recognized that a number of costs are not included in the price of biofuels. First, there are government subsidies to farmers who grow corn for ethanol production. If these subsidies were removed, the price of ethanol would be higher. Second, there are significant hidden environmental costs in producing corn and soybeans for biofuels from prime farmland, such as soil erosion, widespread herbicide and pesticide use and depletion of limited groundwater for irrigation, to mention just a few.[677] If these environmental costs were added to the cost of biofuel production, the per gallon price would be significantly higher. Finally, there are social costs that must be considered. As was indicated in Chapter 6, the rapidly rising demand for agricultural feedstocks to produce biofuels would significantly increase food and feed prices, possibly by more than 50 percent,[678] which could have devastating effects on the poor. According to some estimates, a global increase in food staple prices by only 10 percent would translate into 160 million more hungry people.[679] In summary, if government subsidies were eliminated and all other potential costs that are currently externalized onto the environment and the poor were considered, the true price of biofuels would be much higher than estimated solely on the basis of their production costs.

Regarding the generation of net renewable energy, there has been considerable controversy with respect to ethanol from corn.[680] Superficially, it appears that corn-based ethanol, when powering engines, yields positive net energy—after all, the car is moving! However, this view is too limited because the boundaries of the energy analysis must include not only the energy output of the engine but also the various energy inputs required to produce the biofuel. Thus, to be an effective substitute for fossil fuels, biofuels must have a positive net energy balance in which the energy output from the combustion of the biofuel exceeds all energy inputs in its manufacture. Over the years, more than 20 energy-balance studies have been conducted for corn-based ethanol, with more than a third of them claiming a negative energy balance while the remaining predict various positive energy gains.[681] Most of the discrepancies among these studies can be understood in terms of the selection of system boundaries. In general, as the system boundaries are expanded, the net energy gains become smaller. For example, net energy gains from corn-based ethanol are generally positive if only energy inputs from fertilizer use, farming activities, corn transportation and ethanol processing and distillation are considered. However, if one also accounts for inputs such as the energy embodied in the farming equipment, the ethanol plant or even the food intake by farmers, the net energy balance may become negative,[682] indicating that the manufacture of corn-based ethanol consumes net energy, mostly in the form of fossil fuels, and thereby aggravates climate change.

Regarding the net reduction of greenhouse gas emissions, the benefits of ethanol from corn have until recently been considered positive, although rather limited. For example, while the use of farm equipment and fertilizers, the refining of ethanol and the burning of fuel in vehicles all cause significant greenhouse gas emissions, they are to a large degree compensated for by the photosynthetic uptake of CO_2 from the atmosphere, which occurs during the cultivation of the corn (i.e., about 1.8 metric tons CO_2 equivalents per hectare per year).[683] As a result, compared to gasoline, the use of corn-based ethanol was thought to result in a 20 percent reduction of greenhouse gas emissions.[684] Unfortunately, these previous studies had too narrow a focus because they did not include the effects of land conversion that takes place worldwide as additional forests and grasslands are converted to new agricultural land for food and feed production to replace acreage lost to biomass-for-biofuel plantations. During these land conversions, extremely large amounts of sequestered carbon are released, approximately 604–1,146 metric tons of CO_2 per hectare per year for forests and 304 metric tons of CO_2 per hectare per year for grasslands.[685] The overall result is that, instead of reducing greenhouse gas emissions by 20 percent relative to gasoline, large-scale corn-based ethanol production will double greenhouse gas emissions over the next 30 years and increase greenhouse gases for up to 167 years. Similarly, ethanol from switchgrass, if grown on US corn land, would increase greenhouse gas emissions by 50 percent.[686] Thus, ethanol from corn and switchgrass will exacerbate rather than mitigate global warming. The interest in bio-ethanol would have been much less enthusiastic if, from the very beginning, comprehensive technology assessments that properly estimated real fuel prices, net energy gains and greenhouse gas emissions had been carried out.

Limited Testing of the Effectiveness of Medical Therapies

The majority of medical therapies have never been rigorously tested for their efficacy. As mentioned in Chapter 7, a large fraction, up to about 33 percent, of cures believed to be brought about by modern medical practice, including invasive surgeries, are due instead to the placebo effect and the patient's innate healing response. According to Bernard Dixon, only a small percentage of medical treatments are believed to be effective:

> Various surveys have shown that about 75 to 80 percent of all patients seeking medical help have conditions that will clear up anyway or that cannot be improved even by the most potent of modern pharmaceuticals. In a little over 10 per cent of the cases medical intervention succeeds dramatically, by the adminis-

tration of antibiotics, by surgical maneuvers, or other specific measures. In the remaining 9 per cent the patients' troubles are diagnosed wrongly or treated incorrectly. They end up by becoming cases of iatrogenic illness.[687]

In the absence of the rigorous testing of the usefulness of medical therapies, it is more likely that they will be used inappropriately: they may be (a) unnecessary, because the desired objective can be achieved by simpler means, (b) unsuccessful, because the patient has a condition too advanced to respond to treatment, (c) unsafe, because the risk of complications outweighs the probable benefits, (d) unkind, because the quality of life and the patient's autonomy and dignity are seriously diminished by the treatment and/or (e) unwise, because it diverts resources from activities that would yield greater benefits to other patients.[688] A sad example of an inappropriate medical treatment is the mutilating radical mastectomy surgery that was carried out for more than half a century despite no evidence of its effectiveness in curing breast cancer.[689]

The use of inappropriate and ineffective medical therapies also significantly increases health care expenditures without delivering concomitant benefits to patients. According to a recent analysis by the Midwest Business Group on Health, $390 billion is spent each year in the United States on overuse or inefficient use of medical care (e.g., surgery, tests and medicines), which amounts to one third of all health care expenditures.[690] Clearly, there is an urgent need to more critically assess the effectiveness of medical treatments, if for no other reason than to control spiraling health care costs. As Dr. Richard Taylor pointed out more than 30 years ago in *Medicine Out of Control*,

> Throughout the Western world, medicine and medical care systems are in a state of crisis. There is a dissonance between the escalating costs and the paltry benefit measured in terms of improved health, lower death rates, or increased longevity. Despite the extensive use of science and technology in devising new pharmaceuticals, new diagnostic equipment, and new treatment methods, there has been pathetically little application of the same science and technology in the evaluation of the effectiveness and usefulness of these innovations.[691]

Similarly, Dr. Melvin Konner observes more recently in *The Trouble with Medicine*,

> In fact very little of medicine has been carefully evaluated in
> well designed, well controlled studies. It's really quite amazing,
> but after hundreds of years, I would estimate that only about ten
> to twenty percent of medical practices have been evaluated prop-
> erly. What that means for the patient—and not just the patient
> but for the physician—is that for a large proportion of practices
> we really don't know what the outcomes or what the effects are.[692]

Why has there been so little effort to rigorously test new medical therapies and procedures for their effectiveness and usefulness? Before an-
swering this question, it will be instructive to review the seven typical
"life stages" of a new medical invention. According to John McKinlay,
there are "seven stages in the career of a medical innovation."[693] First,
there is the stage of the promising report on scientific breakthroughs,
newly developed drugs or high-tech treatments, which, after dissemination by the mass media, arouses public and professional interest.
Second, there is the stage of professional and organizational adoption,
in which doctors begin to use the innovative treatment because they
believe that it will deliver improved care and will be viewed by their
patients as more scientific and up-to-date,[694] thereby giving them and
their hospital a competitive financial advantage. Third, there is the stage
of public acceptance and third-party endorsement, where the public
demands that the new treatment should become available to all, which
is generally followed by a formal approval by the state and coverage by
health insurance companies. Fourth, there is the stage of standard procedure and observational reports, in which the new treatment is seen
as the most appropriate therapy and any criticism of its effectiveness
or desirability is likely to invite vigorous retaliation by professional
interests and possibly even public indignation. Fifth, there is the stage
of the randomized controlled trial in which, despite strong objections
by the medical establishment, a study is finally conducted to determine
whether the new treatment is better than doing nothing, for example,
whether a new drug is better than a placebo or a new surgical procedure
is better than a sham operation (see also Chapter 7). If the randomized
control trial shows that the new treatment or therapy is less effective

than existing or otherwise less expensive alternatives, the innovation enters its sixth stage—professional denunciation in which all related clinical data are subjected to intense criticism and vigorous examination. If the negative results of the randomized control trial can withstand professional and public scrutiny or are supported by the independent findings of another trial, the medical innovation enters into the seventh stage—erosion and discreditation, in which the original new treatment may become so discredited that it may even be considered unethical to continue. However, complete discreditation and discontinuation usually occurs only after a "new" replacement treatment or therapy has become available.[695]

John McKinlay concludes his analysis:

> It is reasonable then to argue that the success of an innovation has little to do with its intrinsic worth but is dependent upon the power of the interests that sponsor and maintain it, despite the absence or adequacy of empirical support.... Imagine the potential for harm that could be avoided, and the resources saved, if innovations were routinely evaluated during earlier stages in their careers—certainly well before they become "standard procedure".[696]

Medicine is, of course, not the only enterprise in which new technologies become widely adopted before their usefulness has been satisfactorily demonstrated. As Everett Rogers points out in *Diffusion of Innovations*,[697] sound scientific evidence affects the adoption of new technologies only moderately because a strong and culturally prevalent pro-innovation bias causes advocates to give little attention to negative consequences and to assume that there are only beneficial results.

In the United States, only newly developed drugs are subjected to rigorous testing by the Food and Drug Administration (FDA). However, while the FDA requires that the new drug is safe and more effective than a placebo, it does not demand that the new medicine is more effective than existing, less expensive alternatives.[698] As a result, the market is flooded with new drugs that are more expensive but not necessarily more effective than existing ones. There is little, if any, systematic testing of medical therapies and procedures. As Richard Deyo and Donald Patrick point out in *Hope or Hype*,

> It may be surprising to learn that there are no standard defini-
> tions of experimental and standard care. There is no "Good
> Housekeeping Seal of Approval" that new operations, devices,
> tests, or drugs must receive before they become standard care....
> For medical devices, the FDA's approval process is less rigorous
> than it is for drugs, and for new surgical procedures, there's no
> approval process at all.... There is no standard for judging when
> new medical advances are worth the cost, or when insurance
> should cover them.[699]

Deyo and Patrick explain why medical innovations are adopted so
hastily without being first rigorously tested for their effectiveness and
usefulness:

> We live in a culture that's enamored of new technology in general,
> and doctors are part of that culture.... Doctors love new gadgets,
> new drugs, and new surgical techniques as much as everyone
> else—maybe more. Though most doctors genuinely want the best
> for their patients, they're just as prone as anyone else to assume
> that "newer is better". As we've seen, sometimes doctors have spe-
> cial incentives to love new technology. In some cases, it means
> more patients or more income. Doctors may wish to be the first
> in town to offer a newly touted treatment, proving that they are
> in the vanguard. The practice of "defensive medicine"—the effort
> to avoid malpractice suits at any cost—often means providing
> high-tech tests and latest treatments. And often, doctors are the
> ones who invent new technology, in which case they have both
> an intellectual and a financial investment in the new product or
> technique. In each of these situations, it's easy to rationalize pro-
> viding more care, or more high-tech care, as improving quality.
> For all these reasons, doctors may be slow to put on the brakes
> even when new technology hasn't yet proven itself.[700]

In addition, there are financial interests of medical organizations such
as hospitals, medical centers and health clinics, which acquire new technologies to attract patients seeking state-of-the-art care, thereby increasing return on investment.[701] Furthermore, there are profit motives of
the drug and medical device industries. As Milton Silverman and Philip

Lee point out in *Pills, Profits, and Politics,* the pharmaceutical industry's main motive in developing new drugs is not to make more effective ones but to create slightly modified variations of existing drugs, which can be patented to assure market share and profits:

> Since so much depends on novelty, drugs change like women's hemlines, and rapid obsolescence is simply a sign of motion, not progress. With a little luck, proper timing, and a good promotion program, any bag of asafetida with a unique chemical side-chain can be made to look like a wonder drug. The illusion may not last, but it frequently lasts long enough. By the time the doctor learns what the company knew at the beginning, it has two new products to take the place of the old one.... The pharmaceutical industry is unique in that it can make exploitation appear a noble purpose.[702]

Similarly, the medical device industry, which is about half the size of the drug industry in the United States, also enjoys above average profit margins.[703] Finally, given that total health care costs in the United States are approximately 15 percent of GDP, it is clear that a large number of American jobs would be threatened if unproven, ineffective and unnecessary medical therapies were to be discontinued. If both overuse and inefficient use of medical care were eliminated (i.e., $390 billion per year in the US, or equivalently one third of all health care costs, see above[704]), millions of health care workers would lose their livelihood. Consequently, there is likely to be strong resistance against any type of reforms that require rigorous testing of both the efficacy and usefulness of medical innovations.

It would be a mistake, however, to solely blame the profit motive of the medical establishment and health care industries for the hasty adoption of yet unproven high-tech therapies. To some extent, patients also are responsible for this problem. As Leonard Sagan points out, patients with life-threatening illnesses such as cancer are often so desperate that they are willing to try any new treatment, no matter how painful and ineffective:

> Many people will be surprised to find that there is really very little evidence whether cancer treatment prolongs life, shortens

life, or has no effect at all. If that is the case, then why do patients undergo treatment? The answer is that desperate patients seek therapy and doctors provide it, with neither party demanding rigorous proof of effectiveness. Even the slightest chance of success appears to both patients and physicians to be preferable to doing nothing at all.... The majority of cancer patients are both unlikely and unwilling to question the wisdom of therapy. Being desperate, they often seek even the most bizarre and patently unproven treatment.[705]

In conclusion, most medical treatments have not been rigorously tested for their effectiveness and usefulness. As a result, high-technology medicine currently receives much more credit than it deserves. Unfortunately, because the profit motive is the main driving force behind the development of medical innovations (Chapter 10), there has been strong resistance by the health care establishment to the systematic assessment of new drugs, devices, surgeries and therapies. As will be further discussed in Part III, there is an urgent need to assess the effectiveness and long-term effects of all new technologies, including medical innovations, to assure maximum benefits in a world of increasingly limited resources.

Gross Domestic Product (GDP): A Biased Indicator of Economic Progress

As was mentioned in Chapter 5, science and technology have been and will continue to be key contributors to economic growth and increasing material welfare. In the United States, the size of the economy is measured in terms of the gross domestic product (GDP), which is the sum of all financial transactions related to every good produced and service rendered within a given time period, usually one year. Unfortunately, although this system of national financial accounting was originally devised to manage the economy during World War II,[706] it has since been used as an indicator of economic growth, material progress and even national well-being. As John Cavanagh and Jerry Mander observe,

The GDP came into use in the mid-1980s, when it replaced the previous gross national product (GNP).... Both measurements are rooted in the World War II period, when US economists were trying to find a way to measure the speed at which productive capacity in the country was increasing to meet war production needs. It was a useful measure at that time for that purpose, but its continued application is leading to distorted analyses and conclusions.... GDP measures societal performance by one economic standard, the market value of the aggregate of all economic production—that is, the rate at which resources are converted to commodities and sold, the activities that go into that process, and all other paid services and activities in the formal economy. The assumption is that as GDP grows, society is better off: GDP growth brings progress and national well-being.[707]

While there are serious problems with the assumption that increases in per capita GDP and material affluence will automatically lead to increases in well-being and happiness (a topic that will be discussed in the next chapter), there are also serious issues related to the fact that GDP is a biased economic indicator because economic activities are treated as if they were all equally and inherently beneficial. In fact, GDP is a value-neutral statistic that makes no distinction between economic activities that are beneficial and those that are detrimental to human well-being. As Clifford Cobb and Ted Halstead summarize,

> GDP is the statistical distillation of the worldview of conventional economics. It basically assumes that everything produced is good by definition. It is a balance sheet with no cost side of the ledger; it does not differentiate between costs and benefits, between productive and destructive activities, or between sustainable or unsustainable ones. It is a calculating machine that adds but does not subtract. It treats everything that happens in the market as a gain for humanity while ignoring everything that happens outside the realm of monetized exchange regardless of its importance to well-being.[708]

In many cases, even the costs of recovery from the negative consequences of economic and industrial activities are figured as contributions to beneficial GDP growth.[709] Examples are the costs of remediating Superfund and other sites of major contamination, waste disposal, dealing with the aftermath of oil spills, providing health care for victims of environmentally caused illnesses, rebuilding after floods caused by deforestation and financing pollution control devices.[710] In addition, the GDP is "improved" by the financial costs of social dysfunction, such as expenditures to combat crime and maintain prison systems, and expenditures on medical procedures that address the consequences of unhealthy lifestyle choices such as smoking, drug addiction and meat-based diets.[711] Property damage, death and injuries caused by car accidents also increase the GDP. As Ralph Nader observed many years ago, "You contribute to the GNP when you run your car into someone else's; your contribution is still greater if you hurt people inside."[712] Furthermore, costs related to military activities also contribute to the GDP. John Cavanagh and Jerry Mander write

> Military activity provides the supreme example. All additions to military hardware or personnel salaries or university research on weapons systems add to GDP.... Once this hardware is actually used in war, then great destruction takes place, which requires later redevelopment and reconstruction—yet another plus for GDP.... So a country that becomes warlike may be assumed to be in better economic health.[713]

Finally, the costs of natural resource depletion that are currently transferred to future generations are nevertheless calculated as positive contributions to GDP each year. While the depreciation of human-made capital such as factories and equipment enter into GDP calculations, the depletion and depreciation of natural capital such as loss of topsoil, species extinction and the unsustainable use of both renewable (e.g., forests, fisheries) and non-renewable resources (e.g., fossil fuels and mineral reserves) is ignored in the calculation.[714]

In addition to the problem of counting as benefits the economic costs of detrimental activities and those of counteracting them, GDP also does not account for benefits arising from activities and goods that do not involve formal financial transactions in the marketplace. For example, many activities that give people much pleasure and contentment, such as non-consumptive hobbies such as hiking, singing, bird-watching, and volunteering for charities, as well as child rearing, cooking at home, gardening for self-sufficient food production, and home care of the elderly and ill, do not contribute to GDP because no money changes hands.[715] As more of these activities are shifted from the social economy to the money economy (e.g., to shopping malls, child care centers, restaurants, supermarkets, nursing homes, etc.), the GDP increases but the psychological well-being of the people involved may decrease substantially as a result. GDP is thus a very poor predictor of happiness and psychological well-being (Chapter 9).

In addition, compared to more sustainable local production and consumption, economic globalization adds to GDP because it requires long-distance shipping and delivery, which is inherently unsustainable given the limited supplies of non-renewable fossil fuels.[716] Wastefulness also increases GDP. As Lester Brown observes,

> GDP undervalues qualities a sustainable society strives for, such as durability and resource protection, and overvalues many it does not, such as planned obsolescence and waste.... Shoddy appliances that need frequent repair and fast replacement, for instance, raise the GDP more than a well-crafted product that lasts even though the latter is really more valuable.[717]

In summary, GDP is a heavily biased and inappropriate indicator of national well-being because it (a) overestimates benefits by counting costs as benefits, (b) omits the significant benefits of activities outside the marketplace and (c) does not promote sustainability. Consequently, GDP should never be used as a measure of success or progress. Because of these serious shortcomings, efforts have been made to devise alternative economic indicators that correct GDP by subtracting defensive, destructive and unsustainable activities as costs. As steady-state economist Herman Daly, who pioneered alternative economic indicators such as the "sustainable social net national product," suggests in *Beyond Growth*,

> Let us then define the corrected income concept "sustainable social net national product", as net national product minus both defensive expenditures and depreciation of natural capital.... The category of defensive expenditures can be large or small depending on where the boundaries are drawn.... Five categories have been suggested: 1. Environmental protection activities and damage compensation, 2. Defensive expenditures induced by

spatial concentration, centralization of production, and associated urbanization, such as increased commuting costs, housing,
and recreation costs, 3. Defensive expenditures induced by the
increased risk generated by the maturation of industrial systems,
such as increased expenditures for protection against crime, ac-
cident, sabotage, and technical failure, 4. Defensive expenditures
induced by the negative side effects of car transport, such as traffic accidents with associated repair and medical expenses, and
5. Defensive expenditures arising from unhealthy consumption
and behavioral patterns and from poor working and living conditions, such as costs generated by drug addiction, smoking, and
alcohol.[718]

One of the better known alternatives to GDP is the Genuine Progress Indicator (GPI), also known as the Index of Sustainable Economic Welfare
(ISEW), which was first developed by Herman Daly and John Cobb[719]
and has been used to estimate the extent of "true" economic progress
in a number of nations.[720] The GPI[721] includes many factors that are
currently omitted from GDP, such as resource depletion, environmental pollution, housework and non-market transactions. Positive adjustments are made for leisure time, more equitable income distributions,
greater life spans of durables, and sustainable investments. Negative
adjustments are made for unemployment, increased income disparities
and defensive expenditures (e.g., pollution-control devices and medical
and material costs of auto accidents).[722] When the GPI index is compared to GDP, similar trends appear in all countries that have been studied so far (e.g., the United States, the United Kingdom, Australia, Germany, The Netherlands, Sweden, Austria and Chile). In general, the GPI
rose more or less in tandem with GDP until about the mid-1970s but has
stabilized or declined since then, in spite of continued GDP growth.[723]

These findings clearly indicate that while science and technology
have contributed significantly to economic growth in industrialized
nations, most of the growth during the past 30 years has not been beneficial in terms of enhancing human well-being. A stabilization of GPI
indicates that the total of all economic activities yields no net benefit
(i.e., total benefits equal total costs) while a declining GPI is a warning

that continued economic growth results in net costs (i.e., total costs exceed total benefits). Economic activities currently considered to be beneficial may result in consequences that have negative implications for human well-being, a fact clearly demonstrated by GPI data. Neither has economic growth in industrial nations increased personal happiness or psychological well-being, a topic discussed in the next chapter.

CHAPTER 9

Happiness

A S DISCUSSED IN CHAPTER 5, continuous development in science and technology during the past 200 years has played a key role in contributing to rapid economic (GDP) growth and rising living standards (i.e., per capita GDP) in the industrialized nations. Advances in science and technology have increased material affluence by (a) substituting capital and fossil energy for labor, thereby greatly increasing labor productivities and thus per capita production and consumption, (b) improving efficiencies, thereby reducing the cost of goods and services and thus stimulating their consumption, and (c) creating new products and services that never existed before, thus opening up new avenues of consumption and generating large demands, often spurred by advertising. Improvements in labor productivities and technological efficiencies were discussed in Chapter 5. Here, the focus will be on technological innovation, specifically on the ways in which the invention and advertising of entirely new products and services stimulate desires that, if repeatedly fulfilled, promote rampant consumerism and materialism.

It is first necessary to understand the nature of human needs and their satisfaction. Probably the best known work on human needs is that of Abraham Maslow,[724] who postulated a hierarchy consisting of the following three basic human needs: first, material needs such as physiological needs (e.g., food, water, shelter) as well as safety and security needs. Second, social needs such as the need for belonging (affection and acceptance) as well as for self-esteem and esteem by others. Third, spiritual growth and self-actualization needs, which are often expressed in the aspiration to carry out meaningful work and to use one's talents in the service of something larger than oneself. According to Maslow, higher-order needs can be fulfilled only after lower-order ones have been sufficiently met. More recently, Max-Neef[725] asserted that human needs, unlike wants and desires, are not only few but are virtually the same in all cultures throughout history. According to Max-Neef, there are just nine fundamental needs: subsistence, protection, affection, understanding, participation, recreation, creativity, identity and freedom. It is interesting to note, specifically within the view of creating a more sustainable and happy society (see Part III), that each culture adopts different ways of satisfying these needs, with some approaches being more successful than others.[726]

Compared to these few universal human needs, individual wants and desires appear to be unlimited in number and constantly changing. As Joseph Des Jardins notes,

> Wants seem to be a matter of individual psychology. They are those factors that provide a motive for acting. Wants come to be developed, chosen, learned, created by advertising, and so on.... They are superficial and temporary in the sense that they depend on a person's personal background, history, and culture.[727]

Apparently, desires can be aroused in humans for almost anything. Wants may be stimulated by commercial advertising, which publicizes a new product and claims that happiness will be gained by buying and owning it. A wide range of seemingly frivolous desires may also result from psychological abnormalities such as neurosis or deprivation—material or emotional or both—in early childhood (see below). Satisfaction of these desires may be an attempt to compensate, generally without success, for these deficiencies. Wants and desires are sometimes considered to be "false" or "artificial" needs.[728]

Technological Innovation, Consumerism and Materialism

What, then, is the role of technology and technological innovation in satisfying needs and wants? Throughout human history, various technologies, ranging from the most primitive to the most advanced, have been used to satisfy basic material needs for food, housing, clothing and defense. More recently, many technological innovations have been employed to satisfy a variety of non-material needs. For example, electronic communications technologies (i.e., radio, television, telephone and the Internet) mediate and enable the satisfaction of social needs for participation and affection. Similarly, the industrial mass production of ever-changing fashionable clothing attempts to address the need for identity and affection. And, as Jackson and Marks point out, the automobile has been used to satisfy both material and non-material needs:

> By association, material commodities and economic goods in general may be adopted as attempted satisfiers for a wide range of underlying needs. This is clearly demonstrated in the case of automobiles. In their most fundamental capacity, cars provide mobility. But mobility itself is neither a satisfier nor a need. Rather it is a structural element within the attempted satisfaction of many needs. Mobility allows us to travel to work where we can earn a living (subsistence) and to shop so that we can buy (for example) food and clothing (subsistence and protection). But…cars are associated in the prevailing western culture (and increasingly in other cultures) with social status (participation and identity), with sexual success (affection), with personal power (identity), with recreation and leisure (participation, idleness), with freedom and creativity.[729]

Technological innovation is used to create constantly changing products and services, and advertising is employed to stimulate large-scale public demand. It is important to note that the public's desires for these products and services arise not because there is a fundamental need for them but because advertising has created a "need" and technology delivers the possibility of its satisfaction. Desire can be aroused by almost any new product or service, no matter how useless or trivial. As Langdon Winner observes in *Autonomous Technology*, "Evidently technological accomplishment has become a temptation that no person can reasonably be expected to resist."[730]

Lewis Mumford in *Technics and Civilization* states that our current economic system is based on the ethics of utilitarianism, in which it

is implicitly understood that happiness is directly related to material
consumption:

> Happiness was the true end of man and it consisted in achiev-
> ing the greatest good for the greatest number. The essence of
> happiness was to avoid pain and seek pleasure: the quantity of
> happiness, and ultimately the perfection of human institutions,
> could be reckoned roughly by the amount of goods a society was
> capable of producing: expanding wants, expanding markets,
> expanding enterprises, an expanding body of consumers. The
> machine made this possible and guaranteed its success. To cry
> enough or to call a limit was treason. Happiness and expanding
> production were one.[731]

For the first time in human history, the Industrial Revolution enabled
the inexpensive mass-production of almost anything. According to the
technological imperative, what can be mass-produced will be mass-
produced. But there is no profit in mass production without mass consumption. Thus, mass advertising was required, employing communications technologies, first the printing press, then radio and television,
and finally the Internet, to arouse public desire for new products. Thus,
mass production requires mass consumption which, in turn, requires
relentless advertising by the mass media.[732] Our present materialistic
consumer society is a direct consequence of the success of mass production and of mass media technologies.

As Ernest Braun indicates, technological innovation does not necessarily occur in response to the wants and needs of the public but rather
reflects the desires and goals of those involved in the development of
new technologies:

> We are told constantly that innovation occurs only in response
> to human needs, that the desires for improved technology are the
> desires of the great buying public. In my view, these statements
> are humbug. The desires are those of the engineers and scientists,
> ambitious to achieve ever more elegant solutions to self-imposed
> problems. The desires are also those of the entrepreneurs, eager
> to carve out a niche for themselves and make a good profit. The
> desires are those of manufacturers, eager to stimulate new waves
> of purchases for new products when markets are saturated.[733]

The desire for profit is probably the strongest and most important driving force behind the development of new products and services (Chapter 10) and the subsequent mass-marketing by corporations. Indeed, as David Korten writes in *When Corporations Rule the World*, the American consumer culture was consciously created by businesses with the goal of maximizing their sales and profits:

> The consumer culture emerged largely as a consequence of concerted efforts by the retailing giants of the late 19th and early 20th centuries to create an ever-growing demand for the goods they offered for sale. American historian William Leach has documented in *Land of Desire: Merchants, Power, and Rise of a New American Culture* how they successfully turned a spiritually oriented culture of frugality and thrift into a material culture of self-indulgence. Leach finds the claim that the market simply responds to consumer desires to be nothing more than a self-serving fabrication of those who make their living manipulating reality to convince consumers to buy what corporations find it profitable to sell.[734]

Following the rise of US power after World War II, the American consumer culture first spread to Europe and Japan and, with the more recent advent of economic globalization, is now being adopted as the dominant lifestyle in many developing nations.[735]

To maintain sales and generate continued profits, corporations have to avoid market saturation. One common way of avoiding this problem is to make minor improvements and stylistic changes in basically the same product and then convince consumers that this new, improved version is more desirable than the previous one, which is now considered outdated although still fully functional. This concept of "rapid obsolescence" was first systematically applied on a large scale to automobiles in the United States:

> By about 1925 the engineering approach to building cars exemplified by Ford—unchanging, universal styles—was being replaced by General Motor's Sloanism—rapid obsolescence supported by annual model changes, driven by sophisticated psychological advertising. This evolution was clearly motivated by the fact that the market for transportation alone—exemplified by the Ford—was

becoming saturated. Sloan realized that continued profit growth, his primary goal, could only be achieved by a fundamental shift: sell a dream, make cosmetic changes annually, and make consumers believe they had to keep pace.[736]

Rapid obsolescence is used not only to promote the sale of cars but also many other consumer products. A most striking example is the stimulation of public demand for fashionable clothing, the style of which changes so rapidly that many feel compelled to buy new clothes even when the old ones are still perfectly functional. Clearly, rapid obsolescence not only increases material consumption but also waste and environmental pollution.

The continued stimulation of the desire for new products and services by various means, including rapid obsolescence, requires large-scale advertising by the mass media. Here again, advanced communications technologies, specifically radio, television and the Internet, are employed by corporations peddling their wares. As was pointed out in Chapter 3, modern mass communication technologies, especially television, are used to manipulate and control the public not only with political propaganda but also with incessant commercial advertising. Because most mass media, for their very survival, depend financially on broadcasting advertisements for their corporate clients, the primary goal is to sell advertising, with movies and entertaining stories serving as filler material to engage and maintain the public's attention. Total US consumption for all mass media—newsprint, radio, TV and Internet—was about 3,200 hours per person in 2003, which amounts to more than eight hours per day.[737] In the United States, adults watch about 4.5 hours of TV each day on average, or about 1,600 hours annually, being bombarded with about 28,000 invasive commercials per year.[738] By the time the average person reaches the age of 20, he or she will have been exposed to more than half a million TV commercials, most of them carrying an identical message: "Buy something—do it now."[739]

Advertising achieves the objective of increasing consumption in the following way. First, the stage is set by creating dissatisfaction with current conditions. As Robert Lane points out in *The Loss of Happiness in Market Democracies,*

It is not true that the function of advertising is to maximize satisfaction; rather, its function is to increase people's dissatisfaction with any current state of affairs, to create wants, and to exploit the dissatisfaction of the present. Advertising must use dissatisfaction to achieve its purpose.[740]

Then, information about a newly available product or service is conveyed with the goal of creating desires that formerly did not exist. It is important to recognize that advertising exists only to sell things people don't need. As was the case in all pre-industrial societies, whatever people do need they will readily seek without the need for advertising.[741] Most importantly, the advertised product or service is portrayed as the solution to the dissatisfaction previously created. The advertisement promises that the purchase of the new product or service will lead to more convenience, higher social status and always greater happiness. By incessantly stimulating wants, greed and envy, the overall effect of advertising is not only an increase in consumption but also a profound strengthening of materialistic values in society.[742]

The use of technological innovation to continuously stimulate material wants has had several other unfortunate effects. First, enormous amounts of resources, in terms of money, labor and ingenuity are wasted on the satisfaction of the rather trivial desires of individuals in industrialized nations while many basic needs of the poor in both developed and developing countries go unmet. For example, massive efforts are devoted to the invention, production and advertising of sugary breakfast cereals, unhealthy snacks and prepared foods in wealthy nations while those in poor countries hardly have the basic staples necessary for survival. Similarly, many scientists and engineers in industrialized nations employ their valuable talents for nothing more than inventing and producing consumer products that are designed to become rapidly obsolete while at the same time their abilities could be used in much more constructive ways.

Another unfortunate effect is that many induced wants become actual needs over a period of time. The automobile serves as an example: initially automobiles were desired because they offered the luxury of greater speed and mobility, but they are now required for driving to work and shopping for basic necessities. The United States could not now function without them. Similarly, many other technological inventions, such as the telephone, washing machine, electric lighting, personal computer and the Internet, began as luxuries but soon became necessities following their widespread adoption. Industrialized nations have become so dependent on such a large number of complex technological devices and infrastructures that it would be extremely difficult for the current lifestyle to continue without them.

In summary, technological innovation has played and continues to play a crucial role in the promotion of consumerism and materialism, with the implicit goal of increasing personal happiness and well-being. Whether these technology-induced increases in material affluence have indeed resulted in greater happiness will be examined next.

Material Affluence and Happiness

Many, if not most, people believe that greater material affluence will increase their happiness and psychological well-being. After all, a higher income provides more personal choices, enables conspicuous consumption to raise one's relative social status, and eliminates concerns about meeting basic needs. Indeed, a large body of empirical literature confirms that within a given nation at a specific point in time, subjective well-being increases with income,[743] particularly if the baseline income is below the level necessary to meet basic needs.[744] However, increases in income above this minimum level generally result in rather insignificant improvements in psychological well-being.[745]

Similar relationships between income and psychological well-being have also been observed in comparisons between different nations. In a study of 51 countries, it was found that overall life satisfaction increased with rising national per capita income. However, once an income threshold of about $10,000 was reached, any further increase in material affluence did not result in greater happiness.[746] Thus, while more income does make a difference to people who are very poor and lack food, shelter and health care, an increase in per capita income above that which is required to meet basic needs does very little to improve psychological well-being.[747] In this context it is interesting to note that respondents from *Forbes* magazine's list of the 400 richest Americans, many of them billionaires, had an average life satisfaction score of 5.8

out of 7, while the Maasai, a traditional herding people in East Africa who have no electricity or running water and who live in huts made of dung, had only a slightly lower score of 5.7.[748] Moreover, studies of human happiness have shown that some people even in very poor countries, such as Bangladesh and Azerbaijan, are among the happiest in the world.[749]

While at any given time a larger income correlates to some extent with greater psychological well-being among individuals and even nations, an increase in per capita national income over time appears to do very little to improve happiness. For example, while the per capita income in the United States, Europe and Japan has risen sharply in recent decades, the average happiness has remained virtually constant or has declined over the same period.[750] Consider, for example, that the per capita income in the United States between 1946 and 1991 rose by a factor of 2.5, from approximately $11,000 to $27,000 (in 1996 $US), but happiness on average remained constant during this period.[751] Even more pronounced, despite the fact that per capita income in Japan rose sixfold from 1958 to 1991, the average life satisfaction rating remained constant.[752] Similar results have been reported for other industrialized nations.[753]

While economic and industrial growth and the resulting gains in material affluence have failed to increase people's happiness, at the same time an epidemic of psychological disorders, particularly depression and anxiety, has occurred.[754] In the United States, depression rates have increased tenfold over the past 50 years,[755] and suicide is the third most common cause of death among young adults in North America.[756] Furthermore, 15 percent of Americans have clinical anxiety disorder,[757] and the average American child in the 1980s reported greater anxiety than the average child receiving psychiatric treatment in the 1950s.[758] Depression, though generally not a life-threatening illness except in the cases of attempted suicide, is nevertheless the third leading cause, after arthritis and heart disease, of loss in quality-adjusted life years, ranking above cancer, stroke, diabetes and obstructive lung disease.[759] Unfortunately, current statistics indicate that rates of depression are increasing, and it is estimated that by 2020, depression will be the second leading cause in the world for disability-adjusted life years.[760] As will be discussed further, the incidence of depression is directly correlated with economic development, modernization and the associated materialistic lifestyle.[761] Consider, for example, that the Amish, a Christian sect that rejects modernity, high-technology and materialism, have only one fifth to one tenth the risk of unipolar depression of their average American neighbors who are, by contrast, totally submerged in today's complex high-tech consumer culture.[762]

Clearly, the key utilitarian assumption upon which our economic system is based (i.e., that economic growth and ever-rising affluence will increase psychological well-being and happiness) is false and in serious need of revision.[763] Unfortunately, without exception, all of the world's nations are nonetheless committed to the global goal of continuous economic development and expansion, and it appears that most economists, policy-makers and politicians are still in denial about the failure of economic growth to improve psychological well-being. As Clive Hamilton points out in *Growth Fetish*,

> After sustained attacks from outside the profession, a few economists have at least acknowledged that the link between growth and well-being may not be self-evident. But no serious consideration is given to the critique; after all, conceding that more economic growth may not make people any better off would be a fatal blow to economists—attacking their credibility, their influence, and their jealously guarded place as conjurors of cargo.[764]

Explaining the Paradox

As Aristotle pointed out more than 2,000 years ago, human desires, in contrast to human needs, are by their very nature insatiable.[765] Indeed, *pleonexia*, the "insatiable desire from more," was regarded by Greek philosophers as a human failing and a serious obstacle for attaining the "good life."[766] After a material desire has been fulfilled, a sense of dissatisfaction soon develops, resulting in further craving. The pleasures of a new material acquisition are quickly forgotten and are replaced with a desire for more, leading inevitably to an endless cycle of dissatisfaction and discontent.[767] As Paul Wachtel succinctly observes in *The Poverty of Affluence*, "Wanting more remains a constant, regardless of what we have."[768]

This never-ending cycle of dissatisfaction can also be explained in terms of fundamental economic theory:[769] as long as an object is unavailable or scarce, it has high value and we strive for it, expecting to gain happiness from its possession. However, as soon as we have obtained the desired object, it is no longer scarce, and thereby becomes less valuable and less able to generate happiness. We then repeat this cycle by selecting another scarce item for which we strive, unfortunately without ever finding lasting fulfillment or contentment. We are chasing a mirage, thereby remaining forever dissatisfied and unhappy.

A related explanation of the fact that greater material affluence fails to increase happiness is provided by the psychological theory of hedonic adaptation. According to this well-established theory, satisfaction is determined by the gap between aspirations and achievement,[770] with people being most satisfied when they have attained their expected level of material affluence. Unfortunately, people become quickly accustomed to their higher standard of living while at the same time their expectations for greater affluence continue to rise, primarily because they are continually bombarded by mass-media advertisements that forcefully stimulate their desires for additional material goods and services.[771] This growing gap between expectations and achievement creates ever greater frustration and unhappiness, which people then attempt to ameliorate by reducing this gap through increased material consumption. As a result, people are forever caught on a hedonic treadmill,[772] as Martin Seligman writes in *Authentic Happiness*:

> Another barrier to raising your level of happiness is the "hedonic treadmill," which causes you to rapidly and inevitably adapt to good things by taking them for granted. As you accumulate more material possessions and accomplishments, your expectations rise. The deeds and things you worked so hard for no longer make you happy; you need to get something even better to boost your level of happiness into the upper reaches of its set range. But once you get the next passion or achievement, you adapt to it as well, and so on…. Good things and high accomplishments, studies have shown, have astonishingly little power to raise happiness more than transiently.[773]

The unfortunate end result of this hedonic treadmill is that consumption increases without raising the level of psychological well-being or happiness,[774] a finding that has been confirmed by extensive survey data.

Often, greater material affluence is sought because it creates the illusion of higher social status, which often results in positive feelings of accomplishment and superiority. Unfortunately, the expression of social superiority with the help of materialistic status symbols leads to a "positional treadmill."[775] Pre-industrial societies consisted of small groups (extended families, clans, tribes, etc.) in which everyone and their status was known through frequent direct face-to-face contacts. By contrast, in large modern industrial societies, the individual is surrounded by strangers. Since it is virtually impossible to determine the social status of strangers by direct face-to-face contact, material symbols are used to communicate relative social status.[776] Unfortunately, rather than using simple, environmentally friendly markers of rank, as is done through uniforms with striped appliqués in the military, the public is manipulated by mass-media advertising to express social status through the consumption of expensive, material-intensive products and services such as fashion clothing, elaborate cars, large houses and exotic vacations. Many material status symbols, particularly clothing and cars, are designed for rapid obsolescence, which causes not only rapid disappointment but also the urgent need to continually keep up by purchasing the latest models. It has been estimated that in industrialized nations, a large fraction of household income, 50 percent or more,[777] is spent on competitive consumption, or in Thorstein Veblen's words, "conspicuous consumption."[778]

While positional consumption raises perceived social status and illusions of superiority, its benefits to psychological well-being are regrettably short-lived. The consumption of positional goods loses the ability to raise social status as soon as many others are able to purchase the same status symbols.[779] As Tim Jackson observes,

> Once enough people possess these goods, moreover, their value
> in positioning us ahead of the crowd declines, and those wishing
> to stay ahead must engage in a search for new goods with social
> scarcity....It is a case of everyone in the crowd standing on tiptoe
> and no-one getting a better view.[780]

In conclusion, it is relative rather than absolute income that determines social status and transitory feelings of psychological well-being. However, as long as a nation's relative income distribution remains more or less constant, an increase in average income and material affluence (i.e., per capita GDP) will not enhance an individual's social position or the sense of well-being derived from it. The unfortunate end result is that people, like addicts, are caught on a hedonic, positional treadmill,[781] acquiring ever more material status symbols, thereby raising total consumption of goods and services without obtaining greater happiness.

Furthermore, material consumption does not provide lasting fulfillment because individuals attempt to satisfy their basic non-material needs, such as the need for acceptance, esteem, self-actualization and spiritual growth, by material means.[782] Instead of directly addressing these needs, we have been manipulated by corporate-controlled mass-media to accept material substitutes that are inherently unable to provide lasting satisfaction. While material consumption may provide temporary compensatory relief for some social and psychological needs, it is intrinsically incapable of fulfilling the more complex self-actualization and spiritual needs. As philosopher Alan Drengson notes,

> In our technological society this natural human desire for completion and liberation, which is a deep spiritual need, is not perceived as such. The desire for fulfillment of this need is misplaced as a quest for completion through material accumulation and power.... Humans, and all other beings, have the need, desire, and drive to realize and complete themselves, to fulfill their destiny. This need is expressed in our restless search for something that will give meaning and genuine satisfaction.... When this core need is frustrated and we are unaware of it, we seek compensatory satisfactions.[783]

Similarly, steady-state economist Herman Daly observes,

> Humanity, craving for the infinite, has been corrupted by the temptation to satisfy an insatiable hunger in the material realm.... The infinite hunger of man, his moral and spiritual hunger, is not to be satisfied, is indeed exacerbated, by the current demonic madness of producing more and more things for more

and more people. Afflicted with an infinite itch, modern man is scratching in the wrong place.[784]

Since our basic psychological and spiritual needs remain unfulfilled by material consumption and other techno-distractions, such as watching television, and we are not aware of any other ways to satisfy them, we obsessively try the same approach over and over again. Thus, the addictive process is initiated, which is a natural response to the failure of fulfilling primary needs and the subsequent attempt to satisfy these instead through secondary sources.[785] The addictive and compulsive behavior continues as long as the underlying primary needs are not properly satisfied. The artificial life in technological society promotes an illusory independence of nature as well as mutual alienation between people, thus obscuring opportunities to directly satisfy primary needs:

> The technological construct erodes primary sources of satisfaction once found routinely in life in the wilds, such as physical nourishment, vital community, fresh food, continuity between work and meaning, unhindered participation in life experiences, personal choices, community decisions, and spiritual connection with the natural world. These are needs we were born to have satisfied. In the absence,…the psyche finds some temporary satisfaction in pursuing secondary sources like drugs, violence, sex, material possessions, and machines. While these stimulants may satisfy in the moment, they can never truly fulfill primary needs. And so the addictive process is born. We become obsessed with secondary sources as if our lives depended on them.[786]

Since the fulfillment of fundamental psychological and spiritual needs is often difficult in modern industrialized societies, addictive but ineffective solutions are commonplace. Conspicuous material consumption could easily be considered an addiction, and has, indeed, been compared to substance addiction.[787] Similarly, there is clear evidence that many people are addicted to watching television: they continue watching even though it fails to increase their sense of well-being and often causes passivity and depression.[788] Even when a conscious attempt is made to stop watching, most are, like drug addicts, unable to do so because of unpleasant withdrawal symptoms.[789]

One could argue that, as a culture, we are addicted to technology for the purpose of providing illusory solutions to our problems (Chapter 7), even if these problems are fundamentally social, psychological or spiritual in nature. Since all addictions involve denial,[790] it is possible that our collective "techno-addiction" is one of the main reasons we are unwilling to critically question the negative aspects of technology in society and in our lives. As will be discussed further in Part III, the necessary paradigm shift toward a more sane and sustainable world is possible only if we first acknowledge our collective denial in this regard.

Finally, the rise of intense materialism, defined here as placing high value on the acquisition of material objects,[791] has been shown to have negative effects on character, resulting in a decline in psychological well-being. Materialism may have always been the compensatory response of those who, because of economic or emotional deprivations in early childhood, are insecure about matters of love, self-esteem, competence or control.[792] However, prior to the Industrial Revolution, there was little opportunity to compensate for insecurities through the purchase of material objects. This has dramatically changed during the past century as industrial-scale mass production, combined with incessant advertising, has flooded free-market economies with inexpensive goods and services for mass-consumption. As a result of advertising, which continuously stimulates compensatory desires, citizens of industrialized societies have become not only more materialistic but also more individualistic, egotistic, self-centered, greedy and competitive. Unfortunately, as Richard Ryan observes in *The High Price of Materialism*, selfishness and materialism are becoming a global phenomena:

> Most of the world's population is now growing up in winner-take-all economies, where the main goal of individuals is to get whatever they can for themselves: to each according to his greed. Within this economic landscape, selfishness and materialism are no longer being seen as moral problems, but as cardinal goals of life. This global reality exists, however, only because people, and I mean each one of us, can so readily be converted to the religion of consumerism and materialism. Indeed, such mass conversions seem already to have occurred.[793]

While enormous numbers of individuals strive for happiness through the consumption of material objects, a large number of psychological studies have demonstrated that a materialistic value orientation is inversely related to virtually all factors known to enhance a sense of well-being. Materialistic individuals generally have character traits that are not conducive to harmonious social interactions, thereby increasing the probability of interpersonal conflicts and thus unhappiness. For example, compared to their less materialistic counterparts, individuals exhibiting strong materialistic values have been shown to be more possessive,[794] abrasive, angry, narcissistic,[795] self-centered, envious,[796] controlling, manipulative and power-hungry,[797] while at the same time being less conscientious, agreeable, friendly,[798] empathetic[799] and generous.[800] Materialistic individuals also have a greater probability of being afflicted with a wide range of psychological disorders, many of which interfere with achieving happiness. For example, materialism has been found to be positively correlated with anxiety, stress, neuroticism, depression, behavioral and attention deficit disorders, paranoia, low self-esteem, as well as antisocial and addictive behavior.[801] Materialistic people place less priority on family and social relationships and thus deprive themselves of opportunities for social interactions known to increase psychological well-being.[802] Materialists are also generally more competitive and individualistic, traits that often inhibit close companionship and any kind of happiness to be derived from it.[803] In addition, as Tim Kasser points out in *The High Price of Materialism*, materialistic values are directly opposed to the two social values of benevolence and universalism:

> Benevolence is concerned with the "preservation and enhancement of the welfare of people with whom one is in frequent personal contact," and includes valuing the characteristics of being loyal, responsible, honest, forgiving, and helpful, and of desiring true friendship and mature love. Universalism involves "understanding, appreciation, tolerance, and protection for the welfare of all people and for nature" and includes social justice, world at peace, equality, and being broad-minded. Schwartz's cross-cultural evidence, compiled from thousands of individuals sampled in most parts of the globe, shows that something about

materialism conflicts with valuing the characteristics of strong relationships (loyalty, helpfulness, love), and with caring about the broader community (peace, justice, equality).[804]

In short, compared to those who are not materialistic, individuals who hold strong materialistic values have shorter, more conflicted or impoverished personal relationships and feel alienated and disconnected from others in society,[805] all to the detriment of their psychological well-being.

Materialists are more likely to feel alienated from nature, have a negative attitude toward the environment and have little appreciation or interest in animals and plants.[806] Given that many find solace in the beauty and tranquility of nature, materialistic people are less able to benefit because of their psychological disconnectedness from the natural environment.

Because materialistic values are directly opposed to religious, spiritual and moral values,[807] materialists often become conflicted as they try to balance their many consumerist desires with their religious beliefs and moral convictions. The ongoing value conflict and resulting tensions are likely to decrease overall life satisfaction and can also lead to psychological problems such as depression and neuroticism.[808] Finally, a materialistic focus on the pursuit of wealth, fame and image undermines the satisfactions that may be obtained from autonomy and authenticity. Materialists are often externally motivated by the rewards, praise and opinions of others and thus may lose their self-direction, while those who value autonomy and self-expression are internally motivated by interest, enjoyment, challenge and doing things for their intrinsic value.[809] As Martin Seligman has pointed out in *Authentic Happiness*,[810] the pursuit of self-selected interesting activities that fully engage one's attention and advance self-actualization are major sources of psychological well-being that the average person has control over, and these are generally not chosen by materialistic individuals because of their orientation toward external motivation.

In conclusion, there is much research and plausible theory to indicate that increasing material affluence, beyond levels required to meet basic needs, fails to increase happiness and psychological well-being.

Sources of Happiness

If ever-rising material consumption is unable to increase happiness in any lasting way, is there anything that can? It must first be recognized that possibly about 50 percent of a person's "happiness temperament" may be determined by one's genetic constitution and therefore beyond one's own control.[811] For example, certain personality traits that are positively correlated with psychological well-being, such as extroversion, are strongly influenced by genetic factors.[812] Nevertheless, despite these inherited limitations, people still have significant personal control over the way they choose to manage their internal emotions and thoughts and how they interact with their external environment. Regarding the importance of cognitive factors, there is a large body of evidence indicating that people's thoughts have a direct influence on the quality of their feelings.[813] For example, positive affirmations, optimism, humor, gratitude, forgiveness and detached self-reflection all have been shown to consciously enable people to increase their happiness and psychological well-being.[814] Indeed, cognitive therapy, which involves instruction in transforming irrational negative thoughts into rational positive ones, is extremely effective in permanently addressing depression and neuroticism.[815] Clearly, there is great power in simple rational thought.[816]

Greater freedom, defined negatively as the absence of external restraints, and the accompanying sense of personal control have been positively correlated with psychological well-being.[817] Certain types of freedoms are of particular importance: (a) political freedom, which enables citizens to take part in the democratic process and to exercise both civil and political rights; (b) economic freedom, which gives individuals the opportunity to engage in the free exchange of goods, services and labor and (c) personal freedom, which allows people to do many rewarding things including practice their religion, travel freely or marry without interference from authorities.[818] Since all three—political, economic and personal freedoms—have strong, statistically significant correlations with psychological well-being, it follows that people living in constitutional, free-market democracies are on average happier than those residing in economically mismanaged dictatorships.[819] It is particularly interesting to note that decentralized local control of government and direct democratic participation, as exists and has been studied in Switzerland, is a larger contributor to happiness than material affluence.[820] Thus, while people in industrialized nations have only limited control over the choice of country in which to reside, they can nevertheless work actively to transform their political system to become more participatory. If successful, this direct participation in the democratic process is likely to have a greater effect on happiness than increasing material affluence.

One of the strongest predictors of happiness is the quality of social relationships, which is directly related to the ability to live in harmony with family, friends, neighbors, society and nature.[821] A happy marriage is by far the best predictor of happiness with life as a whole.[822] In addition, a harmonious family life, warm interpersonal relations and a supportive social network all contribute significantly to psychological well-being. This occurs because our daily feelings and moods are often strongly influenced by other people's attitudes, whether they like or dislike us, have a good or bad opinion of us, or accept or reject us.[823] In addition, social relationships enable us to engage in many selfless, kind and caring activities toward each other, which is a proven recipe for creating happiness.[824] As Michael Arygle summarizes in *The Psychology of Happiness*,

> Social relationships are a major source of happiness, relief from distress, and health. The greatest benefits come from marriage and other close, confiding and supportive relationships. Relationships increase happiness by generating joy, providing help, and through shared enjoyable activities. They buffer the effects of stress by increasing self-esteem, suppressing negative emotions and providing help to solve problems.[825]

Religion also appears to increase people's happiness, most likely for the following reasons. Devoutly religious people may have fewer vices, such as excessive drinking, smoking, gambling and promiscuous sex, thereby protecting themselves from the many unpleasant and painful consequences associated with these activities. They may have more peace of mind, because many beliefs diminish people's innate fear of death. Finally, and possibly most importantly, the religious fellowship provides an automatic and instant social support group and thus buffers individuals against stressors such as unemployment, low income and widowhood.[826]

Having satisfying, engaging and worthwhile work that allows maximum autonomy to exercise skills while providing an adequate income is also an important source of happiness for many. As Robert Lane observes in *The Loss of Happiness in Market Democracies,*

> On the job, lack of close supervision, absence of routine and repetitive work, and the presence of substantive complexity in what one does contribute most to work satisfaction.... The intrinsic features of work (largely unpriced by the market) make the most important contributions to satisfaction with one's work: pleasure in what one actually does, the congeniality of colleagues, and the sense of making a difference.[827]

The last point, the sense of making a difference, may very well be one of the most important contributors to happiness. Indeed, research has shown that a sense of meaning and purpose is the single attitude most strongly associated with overall life satisfaction.[828] Individuals who have discovered a higher purpose in their lives often use their talents and energy to engage in meaningful work, thereby satisfying their self-actualization needs. The resulting combination of authentic self-expression, the experience of deep, effortless involvement[829] and the ability to lose one's over-concern with self by contributing to a greater good constitute probably one of the most powerful and effective ways to create long-lasting happiness and fulfillment.[830]

The Destruction of Traditional Sources of Happiness

In Chapter 2, we observed that the application of innovative science and technology often results in a wide range of unintended negative consequences. One of these unexpected consequences is the destruction of traditional sources of happiness, such as strong social relationships, satisfying and meaningful work, closeness to nature, as well as religiosity, spirituality and morality. Before continuing, it will be useful to briefly review how the introduction of new technologies brings about "change," not only social change but even more importantly change in values and traditions, often with devastating effects on psychological well-being.[831]

Technology has a direct impact on values by creating new opportunities. By making possible that which was not possible before, technologies offer individuals and society new options to choose from. By pursuing new goals, social priorities and organizations change in order to take advantage of new technological opportunities, which automatically means that the function of existing social structures will be disturbed. The end result is that traditional values, priorities and social organizations will be replaced by new ones, generally not without causing great distress among those affected by these changes.[832] Stephen Cotgrove summarizes the situation in *Catastrophe or Cornucopia*:

> Industrialization faces society with choosing some options in preference to others. And choice implies criteria—values. Industrialization then maximizes some values, but at the expense of others.... By doing so, it has brought about profound transformations in the nature of society.... In place of the traditional respect for status and position, modern societies replace contract and achievement. Individuals are valued not for the kind of people they are, but by what they have achieved. In place of personal bonds of kinship and community, modern societies substitute impersonal bureaucratic relations, and the cash nexus of the market place.... Individuals become instruments of production in work which maximizes output at the expense of human satisfaction. ...and the speed of technological change leaves little time for reflection.... We have paid a high price for material well-being.[833]

Prior to the Industrial Revolution, most people either worked on family farms or in small workshops in their own homes or nearby. In addition, because there was no efficient and affordable mode of transportation, people in pre-industrial societies rarely moved far away from their place of birth. As a result, an entire lifetime was generally spent in close proximity to family members, long-time personal friends, neighbors and co-workers. This rather stable environment offered excellent conditions for developing strong social relationships and communal bonds, which, in turn, provided many sources of happiness.

Unfortunately, this situation changed profoundly following the Industrial Revolution as people were forced to abandon their traditional way of life and instead work in factories, far away and in isolation from their families and local communities. Thus, technological innovation and the ensuing industrialization resulted not only in the dislocation

and uprooting of people but also caused profound changes in values
and social organization: community spirit was replaced by individualism, common interest by self interest, cooperation by competition,
mutual dependence by personal independence, stability by mobility.[834]
Because these societal changes were so revolutionary, this era has been
appropriately called the Industrial *Revolution*, which, in fact, is still in
progress, particularly in "developing" countries. Given that individualism, self-interest, competition, independence and mobility are all detrimental to the development of strong communities, it is not surprising
that since the Industrial Revolution, personal and communal bonds
have been weakened, social connectedness has been eroded and people
have become increasingly isolated, all with devastating effects on their
happiness and psychological well-being. In his visionary book, *The Conserver Society*, Ted Trainer gives a somber assessment of community life,
or rather the lack thereof, in industrialized nations:

> A most important factor in a high quality of life is community.
> Our industrialized, affluent way of life does not rate at all well on
> this score. We tend to live very privately and to be concerned pri-
> marily with our own welfare rather than that of the group. Most of
> us have little involvement in the public affairs of our region. We
> move frequently. The typical American family relocates every 5
> years, making it difficult to establish strong bonds with people or
> locality.... Many people are quite isolated and lonely and without
> access to sources of emotional support or mutual aid.... But most
> serious is the vast intangible cost in terms of social impoverish-
> ment, of living without much experience of friends, of belonging,
> or social cohesion and of making a worthwhile contribution.[835]

While industrialization in general has contributed to a decline in the
quality of human relationships and community spirit, two technological innovations have been particularly destructive: the automobile and
the television. Since the automobile embodies the values of individualism, independence and mobility, it is inherently damaging to community life. As was indicated in Chapter 2, cities in the United States and
elsewhere have been designed to accommodate automobile traffic rather
than pedestrians, thereby becoming much less suitable for frequent and
relaxed contact among people. At the same time, the car facilitated the

development of low-density suburban housing which, because of job mobility, never produced close-knit, lasting communities. The automobile is essentially designed for the independent and self-directed personal transportation of one individual and possibly a few passengers, in complete isolation from other people. As Floyd Allport reminisced in 1931,

> On the highway we used to meet men and women. Whether on foot, on horseback, in buggies, or even on bicycles, it was always persons whom we encountered. Now we do not meet individuals, but automobiles—grim, impersonal machines which we try to crowd past, unmindful of our fellow being concealed behind the glass and metal. Courtesies and amenities, though natural between men, are lost upon machines. Even such social life as we have preserved has been despoiled of much of its earlier value.[836]

As mentioned previously, television has been used to manipulate and transform citizens into docile consumers. It has also become extremely destructive in terms of human relationships and community. Because television content is controlled by corporations interested in maximizing profit, it is used to broadcast messages that promote individualism, materialism, conspicuous consumption, greed, competition, promiscuous sex and violence, all of which are not conducive to the establishment of healthy social relationships or healthy societies.[837] Also, because television viewing is basically an individual activity involving only interaction with the image on the screen, television tends to isolate persons from one another as well as from their surroundings. This television-induced social isolation is particularly damaging given that people in industrialized nations spend four to five hours per day watching TV,[838] valuable time that is no longer available for establishing and nurturing relationships among family, friends and community members. Thus, as long as people spend a large part of their waking hours watching television, Internet videos, movies and other visual mass media, it will be impossible for them to develop and maintain good social relationships and strong community spirit, which are essential for happiness and psychological well-being.[839] Even watching television together with close family members is likely to be harmful to marriage and family life, as Jacques Ellul writes in *The Technological Society*:

> Television doubtlessly facilitates the material reunion of the family. The members of the family are indeed all present materially, but centered on the television set, they are unaware of one another.... Television, because of its power of fascination and its capacity of visual and auditory penetration, is probably the technical instrument which is most destructive of personality and human relations. What man seeks is evidently an absolute distraction, a total obliviousness of himself and his problems.[840]

The critical reader might object and refer to the numerous modern communication technologies, such as telegraph, telephone, cell phone, electronic mail and Internet, that have facilitated human interactions at a distance. While this is true, long-distance communications are a very poor substitute for face-to-face interactions and will never be sufficient for the establishment of close relationships and strong communities.[841] Indeed, one could consider most modern communications technologies to be counter-technologies that attempt to compensate for the unintended separation of people brought about by transportation technologies. However, as was demonstrated in Chapter 4, counter-technologies are generally not very effective in neutralizing the damage brought about by other technologies. This is also true for modern communications technologies.

As mentioned earlier, since the Industrial Revolution, machines have been used to exploit and control workers (Chapter 3), and the basic aim of industrialization has been to continuously increase technological efficiencies and labor productivities (Chapter 5). While technological development has to a large degree eliminated heavy physical labor, the resulting division and automation of labor has not only created highly repetitive and boring work but has also eliminated nearly every trace of previously enjoyed creativity, independence and autonomy. Consider this account of an assembly-line worker at an automobile plant:

> I don't do nothing but press these two buttons. Sometimes I use my thumb (to push the buttons), sometimes I use my wrists and sometimes I lay my whole arm across. The only time I sweat on the job anymore is when the sun is 100 and something outside.[842]

Extreme alienation and boredom are not only present in many factories but to a lesser extent also in offices, where people tend computers instead of assembly-line machinery. It is therefore not surprising that large numbers of people in industrialized societies, with the possible exception of highly educated, independent professionals, artists, writers and artisans, have become alienated from their work and do not derive satisfaction from it except for receiving their paycheck.[843] This is most unfortunate because, as was discussed earlier, meaningful and enjoyable work is one of the main sources of happiness. In today's industrialized societies, efficiency, industrial output and profits are much more highly valued than meaningful, satisfying work. As E.F. Schumacher writes,

> Dante, when composing his visions of hell, might well have included the mindless, repetitive boredom of working on a factory assembly line. It destroys initiative and rots brains, yet millions of…workers are committed to it for most of their lives.… A person's work is unquestionably one of the most formative influences on his character and personality.… The question of what the work does to the worker is hardly ever asked, not to mention the question of whether the real task might not be to adapt the work to the needs of the worker rather that demanding of the worker to adapt himself to the needs of the work—which means, of course, primarily: to the needs of the machine.[844]

Thus, unless the needs of the worker are given higher priority than the "needs" of the machine or production process, which are really the "needs" of the owners of income-producing property to maximize their profits, work will remain meaningless, boring and unsatisfying for most people. As a result of feeling like "a cog in the machine" and being denied work-related satisfaction and happiness, people search instead for compensatory pleasures in the form of increased material consumption, unfortunately with little success.

In addition to causing alienation at work, modern technologies also lead to separation and alienation from the natural world, depriving many of the opportunity to find inspiration and solace in the beauty of nature. This alienation from nature is a direct consequence of attempts to exploit and control it using various technological means. As was discussed in Chapter 3, the exploiter *must* feel separate from the object of exploitation in order for the abusive acts to appear morally acceptable. Thus, the more we exploit and control nature, the more alienated from it we become.[845]

Technologies that increase our sense of separation from nature abound. Many of us live almost entirely in artificial environments: inside climate-controlled houses, located in air-polluted, noisy, crowded mega-cities made of concrete, glass and steel, far away from wilderness. We drive at high speeds in automobiles, totally enclosed by metal and glass, unable to experience the nuanced beauty of the natural environment we traverse. We wear clothes made of synthetic fabrics instead of natural fibers, we eat food grown thousands of miles away on farmland we have never seen, processed in distant factories and placed in synthetic packages on the shelves of our local supermarkets. Our electronic communications and mass-media technologies such as television, video and Internet have, over time, persuaded us to prefer fantasy over reality to the point that many people have become content to see pictures and movies of nature instead of ever experiencing it themselves.[846] This degree of pathological alienation from nature can only have devastating consequences for psychological well-being and happiness as well as for any interest in preserving the life-support systems of our planet.

As was discussed in Chapter 7, since the Enlightenment, belief in human reason and scientific and technological progress has progressively replaced earlier religious beliefs. The resulting decline in spirituality, in the sense of belief in something greater than oneself, and the increasing acceptance of an utterly materialistic worldview devoid of meaning and purpose has caused much existential suffering for millions, if not billions, of people. Science and technology have been used to bring about an increased emphasis on materialism and consumerism, thereby intensifying the conflict between materialistic and spiritual values. Indeed, it could be argued that excessive emphasis on technological progress and materialism interferes with the development of spirituality and its traditional virtues of charity and compassion.

The world's religions and spiritual traditions generally promote the practices of prayer or meditation; however, these exercises become increasingly difficult in modern fast-paced industrialized societies. Compared to life in traditional agricultural societies, life in industrialized societies is extremely complicated and unnatural, creating significant stress in the absence of traditional and harmless ways of relieving it. Most importantly, modern technologies, including the many consumer products they generate, provide an endless stream of distractions which, as mentioned above, feed a growing egocentrism. Lewis Mumford observed more than 75 years ago in *Technics and Civilization*: "Contemplation, one of the great contributions of western religions, has become an ever remoter possibility."[847] This is a most unfortunate consequence of technological modernization because true religiosity and spirituality provide one of the few sources of lasting fulfillment.

In addition to undermining spiritual development, modern technologies can, under certain circumstances, promote unethical and immoral behavior. As was mentioned above, industrialization has to a large degree destroyed the sense of community, something vitally important for the development of ethical sensitivities. Ethics is concerned with the just and proper treatment of other members of society. If local communities have been destroyed, there will be fewer moral qualms about treating others badly, the "out group" becomes very large indeed. As discussed in Chapter 3, science and technology have greatly increased our knowledge and power to control and exploit. The power to control and exploit is, for many, too large a temptation to resist, generally resulting in unethical actions with serious consequences for the victims. Technologies often facilitate this exploitation and control by creating a safe distance between exploiters and exploited. As discussed in Chapter 3, modern warfare technologies facilitate the act of killing other human beings by blunting normal ethical sensitivities through the creation of safe physical and psychological distances. Even when there are no specific motives for exploitation, the introduction of a technology always has both positive and negative effects (Chapter 1), which accrue differentially to some and not to others. Generally little effort is made to help those who suffer the negative effects, resulting in unethical situations and unresolved issues of distributive justice (Chapter 8).

Finally, modern technologies and associated consumer products may remove immediate negative consequences that, in earlier times, caused people to carefully consider their actions and to restrain themselves from unwise behavior. An example is the oral contraceptive, which is clearly an effective and important family planning tool. Unfortunately, it has given people in uncommitted relationships the opportunity to engage in casual and promiscuous sex, primarily because the previous fear of unwanted pregnancy has been removed. In short, the oral contraceptive has substantially weakened former moral restraints with respect to sexual relations. As early as 1968, Pope Paul VI warned in his encyclical, *Humanae Vitae*, that modern contraceptives would promote hedonism and lower moral standards:

> Not much experience is needed to be fully aware of human weakness and to understand that human beings—and especially the young, who are so exposed to temptation—need incentives to keep the moral law, and it is an evil thing to make it easy for them to break that law.[848]

A more secular statement to this effect is that it is unfortunate when technology facilitates or encourages actions that have serious negative effects on psychological well-being as well as society. Given the present overpopulated state of the world, however, the widespread use of contraceptives should not be discouraged. It may be seen as a necessary counter-technology (Chapter 4).

In summary, the role of technology in increasing happiness is problematic. Although technological innovation promotes consumerism and materialism, the resulting economic growth and material affluence have failed to improve psychological well-being and happiness. In addition, industrialization and modernization have diminished or destroyed many traditional sources of happiness, such as strong social relationships, satisfying and meaningful work, closeness to nature as well as spirituality and ethics. Technology has not delivered on the promise of continual progress and increased happiness. Despite serious failures of industrialization and modernization, the role of modern technology is not questioned but rather passively accepted or even applauded. This widespread lack of critical thinking will be discussed in the next chapter.

CHAPTER 10

The Uncritical Acceptance
of New Technologies

"I NEVER CEASE TO BE ASTONISHED at the docility with which people accept technology uncritically, as if technology were a part of Natural Law."[849] Although E. F. Schumacher wrote this more than 30 years ago, it is even more true today.

Several factors have led to the uncritical acceptance of new technologies. Among these are the beliefs that technologies are intrinsically value neutral, evolve autonomously with little or no human guidance and that their existence and trajectories are inevitable.

There are also factors that discourage public influence over the direction of technological change. These are technological dependency and undemocratic decision making. Citizens of industrialized nations may feel trapped within technological megasystems and believe that they have no other options. Furthermore, the direction of technological change is not determined in a democratic manner: the public is regularly excluded from any meaningful debate regarding the desirability of new technologies. We now examine the above in more detail.

The Myth of Value-Neutrality

According to the myth of value-neutrality, technologies are simply neutral tools that can be used for good or evil purposes depending on who controls them[850] (i.e., "Machines don't cause problems, people do"[851]). Technological change appears to arise in an objective and rational manner, thereby creating the illusion of inevitability. As Jerry Mander indicates, the belief that technology is value neutral and inherently unbiased is widespread:

> No notion more completely confirms our technological somnambulism than the idea that technology contains no inherent political bias. From the political Right and Left, from the corporate world and the world of community activism, one hears the same homily: "The problem is not with technology itself, but with how we use it, and who controls it." This idea would be merely preposterous if it were not so widely accepted, and so dangerous. In believing this, however, we allow technology to develop without analyzing its actual bias. And then we are surprised when certain technologies turn out to be useful or beneficial only for certain segments of society.[852]

Let us analyze then how specific values are inherently embedded in a particular technology, making it politically and socially biased independently of how people choose to use it. As anyone who has been involved in creative activities is aware, myriad choices have to be made during the process of creation, and these choices generally reflect specific value orientations. The same is true for the engineering design process in which the final technological product is created within the constraints of specific goals and objectives, which, in turn, reflect the values of the engineer, the corporate or government funding agency or society at large. As Stephen Cotgrove points out,

> The conviction that technology is a neutral and objective force springs from the fact that the values which are incorporated in technological products and which guide and inform the actions of technologists and those who direct their work, are either unrecognized, or simply taken for granted. In the practice of engineering, it is constantly stressed that engineering solutions must be cost-effective and efficient, and that products must be marketable.[853]

In addition to cost-effectiveness, efficiency and marketability, a wide range of other values are commonly embedded in many modern technologies. These are, for example, power, control, exploitation and violence, with respect to both people and nature (Chapter 3), profit-maximization, speed, mass-production, uniformity, repetitiveness, quantification, precision, standardization, dependency, materialism, consumerism and individualism.[854] In general, the more specifically a technology has been designed for a particular use, the more completely it will embody the values of the designers, and the less it will be of use for purposes reflecting different values. Thus, the more specific the design, the less human choice and control there will be regarding the final use of the respective technology. As Robert Proctor writes in *Value-Free Science? Purity and Power in Modern Knowledge,*

> Tools, we realize, have alternative uses; the knife bought for cooking might be used for killing. Yet knives or levers are not what modern science-based technology is all about. A nuclear power plant, cruise missile, or linear accelerator can hardly be used for ends other than those for which they are designed. Science-based technologies are increasingly end-specific: the means constrain the ends; it is no longer so easy to separate the origins of a tool from its intended use.[855]

Indeed, military technologies are excellent examples of how a unique purpose, which is the use of violent means for conflict resolution, is built into their design. The very presence of military technologies, ranging from guns to nuclear weapons, increases the likelihood that they, instead of diplomacy, will be used to resolve conflicts among adversaries. Military technologies have a built-in bias toward violence, control and exploitation, and they undermine alternative courses of action that reflect other values, such as negotiation, non-violence, cooperation and justice. Similarly, as was discussed in Chapter 3, many modern technologies, not only military, embody the values of control, exploitation and violence, often with respect to both people and the environment. Consider, for example, the difference between the generation of nuclear or fossil energy and renewable solar energy. Nuclear energy can be generated only in highly complex and centralized power plants that pose potential radiation hazards to the surrounding communities and even to very large geographic regions. Furthermore, the very serious and extremely long-term risks related to radioactive waste management are transferred to future generations. Similarly, the generation of electricity from fossil fuels (coal, petroleum, natural gas) involves complex and centralized power plants that emit carbon dioxide, which, in turn, accelerates global warming and its dire consequences that will also be visited upon future generations. These two conventional energy-production technologies embody violence against both people and the environment as well as centralized authoritarian control, the latter of which is incompatible with the ideal of democracy. By contrast, many renewable energy technologies are more environmentally friendly and more amenable to local control, particularly if deployed on a small scale near the location of energy end use. Solar-energy generation has an inherent ideological bias toward environmental compatibility, decentralization and democratic control.[856] The design of centralized installation of acres of solar panels as well as wind farms are obvious attempts to maintain corporate control even of technologies that lend themselves to decentralization and individual use. Small solar panels placed on the roof of one's home would be much less destructive than vast arrays of solar collectors in sensitive desert ecosystems. The public is not being alerted to the implications of these two types of application of the same technology. Control is enhanced by centralization. What should be simple, local, environmentally friendly technologies are being transmuted into corporate-controlled, centralized, mega-profit-generating enterprises.

Examples of the value-laden character of technology abound—indeed, a distinct value-orientation can be demonstrated for any technology. Most high-tech medical technologies have a bias toward saving life at any cost, even if this causes severe physical and emotional distress to the patient and family.[857] Genetic engineering technologies commodify plants and animals for the maximization of profits.[858] The automobile reflects the values of speed, mobility and individualism as well as violence against the environment (i.e., noise, pollution, road kill, etc.) and even people (those unfortunate millions who are killed or maimed in car accidents every year). Television, as was discussed in Chapter 9, promotes individualism, consumerism and materialism, thereby contributing to the destruction of family life and community participation. It is important to recognize that, independent of content, television viewing is harmful to family and community life because it is largely an activity that precludes conversation and social interaction, thus isolating individuals from one another. Furthermore, as Jerry Mander writes in

Four Arguments for the Elimination of Television, because the electronic medium is constrained by technological factors to convey a view of reality on a confined two-dimensional screen, television and related video technologies are inherently biased toward images that

> tend to the larger as opposed to the smaller, to the broad as opposed to the detailed, to the simple as opposed to the complex, to the obvious rather than the subtle. Because of these tendencies, inexorably imposed by the technology itself, the communicable content of all programs is affected.... Subtle feelings are more difficult to transmit through television than the larger emotions.... The human relationships which are shown on television, therefore, tend to be those that can be shown on television. These dwell on the grosser end of the human emotional spectrum. The more subtle expressions, those which express intimate, deeply personal feelings, are lost in the blur.[859]

Thus, it is not surprising that television programming is heavily biased toward more gross and active content such as sports, violence, police action, conflict, war, drama, game shows, soap operas and flashy product advertising. It is difficult to broadcast more subtle, emotional content such as social interaction, being within and surrounded by nature, or a profound appreciation of other cultures or religions.[860] Even if the message on TV is pro-environmental, TV viewing is intrinsically anti-environmental because it provides a substitute for experiencing nature first hand and because it encourages passivity, thereby undermining interest in environmental activism.[861] In addition, because electronic media are able to make consumer products seem "more alive than people,"[862] it is inherently biased toward materialistic consumerism, which is one of the causes of the environmental crisis. In summary, television and related video technologies are intrinsically biased toward values destructive of society and the environment. It would be extremely difficult, if not impossible, for electronic media to play a more constructive role in society.

Finally, because all technologies are value laden, their embedded values move with them as they are transferred from one culture to another.[863] Knowingly or not, industrialized nations often use the avenue of technology transfer to impose their own cultural values and ideologies on unsuspecting developing nations.[864] As Robert William Stevens writes,

> Built into a technology are the values and ideals of a society that developed it. So when we export technology, we export a whole system of values: a certain attitude toward nature, toward society, toward work, and toward efficiency. As yet, no developing society has been able to withstand the effects of this onslaught. The result has been that local societies always change to meet the incessant demands of the new technology. The end of this process could well be a global uniformity of cultures, all perfectly adapted to high technology, and everywhere the same.[865]

Despite the fact that all technologies are value laden, the myth of value-neutrality continues to be perpetuated. There are at least four reasons for the promotion of this myth. First, by giving technology the mask of objectivity, the fact is hidden that many modern technologies are used for the exploitation and control of both people and the environment (Chapter 3).[866] As Robert Proctor writes, "Systems of domination have always sought to clothe themselves in the guise of neutral and objective legality—so that power appears as justice, force as law."[867] Second, by promoting the view that social and environmental problems are solely caused by the misuse of technology rather than by the inherent characteristics of the technology itself, the myth of value-neutrality isolates technologies from criticism.[868] Consequently, both the technology itself and those who designed it abdicate their responsibility and escape blame.[869] As Jeremy Rifkin observes,

> By vesting every new technology with neutrality and inevitability, the many special interests who have so much to gain from the speedy introduction and acceptance of their "inventions" free themselves of any responsibility for having to ponder the merit, wisdom, or appropriateness of their "contributions." Technologies, however, are not value-free, nor are they inevitable.[870]

Third, by making technologies and industrialization appear to be value neutral, political decisions regarding the development and deployment of new technologies are given the mask of legitimate objectivity, thereby removing citizens from political participation in regard to these innovations.[871] Consequently, as will be discussed later in this chapter, because people are denied these basic democratic decision-making rights, they often feel that they have no choice but to accept any new technology, no matter how destructive to health, society or the environment. Finally, and probably most importantly, the myth of value-neutrality hides the fact that technological development is consciously directed by special interests and powerful social classes whose primary objective is the maximization of profits, a topic that will be discussed in greater detail in the next chapter.[872]

The Myth of Autonomous Technology

This is the belief that technology has autonomy and therefore is an independent, uncontrollable force.[873] French theologian Jacques Ellul introduced the concept of autonomous technology (technique) in *The Technological Society*: "Technique…is truly autonomous.… The autonomy of technique forbids the man of today to choose his destiny."[874]

Similarly, Langdon Winner, professor of political science at MIT, writes in his book *Autonomous Technology*,

> The same technologies that have extended man's control over the world are themselves difficult to control.… Technology seems to go forward at its own inertia, is self-sustaining, and appears autonomous while at the same time humans have free will to make decisions.… technology moves along its own evolution.[875]

The belief in autonomous technology is related to the concept of technological determinism, the idea that technology develops independently of society and is subsequently imposed upon it. For example, according to this view, the Industrial Revolution and the social upheaval that followed it were solely the result of major technological developments such as the invention and introduction of steam power. As David Dickson writes in *The Politics of Alternative Technology*, the concept that technological innovation is autonomous and determines the fate of society is still the most common model of technology today.[876] However, as we will see in the next chapter, technological change is anything but autonomous because it is consciously directed by certain influential players.

One reason technological progress may appear autonomous is that many technological systems, because of their size, complexity and mutual interdependence, have a rigidity and inertia that is difficult to overcome.[877] As technological choices become permanently "locked in" via very large capital investments and major adjustments in social habits, it becomes difficult to change the direction of technological development.[878] As Richard Sclove explains in *Democracy and Technology*,

> The flexibility associated with a given technology, or with other social structures, tends to diminish with time. After a society has habituated itself to one technology, alternatives tend to become less accessible. Once designed and deployed, a technology, like a law or a political institution tends—if it is going to endure—gradually to become integrated into large systems of functionally interdependent artifacts and organizations and then to influence the design of subsequent technologies, laws, and institutions.... Thus, owing to the accompanying evolution of supporting custom, entrenched interest, and various sunk costs, it is often difficult to achieve radical design alterations once an initial decision has been implemented.[879]

An excellent example of this technological "lock-in" and resulting perception of technology as an autonomous force is the worldwide dissemination of the car culture. Following the invention of the internal combustion engine more than a century ago, the first automobiles were built and tested. Shortly thereafter, new industrial manufacturing technologies such as the assembly line made possible the mass production of automobiles. After millions of cars flooded the market, an extensive infrastructure consisting of road and highway networks, bridges, tunnels, parking lots and repair shops came into existence to accommodate the needs of the automobile. Furthermore, a highly complex fuel-delivery network was built, consisting of petroleum extraction facilities across the globe, tanker transport, oil refineries, pipelines, truck and train delivery systems and gas stations.[880] With increasing use of the automobile worldwide, the components of this technological megasystem not only grew enormously in size but also became more and more interlocking, making it virtually impossible to stop the global automobile juggernaut.

As people begin to feel trapped by the various entrenched technological megasystems, they often blame the technologies themselves, claiming that humans have been taken over by self-directed, autonomous technologies. This view is seriously incorrect because humans, and not machines, made the decisions to invent, develop, build and produce these technologies and associated systems. In the case of the automobile, conscious decisions were made to invent the internal combustion engine, to mass produce cars in assembly lines and to build the supporting road networks as well as fuel-production and delivery infrastructures. As Richard Sclove warns, the myth of autonomous technology is dangerous because it diverts attention from the need for more social and democratic control of technology:

> One of the most serious and prevalent misconceptions about technologies is that they are natural or inevitable rather than the result of contingent social choices. This idea is dangerously false, because it hampers the establishment of a strongly democratic politics of technology.[881]

The Technological Imperative

The technological imperative may be expressed as "Whatever can be done technically, will be done and should be done." This highly deterministic view of technology has also been expressed as "Whatever becomes technologically possible—within certain economic limits—must be done. Society must adapt itself to it."[882] "What it is possible for science to know science must know. What it is possible for technology to do technology will have done."[883] "You can't stop progress....Once the genie is out of the bottle you cannot put it back."[884] "If a discovery or a technology can be used for evil purposes, it will be so used."[885] "Technological possibilities are irresistible to man. If man can go to the moon, he will. If he can control the climate, he will."[886] "Technological accomplishment has become a temptation that no person can reasonably be expected to resist."[887] Even the world-renowned physicist Werner Heisenberg was under the spell of the technological imperative when he proclaimed, "The worst thing about it all [the building and use of the atomic bomb] is precisely the realization that it was all so unavoidable."[888] Similarly, J. Robert Oppenheimer, a physicist in charge of the Manhattan Project (designing and building the world's first nuclear weapon), said in 1945,

But when you come right down to it the reason that we did this job…is because it was an organic necessity. If you are a scientist you cannot stop such a thing. If you are a scientist you believe that it is good to find out how the world works; that it is good to find out what the realities are; that it is good to turn over to mankind at large the greatest possible power to control the world.[889]

As Lewis Mumford observes,

Western society has accepted as unquestionable a technological imperative that is quite as arbitrary as the most primitive taboo: not merely the duty to foster invention and constantly to create technological novelties, but equally the duty to surrender to these novelties unconditionally just because they are offered, without respect to their human consequences. One may without exaggeration now speak of technological impulsiveness: a condition under which society meekly submits to every new technological demand and utilizes without question every new product, whether it is an actual improvement or not.[890]

Examples of the technological imperative abound: if it is possible to pump oil out of the ground, it must be done even if it causes pollution and global warming. If it is possible to create new synthetic organic chemicals, they must be mass produced even if this causes harm to the environment and human health. If it is possible to dam rivers, it must be done even if it causes ecological disaster. If it is possible to genetically engineer plants and animals, it must be done even if it may cause untold harm to the environment. If it is possible to extend human life by a few months using heroic high-tech medical procedures, it must be done no matter how much suffering and loss of dignity will result. If it is possible to build nuclear weapons, they must be mass-produced even if their deployment could result in the end of civilization.

Anyone who allows the technological imperative to guide his actions has, in fact, given up any consideration of ethics in his decision making. Instead of carefully considering whether a new technology, if fully developed and deployed on a large scale, is useful or useless, constructive or destructive, harmless or harmful, humane or inhumane, right or wrong, good or evil, the person who believes in the technological imperative will endorse and promote new technologies simply because they are new. Being able to accomplish something is never a good reason for doing it. A professor of our acquaintance, when explaining a certain action he had taken, declared, "I like it. Why not?" We felt somewhat uneasy about his witty rule of conduct, which, in effect, like the technological imperative, omitted recognition of any ethical limits whatsoever. As Erich Fromm writes,

> Something ought to be done because it is technically possible to do it. This principle means the negation of all values which the humanist tradition has developed. This tradition said that something should be done because it is needed for man, for his growth, joy, and reason, because it is good, beautiful, or true. Once the principle is accepted that something ought to be done because it is technically possible to do it, all other values are dethroned, and technological development becomes the foundation of ethics.[891]

The technological imperative serves at least three purposes, all of them dubious: First, it is used as an excuse to avoid democratic deliberation as well as ethical decision making with respect to the development and acceptance of innovative technologies. Second, as was the case for the myth of value-neutrality, the technological imperative makes new technologies appear inevitable by hiding the fact that the course of technological development is directed by special-interest groups and powerful social classes (Chapter 11). Third, as a result of this concealment, belief in the technological imperative discourages not only critical thinking but also promotes a culture-wide passive acceptance of any new technology, no matter how destructive.[892] As will be discussed in Part III, critical ethical reasoning rather than submission to the technological imperative is needed to consciously direct both science and technology in creating a world that is both sustainable and humane.

Technological Dependency and Loss of Freedom

It is claimed that technology increases freedom in two ways. First, it provides freedom from a wide range of former external constraints imposed by a hostile natural environment. It has also been pointed out that technological innovations have provided many new opportunities to indulge in new types of positive freedoms, the freedom to choose new ways of working, travel, communication and even entertainment. As Emmanuel Mesthene optimistically proclaims,

> We have the power to create new possibilities, and the will to do so. By creating new possibilities, we give ourselves more choices. With more choices, wé have more opportunities. With more opportunities, we can have more freedom, and with more freedom we can be more human. That, I think, is what is new about our age. We are recognizing that our technical prowess literally bursts with the promise of new freedom, enhanced human dignity, and unfettered aspiration.[893]

Although a number of technological innovations have, indeed, had a profound impact on freedom, such as the freedom to work at night, to travel at high speeds for great distances and to quickly access large amounts of information, for the most part, technology has provided only greater freedom to choose among an enormous array of competing consumer products and services, including vapid, if not degrading, entertainment.[894] For example, we are now free to choose among many different brands of toothpaste, soft drinks, cigarettes, automobiles, diapers, toilet paper and TV programs. This vast but trivial array of consumer choices creates an illusion of freedom within the constraints of the industrialized society. There is, however, a largely unnoticed dark side: the loss of important fundamental freedoms and increasing technological dependency.

What kinds of freedoms have been lost by subjecting ourselves to life in industrialized society? E.F. Schumacher attempts to provide an answer:

> What is freedom? Instead of going into long philosophical disquisitions, let us ask the more or less rebellious young what they are looking for. Their negations are such as these: I don't want to join the rat race. Not to be enslaved by machines, bureaucracies, boredom, ugliness. I don't want to become a moron, robot, commuter. I don't want to become a fragment of a person. I want to do my own thing. I want to live relatively simply. I want to deal with other people, not masks. People matter. Nature matters. Beauty

matters. Wholeness matters. I want to be able to care. All this I call
a longing for freedom. Why has so much freedom been lost?[895]

There are several reasons for the loss of so much real freedom. As was
discussed in Chapter 3, to the extent that technology has provided a partial escape from the direct forces of nature, it has required submission
to many constraints imposed by technological structures. Many people
are willing to sacrifice freedom in exchange for security and material
comfort.[896] Hundreds of millions of people in industrialized societies
relinquish their personal freedom by spending most of their life following highly regimented schedules operating machines, computers and
automobiles, acting as a mere cog in the technological system, rewarded
only by the opportunity to purchase more consumer products. Dependence on mega-technological systems has increased to the point that
many feel trapped inside an increasingly coercive and invasive technological structure. As Theodore Roszak observes,

> The thousand devices and organizational structures on which
> our daily survival depends are far more than an accumulation
> of technical appendages that can be scaled down by simple sub-
> traction. They are an interlocking whole from which nothing can
> easily be dropped. How many of us could tolerate the condition
> of our lives if but a single "convenience" were taken away... the
> telephone...the automobile...air conditioner...refrigerator...
> computerized checking account and credit card...? Each ele-
> ment is wedded to the total pattern of our existence; remove any
> one and chaos seems to impend. As a society, we are addicted to
> the increase of environmental artificiality; the agonies of even
> partial withdrawal are more than most of us dare contemplate.[897]

Collective dependence, actual or supposed, on high technology is extreme, as the following list of randomly selected examples illustrates. We
are told that without industrialized agriculture, massive irrigation infrastructure, fossil-fuel based nitrogen fertilizers, pesticides and herbicides, heavy machinery and more recently genetically modified organisms, farming cannot provide sufficient food for all, placing billions of
people at risk of starvation. Appliances of every sort, which incorporate

the latest techno-features, are "required" for a comfortable existence. We depend on physicians and medications credited with keeping us alive and pain-free while we suffer from self-inflicted conditions such as heart disease, cancer and other degenerative diseases that are to a great extent the result of poor lifestyle choices (encouraged by advertising) or environmental pollution (caused by other forms of technological "progress")[898] We depend on the Internet and television instead of face-to-face communication. Instead of receiving important and relevant information directly from members of our local community, we rely on the mass media to provide us with the latest "news," most of which is a form of perverse entertainment and totally irrelevant to our lives. And, of course, the car, fully equipped with state-of-the-art techno-gadgets, has been become a necessity.

While many people in industrialized society suffer loss of freedom and increased technological dependency, there are some who benefit. Scientists and engineers, because of their special knowledge and skills, are rewarded for designing "technologies of the future." Large corporations destroy local self-sufficiency and subsequently increase dependency on the new products and services they provide. Efforts designed to empower people by increasing self-sufficiency and decreasing technological dependency are vigorously and sometimes violently opposed.

The Undemocratic Control of Technology

Most decisions regarding the development of new technologies are made by the directors of corporations and government bureaucracies, with ideas, research and technical work being supplied by the scientists and engineers they employ. Many professors are financially dependent on research grants from the same corporate and governmental entities. In most cases, the public is excluded from any meaningful debate; thus an authoritarian technocracy has replaced democracy.[899] For example, no vote was ever taken as to whether more than 100 million tons of untested synthetic organic chemicals, representing over 65,000 different compounds, should be introduced into our environment every year[900] or whether genetically modified plants should be allowed to enter our food supply. As Jerry Mander observes,

The great majority of us have no say at all in choosing or controlling technologies. These choices...are now solely within
the hands of this technical-scientific-industrial-corporate elite
whose power is enhanced by the technology they create. From
our point of view the machines and processes they invent and
disseminate just seem to appear on the scene from nowhere....
We don't get to vote on these things as they are introduced. All
we get to do is pay for them, use them, and then live within their
effects.... We believe ourselves to be living in a democracy because from time to time we get to vote on candidates for public
office. Yet our vote for congressperson or president means very
little in light of our lack of power over technological inventions
that affect the nature of our existence more than any individual
leader has ever done. Without our gaining control over technology, all notions of democracy are a farce. If we cannot even think
of abandoning technology or, thinking of it, effect the ban, then
we are trapped in a state of passivity and impotence hardly to be
distinguished from living under a dictatorship.[901]

Politicians, economists and technocrats generally defend the present arrangement by pointing out that people have nevertheless a democratic
influence over the direction of technological change by exercising
their purchasing power in the free market. According to this view, technologies and technological products survive in the marketplace only if
consumers buy them. But there are at least two problems with this position. First, corporations use billions of dollars each year to manipulate
consumers through advertising campaigns to buy products and services
they do not need and, in the absence of manipulation, would not want
(Chapter 9). Second, the free market, which operates according to the
principle "one dollar–one vote,"[902] is inherently undemocratic in terms
of directing technological change because of the very large inequalities
that currently exist in wealth and income. As US Supreme Court Justice
Louis Brandeis noted more than half a century ago, "We can either have
democracy in this country, or we can have great wealth concentrated
in the hands of a few, but we can't have both."[903] When there is great
inequality, the wealthy have many more "purchasing votes," often

hundreds of times as many as ordinary citizens. Technologies and technological products reflect this bias.

In the early 1990s, the wealthiest one percent of the US population earned 15.7 percent of the nation's annual income and owned 37.2 percent of all privately held wealth, including 49.6 percent of all corporate stocks and 62.4 percent of all bonds.[904] In 2007, the 50 highest-paid private investment fund managers were paid on average $588 million, 19,000 times as much as the average worker.[905] Globally, inequalities are even more severe. Consider, for example, that the richest 2 percent of the world's people now own 51 percent of all of the world's assets while the poorest 50 percent own only 1 percent.[906] The average individual income of the global top 20 percent is approximately 150 times larger than the bottom 20 percent. By the end of the 20th century, the world's three wealthiest individuals had a total wealth greater than that of the poorest 48 nations combined.[907] Similarly, the net worth of a few hundred billionaires is equal to the combined wealth of more than 2 billion of the world's poorest people.[908]

This global inequality ensures that technological innovation in wealthy industrialized nations will be employed almost exclusively to satisfy the trivial desires of well-off "consumers" while hardly any technological ingenuity will be employed to address the basic needs of the world's poor.

Because of its size and complexity, contemporary technological infrastructure is difficult to control in a democratic fashion. Most technological systems have very large capital requirements that can be provided only by wealthy investors or by governments, which are often controlled by the same entities. The average citizen is not able to make major selective capital investments and is thus excluded from influence regarding the direction of technological development.[909] Also, because of the complexity, the average citizen often cannot understand the scientific and technical details related to newly developed technologies and technological products, including their potential long-term environmental and social consequences. This is aggravated by declining scientific literacy and increasing specialization, the latter making it difficult for even highly educated people to comprehend necessary technical details, as mentioned in Chapter 7. The public is effectively excluded from any meaningful debate about the wisdom of most major technological developments.

Giving "experts" as well as business interests the power to determine the direction of technological change on behalf of society poses a number of problems. As was mentioned earlier, the masks of objectivity and value-neutrality hide underlying value conflicts, which should be brought to public attention and scrutiny and be resolved in the political arena. Also, as mentioned previously, most "experts" are ignorant of matters not directly related to their specialized area and generally do not understand or appreciate the larger context. Specifically, as Richard Sclove points out in *Democracy and Technology*, many experts do not represent the values of average citizens:

> Their tendency to acquire specialized competence at the expense of integrative knowledge and experience, render them statistically unrepresentative of the values and outlook of the larger citizenry. Moreover, technical experts—especially the most influential, elite experts—tend to live experientially far removed from the everyday events and concerns of non-expert, non-elite citizens.[910]

Most importantly, nearly all experts represent the interests and values of those who employ them, mainly large corporations and government agencies.[911] Since the primary objective of corporations is the maximization of profits for their shareholders, it is not surprising that most experts have only the choice of resigning or of coming to the defense of new technologies if these promise to generate profits, even if they may cause harm to society or the environment. Thus, in essence, the control of technology is undemocratic because corporations and the governments they influence or control, rather than average citizens, are the key decision-makers in matters of technological innovation. The end result is that the developers of new technologies have almost complete control over the direction of technological change without having to confront a critical public. New technologies are not generally developed with the goal of meeting the needs of the many but rather of maximizing profits for the few, a topic that will be treated in the next chapter.

CHAPTER 11

Profit Motive:
The Main Driver of
Technological Development

A S WAS DISCUSSED in the previous chapter, ordinary citizens have or believe that they have little or no control over the direction of technological change. This is partially the result of the widespread acceptance of several myths: that all technology is value neutral and thus objective and inevitable, that technological development is a powerful, independent, autonomous force beyond human control, and that, in accordance with the technological imperative, whatever can be done technically, will be and should be done. In this chapter, we show that the development and large-scale adoption of new technologies is a social process guided by the principle of profit maximization. We then demonstrate that selection criteria based on profit maximization do not necessarily lead to the development of technologies and products best suited to meet the needs of people in terms of, for example, food, health or security.

Technological Development as a Social Process

As David Dickson writes in *The Politics of Alternative Technology*, "The idea that technology develops independently of society, and is subsequently imposed upon it, is still the most common model used today."[912] According to this view, inventors periodically create new technologies. Most inventions are condemned to failure while a few successful ones are soon widely adopted by society, which, in turn, is transformed by them. Unfortunately, this simplistic idea of technological determinism is not informed by the important historical fact that many technological inventions were ignored or even suppressed by the society in which they were first produced, only to be rediscovered or reinvented many centuries later under different social conditions.[913] For example, both the Greeks and Romans had little use for labor-saving devices because slave labor was abundant. The Roman Emperor Vespasian, when presented with a novel labor-saving device, is said to have replied "Take it away; I have my poor to feed."[914] Similarly, in ancient China, certain technologies were known but were not adopted.[915]

David Dickson asserts that new technology is selected and constructed by the politically dominant social group to maintain power and control:

> My general thesis is that technology plays a political role in society, a role intimately related to the distribution of power and the exercise of social control. It does this, I maintain, in both a material and an ideological fashion, implying that in both senses technological development is essentially a political process. At a material level, technology sustains and promotes the interests of the dominant social group of the society within which it is developed.... The implication of this thesis is that one can only understand the nature of technology developed in any society by relating it to the patterns of production, consumption and general social activity that maintain the interests of the politically dominant section of that society.[916]

Who are then the dominant actors in the arena of technological development? There are basically three: governments, corporations and the public. Governments are primarily responsible for sponsoring the development of military and space technologies. Furthermore, governments also provide significant funding for basic research, particularly in areas that may lead to commercial applications, such as new pharmaceuticals and high-tech medical therapies. Other than through elected representatives, the public has very little input with respect to government-sponsored technological development. This is most obviously the case in all nations for military technologies that are designed and produced under strict secrecy.

Non-government-sponsored technologies and technological products are developed and marketed by private and publicly traded businesses. In free-market economies, companies select for further development only those technological innovations that promise to reduce production costs, increase sales and maximize profits.[917] Earnest Braun writes,

> In a Darwinian analogy, technological inventions may be regarded in the same light as spontaneous mutations of species. Inventions become successful innovations if society selects them.... Only those technologies will be selected that can make money in the marketplace. In the case of process technologies, the selection mechanism is dominated by the ability of the technology to increase productivity, improve the competitive position, and increase profits for its owner.[918]

Understanding the Meaning of Profit

Because profit maximization is a primary criterion that determines whether technological innovations will be adopted, it is important to understand the concept of profit, a term that is often misunderstood and incorrectly used. Profit is simply the difference between income collected from sales of produced goods and services minus all costs incurred during production.[919] There are many types of expenses: wages and salaries; the costs of raw materials, including energy; the depreciation of equipment, rents and so forth. In order for profits to be maximized, expenses must necessarily be minimized. This is generally done by employing more efficient, more productive, and more labor-saving technologies. However, it is often overlooked that profits are also maximized by not compensating workers for the full value of their labor. The same is true for the suppliers of materials and energy. As Jerry Mander succinctly states, "Profit is based on paying less than the actual value for workers and resources.... Profit is based on underpayment.... This is called exploitation."[920]

Who collects the profits? In free-market economies, profits are paid to the owners of income-producing properties (i.e., either to private business owners or to holders of corporate stocks). It is important to recognize that these owners do not have to work to collect profits.[921]

Simply by providing the necessary financial resources to build and operate factories, they expect a return on their investment in the form of profits. Given that these owners do not work themselves other than periodically checking the performance of their investments, profits are never really "earned" through work but rather appropriated through underpayments of various kinds. This upward transfer of wealth is fundamentally exploitative because the owners benefit at others' expense. For example, workers have little choice in the matter as they do not own the income-producing property themselves. Because of its inherently exploitative nature, the collection of profits based solely on the lending of money was considered highly immoral by the Catholic Church and was therefore strictly forbidden prior to the rise of modern free-market economies. It should be noted, however, that it is still considered unethical today in many Islamic countries. Indeed, as Thomas Princen writes in *The Logic of Sufficiency*, the entire concept of profit is a rather recent historical phenomenon:

> The profit motive, we are constantly being told, is as old as man himself. But it is not. The profit motive as we know it is only as old as "modern man." Even today the notion of gain for gain's sake is foreign to a large portion of the world's population, and it has been conspicuous by its absence over most of recorded history.[922]

Several other facets of profit are worth mentioning. First, profits are often reinvested by owners to expand productive capacity. Under normal economic conditions, increasing production automatically leads to increasing consumption. The reinvestment of profits and the resulting upward spiral of production and consumption is one of the key drivers of economic growth in free-market industrial societies. Our economic system's inherent dependence on growth for survival (i.e., more of everything: more markets, more consumers, more raw materials, more energy, more cheap labor, etc.) is the root cause of many environmental problems and is in direct conflict with sustainability, as was discussed in Chapters 2 and 6.

Second, unless counteracted by government regulations, the continuous upward transfer of wealth (e.g., from workers to owners) tends to increase the income gap between rich and poor. Inequality in income and wealth, in turn, has a profound effect upon what types of technological products and services are produced. As was already discussed in the previous chapter, if there is great inequality, the wealthy have vastly more purchasing "votes" in the marketplace than others, so that the development of technologies and products becomes biased in favor of meeting the needs of the comfortable, while ignoring the needs of others.

Given the inherently exploitative nature of free-market economies in which wealth is continuously transferred from workers to owners, where people are constantly in danger of being replaced by machines or cheaper, often foreign, labor, where large inequalities exclude many from purchasing even the basic necessities of life and where there is a constant need for more economic growth which threatens environmental sustainability, one wonders how such an intrinsically immoral system could have been maintained for more than 200 years. Though there is no simple answer, several explanations may be advanced. First, it is most likely that the average citizen does not understand how the economic system functions. Even most economists and business people may not recognize that the collection of profits may be considered inherently unethical. A brief survey of several major economics and business textbooks confirmed that the ethical dimensions of profits, particularly their exploitative character, are not discussed. As David Korten, who received his Ph.D. from Stanford University business school, comments,

> In business school, I learned the art of assessing investment options to maximize financial return. My teachers never mentioned that what we were really learning was to maximize the returns to people who had money, that is, to make rich people richer.[923]

Second, there is a taboo, particularly in the United States, against discussing the topic of class, or worse, class conflict. The widely accepted myth that the United States is a classless society effectively prevents confrontations between the owning and working classes.[924]

Third, because of the positive synergism between free markets and science and technology, industrial economies have been extremely successful in "delivering the goods." Per capita affluence, measured as gross domestic product (GDP) per person, has risen dramatically in all industrialized nations for the past two centuries. For example, per capita GDP rose on average 14-fold between 1820 and 1989 in the industrialized countries.[925] People who have benefited from such a tremendous increase in personal income and wealth are willing to accept the ethical shortcomings of the economic system in which they participate. People who have not yet benefited are kept in a constant state of hope and expectation that they will do so in the very near future. For example, a recent survey in the United States showed that 19 percent of the population thought they were already in the top 1 percent income bracket and another 20 percent thought they would soon be.[926] These statistics clearly reflect a high level of unrealistic hope. Although developed nations already have a very high standard of living, there is a constant quest for more. The expectation of ever-increasing affluence in the future has a pacifying effect on the economically less fortunate who, if economic growth were to stop, might very well demand a more equitable distribution of wealth. Similarly, the poor in so-called underdeveloped countries readily accept the need for Western-style industrial and economic development, believing that they will soon reach the same level of affluence as enjoyed in industrialized societies. As shown in Chapter 6, unlimited economic growth is not environmentally sustainable. And as will be discussed in Chapter 12, a transition to a steady-state economy will be necessary to guarantee environmental sustainability, but this will be complicated by the fact that people would no longer be pacified by the unrealistic hope of never-ending growth in material affluence.

Profit Maximization and the Development of New Technologies

Standard economics and business textbooks claim that profit maximization is necessary for improving the efficiency of resource allocation and optimizing overall utility to society. Consider this usual explanation from the textbook by Michael Baye, *Managerial Economics and Business Strategy*:

> A common misconception is that the firm's goal of maximizing profits is necessarily bad for society. Individuals who want to maximize profits often are considered self-interested, a quality that many people view as undesirable. However, consider Adam Smith's classic line from *The Wealth of Nations*: "It is not out of the benevolence of the butcher, the brewer, or the baker, that we expect our dinner, but from their regard to their own

interest". Smith is saying that by pursuing its self-interest—the goal of maximizing profits—a firm ultimately meets the need of society.[927]

Although it is clearly true that the profit motive has been a powerful driver of invention and the dissemination of technological innovations, including their widespread application for economic gain, there are a number of problems with profit maximization as the primary criterion for guiding technological and economic change in society. First, it is questionable whether profit maximization has actually improved the total welfare of society. Though it is true that material affluence has increased significantly in industrialized nations, psychological well-being and average happiness have not improved (see Chapter 9). Second, the entire concept of profit maximization as a guiding principle for technical and economic development is ethically problematic because profit depends upon exploitation. As discussed in Chapter 3, many technologies are developed and applied with the goal of exploitation and control, the profit motive providing the incentive for these morally dubious activities. Third, it is disappointing to have greed, expressed in the form of desire for maximum profits, to be the primary motive for technological innovation and economic activities. Current free-market economies legitimize and encourage selfishness and greed,[928] one of the seven deadly sins proclaimed by the Catholic Church in earlier times.[929]

Finally, and probably most important from a practical perspective, is that profit maximization often leads to the neglect of important and fundamental social needs. In the conflict between corporate profits and human or environmental welfare, the economic system has a strong bias toward profits. As Jerry Mander observes,

> US corporate law holds that management of publicly held companies must act primarily in the economic interests of shareholders. If not, management can be sued by shareholders and firings would surely occur. So managers are legally obliged to ignore community welfare (e.g., worker health and satisfaction, environmental concerns) if those needs interfere with profitability.[930]

Nobel laureate economist Milton Friedman expresses this lack of social responsibility:

> Few trends could so thoroughly undermine the very foundation of our free society as the acceptance by corporate officials of a social responsibility other than to make as much money for their stockholders as possible. This is a fundamentally subversive doctrine.[931]

Although over the years the government has enacted labor and environmental laws to limit the abusive power of corporations, the primary goal nevertheless remains profit-maximization rather than social responsibility. As David Korten writes in *The Great Turning*,

> Consequently, under current US law, the publicly traded limited-liability corporation is prohibited from exercising the ethical sensibility and moral responsibility normally expected of a natural-born, emotionally mature human adult. If it were a real person rather than an artificial legal construction, we would diagnose it as sociopathic. Unless constrained by rules set and enforced by a public body functioning as a kind of parent surrogate, the publicly traded corporation operates in an ethical vacuum.[932]

Ethical and socially conscious managers who try to manage the corporation in the public interest face the risk of being expelled as soon as profits are threatened.[933] In the final analysis, the conflict between profit and social responsibility reflects differences in interest between the wealthy and the rest of society. The privileged and wealthy appropriate maximum profits so they can live in luxury without working while the majority only expect humane working conditions, food, shelter, entertainment and a clean and safe environment. Because of the great inequality of wealth (see Chapter 10), the interests of the wealthy, who have disproportionate economic and political power, have taken precedence over the rest of society. As a result, workers are laid off as soon as profits are threatened in times of economic recession;[934] the environment continues to be polluted while the needed cleanup and protection is slow if pursued at all; and there is no significant progress toward true sustainability (see Chapter 6). Furthermore, as will be discussed in the remainder of this chapter, the overriding objective of profit maximization has a profound influence on what types of foods are produced, what kind of medical technologies are developed and how military threats are perceived.

Profit Maximization: Agriculture and Food

In much of Europe and elsewhere, traditional farming at one time produced food on family or community land in a way that was environmentally sustainable for generations. Slash-and-burn agriculture was practiced in some parts of the world but by such sparse populations as not to irreversibly degrade the environment. By contrast, modern industrial agriculture attempts to maximize profits for shareholders by maximizing harvests while minimizing costs. The profit maximization strategy is directly or indirectly responsible for many of the negative consequences of high-technology agriculture, discussed in Chapter 2. These are the loss of farm jobs as a result of mechanization, the damming of rivers and drawdown of aquifers for irrigation, the irreversible loss of top soil, the unsustainable use of inorganic and synthetic fertilizers, the widespread application of toxic and environmentally harmful herbicides and pesticides and the use of genetically engineered crops that may pose serious risks to ecosystem integrity. Furthermore, profits to agribusinesses remain guaranteed as long as individual farmers remain dependent on farm inputs such as machinery, fossil fuel, fertilizers, herbicides, pesticides and genetically engineered seeds. In a competitive free-market environment, the average farmer cannot survive unless he adopts these environmentally unsustainable farming practices. An alternative is organic farming, but this is currently possible for only a small number of farmers who serve a rather limited niche food market.

One would think that the large increase in food production that was brought about by the profit-driven industrialization of agriculture would at least have had the benefit of abolishing world hunger. Unfortunately, this has not been the case. As Michael Perelman states in *Farming for Profit in a Hungry World*, the overall objective of agribusiness is to maximize profits rather than pursue socially responsible goals such as providing food to the world's poor.[935] Because profits must be maximized in a free-market economy, food production is geared toward addressing the culinary desires of those who can pay a premium price rather than toward meeting the minimum nutritional needs of the poor.

For example, most of the grain grown in the United States is fed to cattle to provide meat for the well-off, while over a billion people are at the risk of starvation because they cannot afford to purchase grains whose price has been driven up by the excessive demand for animal feed. [936]

Not only has profit-driven industrialized agribusiness not ended world hunger, it has produced and manufactured foods that are distinctly unhealthy for consumers in wealthier nations. To maximize profits, the industry produces highly refined foods, generally too rich in animal products, salt, sugar and fillers, that can be sold at higher prices instead of the basic staples such as grains and vegetables.[937] People who eat this processed food are at higher risk of heart disease, cancer, diabetes, obesity and other degenerative diseases. As Marion Nestle writes in *Food Politics: How the Food Industry Influences Nutrition and Health*,

> In order to make profits, the food industry must always entice people to eat more.... Primary health advice should be "eat less" but this is in direct conflict with the food industry's mission to sell more food to make more profit.... I have become increasingly convinced that many of the nutritional problems of Americans— not least of them obesity—can be traced to the food industry's imperative to encourage people to eat more in order to generate sales and increase income in a highly competitive marketplace.[938]

Efforts at reforming the food system to provide healthy unrefined foods consisting mostly of grains, legumes, vegetables and fruit is strongly resisted because it will threaten corporate profits. Marion Nestle continues,

> Ethical or not, a message to eat less meat, dairy, and processed food is not going to be popular among the producers of such foods.... The message will not be popular with cattle ranchers, meat packers, dairy producers, or milk bottlers; oil seed growers, processors, or transporters; grain producers (most grain is used to feed cattle); makers of soft drinks, candy bars, and snack foods; owners of fast-food outlets and franchise restaurants; media corporations and advertising agencies;...and eventually, drug and health care industries likely to lose business if people stay healthier longer.[939]

In summary, as long as profits and greed are the main driving forces behind food production, the nutritional needs of people, both rich and poor, will not be adequately met, while at the same time the environment will be degraded, sometimes irreparably, by polluting and unsustainable farming practices.

Profit Maximization: Medical Care

Prior to the industrialization of medicine, the needs of the patient were primary. Health-related issues were dealt with by family doctors, who were self-employed and independent professionals. Currently, since the rise of the medical-industrial complex, many doctors are no longer independent but are employees of clinics and hospitals that are operated as corporate entities according to the principle of profit maximization.[940] There are, indeed, many opportunities to optimize profits in medicine: overuse of diagnostic tests, prescription of expensive but marginally effective drugs, unnecessary high-technology treatments, excessive and risky surgery, excessive specialization, and so forth.[941] Just as President Eisenhower warned the nation during his farewell address in 1961 about the dangers of the military-industrial complex, similarly Arnold Relman MD, editor of the *New England Journal of Medicine*, cautioned the medical community almost 30 years ago about the rise of a new medical-industrial complex:

> The most important health-care development of the day is the recent, relatively unheralded rise of a huge new industry that supplies health care services for profit.... This new "medical-industrial" complex may be more efficient than its nonprofit competition, but it creates the problems of overuse and fragmentation of services, overemphasis on technology,...and it may also exercise undue influence on national health policy.... Another danger arises from the tendency of the profit-making sector to emphasize procedures and technology to the exclusion of personal care....The private health-care industry can be expected to ignore relatively inefficient and unprofitable services, regardless of medical or social need.[942]

The explicit goal of profit maximization in medicine has been responsible for the development and application of high-technology diagnostic tests and therapeutic interventions. The application of high-technology medical procedures requiring specialization is far more remunerative than family practice. Many doctors have become highly paid specialists, which increases our health care costs and their income, while at the same time fewer general practitioners are available to attend to preventative and more mundane, but nevertheless very important, health needs of the public.[943] As Richard Deyo and Donald Patrick point out in *Hope or Hype*,

> Historians point out that the success of twentieth-century hospitals and the growth of medical specialties were both mediated through technology. Hospitals are often quick to acquire new technologies in an effort to be at the "cutting edge" and to attract patients seeking state-of-the-art care. Many hospitals have committees devoted to deciding which new imaging devices, surgical implants, or treatment techniques are worthy of investment. Often, these committees are preoccupied with return on investment rather than with clear evidence of health benefits from the new technology.[944]

Physicians are also constantly pressured by salespeople from pharmaceutical companies to prescribe new drugs, many of which are not superior to existing ones but promise greater profits.[945] It is well known that pharmaceutical companies create new molecular structures that are similar to previous ones but which, by being different, can be patented, thereby increasing their competitive advantage and profit potential.[946] Any preparations that are not patentable, such as naturally occurring compounds derived from traditional medicinal plants, are of little interest although they may in fact be more effective, because they would not be sufficiently profitable.[947] As Dr. Dale Console, former director of research at Squibb, told a Senate committee more than 50 years ago,

> With many of these products [pharmaceuticals], it is clear while they are on the drawing board that they promise no utility. They promise sales. It is not a question of pursuing them because something may come of it.... It is pursued simply because there is profit in it.[948]

Finally, the profit motive can also be blamed for the fact that disease prevention has been largely ignored, including prevention of the two major killers in industrialized countries: heart disease and cancer. A first step toward preventing the incidence of these diseases would be to eat less animal fat, but this is likely to be resisted by agricultural and food business interests because it will threaten their profits. In the case of cancer, little real progress has been made since President Nixon declared a "war on cancer" almost 40 years ago.[949] The main reason for this failure is that the war has been fought on the wrong front: on finding a cure for cancer instead of preventing it.[950] As Ralph Moss writes in *The Cancer Industry—The Classic Exposé on the Cancer Establishment*,[951] there is a lot of money to be made in cancer research, and the development, marketing and use of relatively ineffective chemotherapy drugs and mutilating surgeries which have questionable efficacy. If only a fraction of these wasted resources had been used for preventative measures such as the widespread promotion of dietary changes, development of smoking-cessation programs and banning of industrial chemicals that are known or suspected carcinogens, cancer incidence rates would have declined by now. However, the outlawing of carcinogens has been, with few exceptions, almost impossible because of strong resistance by the chemical industry that sees its profits threatened. Similarly, prevention has been only half-heartedly endorsed by the medical-industrial complex because successful disease prevention programs would sooner or later decrease the demand for many medical services. As Robert Proctor points out in *Cancer Wars*, a common way to avoid the prevention approach is to call for more research in order to buy time, to avoid change and to create more confusion.[952] For example, the tobacco industry has for decades funded basic research into the molecular and cellular mechanisms of cancer causation to obfuscate the obvious fact that smoking causes lung cancer.[953] Similarly, the recent focus on developing diagnostic tests for "cancer genes" shifts attention from the problem of environmental and dietary carcinogens, which undoubtedly benefits many corporate interests, and instead "blames the victim" for having "defective" genes.[954] But clearly, the sharp rise in cancer incidence rates over the past 50 years cannot, to any significant extent, be due to patients' genetic constitutions because it is unlikely that the gene pool has changed significantly for many generations.

In conclusion, as long as profit maximization continues to be the primary objective of the medical-industrial complex, the health needs of patients will remain secondary, and the health status of our nation will suffer. Though a profit-driven medical system assures the efficient delivery of high-technology health care to those who can pay for it, it neglects the needs of those who cannot afford it, and it undermines the establishment of more effective disease-prevention programs. Indeed, compared to other industrialized countries, the profit-oriented medical system in the United States must be considered a failure. According to an in-depth analysis by the World Health Organization, the delivery of health care in the United States is the most expensive in the world, now exceeding 15 percent of GDP,[955] while the overall health system performance ranks number 37 in the world, only slightly better than Cuba (ranked 39).[956] By comparison, the national health care system in the United Kingdom, which even guarantees universal coverage, costs only half as much[957] but performs much better, ranking 18.[958] Clearly, it is time to reform the medical system in the United States to improve the overall health of the nation instead of optimizing the profits for the medical-industrial complex.[959]

Profit Maximization: Military Technologies and Foreign Policy

Although governments of the world provide the funding for military defense and offense, privately and publicly owned corporations frequently develop and manufacture the necessary warfare technologies. Defense contractors have gained increasing economic and political power as a result of increased military spending and the resulting rise in sales and profits. President Eisenhower, in his final address, warned the nation, "In the councils of government we must guard against the acquisition of unwarranted influence, whether sought or unsought, by the military industrial complex."[960] Given the magnitude of spending worldwide on military technologies, the profit opportunities are large indeed. World military expenditures in 2005 exceeded a trillion dollars, with the United States spending approximately 48 percent of the world total.[961] In the same year, 57 percent of the 131-billion-dollar United States federal R&D budget was allocated to warfare-related projects.[962] Accrued over time, large military expenditures by any nation become enormous. For example, it has cost the United States 5.5 trillion dollars to build and maintain the nuclear weapons arsenal.[963] The development of such technologies may also contribute to instability and violence worldwide. As Michael Parenti writes,

> Along with the corporations, there exists a whole subsidiary network of subcontractors and politicians who are part of the permanent war economy.... Almost all defense contracts are awarded without competitive bidding and with little or no subsequent supervision, allowing for outrageous profiteering.[964]

As Edward Herman observes,

> Missions are needed to justify weapons acquisition, and they are usually couched in terms of some threat.... The record shows that such threats will be manufactured or artificially stoked if no real threat is available, and that the mass media will not challenge their basis of reality.[965]

High levels of military expenditure worldwide are responsible for the massive diversion of wealth that could be used for more constructive purposes (e.g., universal health care, education and poverty programs). The profit motive also biases the foreign policy of nations toward military rather than diplomatic solutions to domestic as well as international conflicts.

In conclusion, the development and large-scale adoption of new technologies is a process currently guided by the principle of profit maximization. As a result, there is an inherent bias toward technologies and products that maximize profits for owners of income-producing properties, while the vital needs of people and the environment are largely ignored. If our goal is to maximize the well-being of people by meeting their basic needs and to protect the environment for future generations, we need selection criteria other than profit maximization to guide the development of new technologies. A major paradigm shift is required.

PART III

THE NEXT SCIENTIFIC
AND TECHNOLOGICAL REVOLUTION

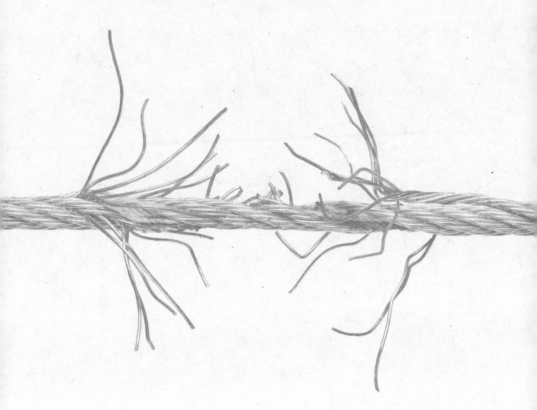

CHAPTER 12

The Need for
a Different Worldview

ALBERT EINSTEIN ONCE REMARKED, "No problem can be solved at the same level of consciousness that created it."[966] For problems caused by ignorance, one must rise above that level of ignorance before workable solutions can be found. This often involves paradigm shifts and a substantial change in worldview, a change in the very perception of reality. In this chapter, we first discuss the importance of paradigms and overall worldviews, and how conflicts between different paradigms and worldviews are generally resolved. We then make the case that a new view of reality is needed to successfully address humanity's most serious problems. A view is needed that is based on inclusiveness rather than separateness. Paradigm shifts are needed in all major areas of human activities: how the economy is organized, how science and technology are applied and how medicine is practiced.

The Power of Worldviews and Paradigms

Because of its enormous complexity, the nature of reality cannot be comprehended in its totality. Instead, to make sense of the world and to provide guidance for personal action, humans construct mental maps that include only certain selected aspects of this complex reality. These mental models are deeply held internal images of how the world works. These paradigms constitute a person's worldview.[967] The construction of a specific worldview is guided by "the set of experiences, beliefs, and values that affect the way an individual perceives reality and responds to that perception."[968] These consist of very fundamental background assumptions of which most people are not consciously aware. As a result, the validity of the underlying beliefs and assumptions is usually not questioned.

Most commonly, worldviews are shared social constructs inculcated by and consistent with the prevailing culture whose dominant values, beliefs and traditions serve as a lens through which the individual perceives reality.[969] As Donella Meadows and colleagues write,

> Everybody has a worldview; it influences where they look and what they see. It functions as a filter; it admits information consistent with their (often subconscious) expectations about the nature of the world; it leads them to disregard information that challenges or disconfirms those expectations. When people look out through a filter, such as a pane of colored glass, they usually see through it, rather than seeing it—and so, too, with worldviews. A worldview doesn't need to be described to people who already share it, and it is difficult to describe to people who don't.[970]

Because of their self-validating nature, paradigms and overall worldviews are transmitted virtually unchanged from generation to generation, unless a conscious effort is made to examine them in an objective way. At the very core of a particular worldview one always finds a set of dominant values and norms that give people meaning and guide their actions and behavior. A worldview provides the necessary normative and cognitive framework for defining problems and finding the corresponding solutions. Because this necessarily offers an inherently limited view of reality, it is common that the realities and very real dangers that lie outside the boundaries of the particular worldview are ignored. For example, prior to the rise of the environmental movement, the concept of "environment" was not an integral part of the dominant worldview. As a result, environmental pollution was not recognized as a problem that required attention.

Despite the inherent shortcomings of any worldview, no society can effectively function without being guided by some worldview and its more or less consistent paradigms. As Kenneth Stunkel and Saliba Sarsar point out,

On either level, individual or social, it is not possible for human beings to live without standards or ideas of meaning and purpose to guide decisions about what is believed, said, or done. All societies have standards and ideas embodied in traditions, rituals, theologies, philosophies, folkways, mores, and public policies.[971]

Many, if not the majority, of the most intractable social and political conflicts are caused by underlying differences in worldviews, paradigms and values. The worldview of a segment of society is often called an ideology. Being a worldview, an ideology is a unique view of reality and a complex set of values, rationales and rationalizations that serve to guide and justify the actions of a particular social group, class or nation.[972] In many cases, an ideology serves to legitimize and justify the actions of the socially dominant group or class. There is some truth to the historical proclamation: "The ruling ideas of any particular age have ever been only the ideas of its ruling class."[973] An ideology is often used to camouflage illegitimate policies and injustices perpetrated by the dominant social class against the less powerful, whether other humans, animals or the environment.[974] Thus, ideologies may become systematic attempts to offer plausible explanations and justifications for social behavior that might otherwise be subject to severe criticism.[975] Ideologies are maintained by repeatedly subjecting the public to narratives that offer simplistic explanations as to how the world works, while carefully omitting anything that might lead to inconvenient questions.[976]

Conflicting Worldviews and Paradigm Shifts

If our goal is to create a society that is more humane and environmentally sustainable, we need to understand that any such attempt will result in a conflict between worldviews and paradigms, between people who believe that the current system is preferable and can be maintained indefinitely and those who believe that it is rapidly approaching collapse and must be changed. Thus, it will be useful to understand the nature of clashes between opposing worldviews and paradigms, and how these conflicts are generally resolved. As mentioned before, irresolvable disagreements among individuals and groups in terms of what constitutes the best course of action are often the result of conflicting worldviews and paradigms. Consider, for example, the typical conflict between growth-oriented developers and conservation-oriented environmental-ists.[977] Although both parties are convinced that their behavior is ethi-cally justified, their respective actions are completely opposite because their underlying views of reality, in this case of nature, are totally differ-ent. As Arne Naess and David Rothenberg explain,

> What a conservationist sees and experiences as reality, the devel-oper typically does not see—and vice versa.... The difference be-tween antagonists is one rather of ontology than of ethics. They may have fundamental ethical prescriptions in common, but ap-ply them differently because they see and experience reality so differently. They both use the single term forest, but referring to different entities.... To the conservationist, the developer seems to suffer from a kind of radical blindness. But one's ethics in en-vironmental questions are based largely on how one sees reality. If the developer could see the wholes, his ethics might change. There is no way of making him eager to save a forest as long has he retains his conception of it as merely a set of trees. His charge that the conservationist is motivated by subjective feelings is firmly based on his view of reality. He considers his own positive feelings towards development to be based on objective reality, not on feelings.[978]

Parties to a conflict may have opposing interpretations of the same facts[979] or even disagree as to what constitutes relevant information.[980] As Thomas Kuhn writes in *The Structure of Scientific Revolutions*, adher-ents to different schools of scientific thought may essentially talk past each other because they are holding competing paradigms:

> To the extent, as significant as it is incomplete, that two scientific schools disagree about what is a problem and what a solution, they will inevitably talk through each other when debating rela-tive merits of their respective paradigms. In the partially circular arguments that regularly result, each paradigm will be shown to satisfy more or less the criteria that it dictates for itself and to fall short of a few of those dictated by its opponent.[981]

How then, are conflicts in worldview and paradigms ultimately resolved, as they ultimately must be? A society can function most effectively if it is guided by an internally consistent worldview and system of values. One way to resolve paradigm conflicts is to provide more convincing facts in support of a particular view of reality. This is the case in science. As the history of science has repeatedly shown, the scientific theory that most accurately describes, explains and predicts natural phenomena will ultimately be adopted as the dominant paradigm by the scientific community.[982] For example, as more and more measurements confirm that Earth temperatures are indeed rising, global climate change is being accepted as real, necessitating drastic mitigation actions. Finally, a conflict between scientific paradigms may also be resolved when a more general scientific theory (e.g., quantum theory) arises that unifies formerly incompatible lower-level theories.[983]

One would think that most conflicts between worldviews and paradigms could be resolved by providing superior factual evidence. Unfortunately, this is not necessarily the case. Often, conflicts become more severe as each party attempts to bolster its position with additional factual evidence. As Daniel Sarewitz points out in an article titled "How Science Makes Environmental Controversies Worse," a debate using scientific and technical facts may not only conceal underlying value conflicts because science is wrongly believed to be value neutral, but the use of selective facts is also unlikely to resolve policy questions about complex environmental issues such as global climate change:

> The point is that, when cause-and-effect relations are not simple or well-established, *all* uses of facts are selective. Since there is no way to "add up" all the facts relevant to a complex problem like global change to yield a "complete" picture of "the problem," choices must be made. Particular sets of facts may stand out as particularly compelling, coherent, and useful in the context of one set of values and interests, yet in another appear irrelevant to the point of triviality.[984]

If factual evidence fails to settle differences in worldviews, the conflict will most likely be resolved by resorting to power. As history has repeatedly shown, conflicts between cultural and religious values are often decided by physical might. In other circumstances, the conflict is resolved by economic power; that is, those who can pay for massive media propaganda for the longest time are able to convert large fractions of the populace to their point of view. This has been the case with economic globalization, in which Western consumer values are broadcast worldwide, destroying local cultures and promoting materialism, with the goal of increasing profits for transnational corporations. Finally, in democratic societies, most disputes over differences in values and worldviews are subjected to political debate,[985] in which each party uses its economic power to mobilize support for its position.[986] As Stephen Cotgrove writes,

> Conflicts over what constitutes the paradigm by which action should be guided and judged to be reasonable is itself a part of the political process. The struggle to universalize a paradigm is part of the struggle for power. Parties to a conflict will draw on all the resources they can muster, and will raid the cultural repertoire for beliefs and values which will authorize and provide acceptable reasons for their actions and support their interests.[987]

With regard to many environmental issues, political debate has been raging for decades. However, as more people feel the negative consequences of the outdated view that unlimited economic growth is possible forever without harming the environment, the balance is slowly shifting toward a new, more accurate view that the environment is vulnerable to human activities and thus must be protected. As David Korten notes, "When the stories a society shares are out of tune with its circumstances, they can become self-limiting, even a threat to survival. That is our current situation."[988] This threat to survival, when finally recognized, is a powerful force for change, for adopting a new worldview, for enabling a paradigm shift on a large scale.

There will always be people who cannot change their mental models, who will resist an impending paradigm shift. In this case, a complete paradigm shift can occur only after those who oppose the new worldview, which generally belong to the older generation, have passed on. As Thomas Kuhn points out in *The Structure of Scientific Revolutions*, even in science, a new paradigm is wholeheartedly embraced by the scientific community only "after the last holdouts have died."[989] This is why Nobel laureate quantum physicist Max Planck wrote in his autobiography, "A new scientific truth does not triumph by convincing its opponents and making them see the light, but rather because its opponents eventually die, and a new generation grows up that is familiar with it."⁹⁹⁰ Thus, to affect a society-wide change in values and worldview, it is probably much more effective to instill in the younger generation the new paradigm as part of their standard education than to waste time debating it with the older generation, who may never be able to change their mental maps.

Finally, it must be noted that after a paradigm shift has been completed, there is no way to return to the previous worldview. Paradigm shifts are the cause of irreversible cultural, religious and scientific revolutions. Western society, which has been operating for centuries on an increasingly outdated paradigm of growth in material affluence and human population, is currently on the brink of either an environmental collapse or a sustainability revolution. In order to avoid impending disaster, a cultural revolution must occur in the very near future. Paradigm shifts must occur in the way we view ourselves, nature, the economy and the role of science and technology, and they must occur more or less simultaneously in order to be effective.

A Different View of Reality

Our current worldview is based on the illusion of separateness, that our interests and destinies are separate from each other and from nature. Such a worldview will have serious negative consequences. Having good intentions and obeying all applicable moral laws is not sufficient to avoid inflicting harm if one's worldview is based on the illusion of separateness. For example, as long as the wealthy feel separate from the poor, they will continue to exploit them and feel no moral obligation to share or to treat them as equals. As long as one race feels superior to another, one nation better than another, the feelings of superiority and separateness will be used as an excuse to carry out deplorable acts that would be considered highly immoral if committed toward members of one's own race or nation. As long as people feel intrinsically separate from nature, they will continue to degrade and exploit it, resulting in massive environmental pollution and, eventually, ecosystem and economic collapse. As long as humans consider themselves separate and comfortably different from other animals, they will kill and eat them with no moral qualms whatever. Therefore, unless we change our current, limited worldview, which is based on the illusion of separateness, to a more inclusive worldview based on the fact of interconnectedness and shared destiny, there will be no chance that our problems with each other and the environment will ever be solved. If we genuinely embrace a worldview that places all life forms and the environment within our circle of ethical concern, our actions will lead necessarily to more constructive and positive outcomes.

The concept of interconnectedness is nothing new but, unfortunately, it has not been recognized by a sufficiently large number of people to result in a general paradigm shift. As mentioned earlier, there is indisputable evidence from many fields of science, ranging from ecology to quantum physics, that there is an underlying unity to observable natural phenomena that superficially appear to be unrelated and separate. Prior to the advent of modern science, humanism and the world's great religions provided ethical maxims based on the view that we all belong to the same family, that our similarities are more important than our differences. Immanuel Kant's "categorical imperative," "the golden rule," Albert Schweitzer's "reverence for life,"[991] the Christian admonition to "love thy neighbor as thyself," the Buddhist concept of "compassion toward all beings," and the Hindu belief in the "unity of life" are all based on the view that there is an underlying communality that should not be violated. As the world's great spiritual traditions have warned repeatedly, anything that violates this unity will surely result in negative consequences.

How would a worldview based on unity rather than separateness change our attitude toward the environment and our treatment of nature? If we consider ourselves as part of nature, we will include it in our ethical circle of concern and thus will treat it as we would our best friend. We would no longer attempt to conquer, control, exploit and destroy. Instead, we would treat all of creation with respect, even reverence, and we would try to live in harmony with nature. Our relationship to creation would be based on compassion and reciprocity, and we would consider all of nature as having intrinsic value. Thus, we would embrace the idea of biological egalitarianism, that all beings have an intrinsic right to flourish.[992] We will therefore, whenever possible, follow Aldo Leopold's land ethic: "A thing is right when it tends to preserve the integrity, stability, and beauty of the biotic community. It is wrong when it tends otherwise."[993] We would preserve biological diversity and try to minimize our impact on the environment. We would consider minimum interference to be the best policy. We would minimize potential risks to the environment by following the precautionary principle whenever possible (see Chapter 13).[994]

Such a change in our view of nature, one that is based on interconnectedness rather than separateness, would necessarily force us to change our view of the economy.

A Different View of the Economy

"Anyone who believes that exponential growth can go on forever is either a madman or an economist," wrote critical economist Kenneth Boulding.[995] Clearly, the belief that economic growth can continue indefinitely on a finite planet is outdated and is in serious need of revision. The historic view that unlimited economic expansion and affluence is possible was more or less justified as long as the size of the human population and economy were small relative to the natural environment. But this situation has changed completely. Almost seven billion people aspire to an energy- and material-intensive affluent lifestyle that will put impossible demands on the environment, in effect threatening the functioning of ecosystems around the world as well as the climate. Clearly, we now live in a world that is no longer largely empty but rather dangerously overfull, so that any additional economic growth is likely to have far more negative than positive consequences (see Chapter 8 on the problems of GDP accounting). Therefore, a transition from the current growth-oriented economy to a steady-state economy will be necessary and unavoidable.[996] In this steady-state economy, the size of the entire human economy, which is the product of the number of people times per capita affluence (i.e., P·A, see IPAT equation in Chapter 6), will have to remain more or less constant over time.

The concept of an economic steady state is nothing new. Ecological economist Herman Daly reminds us that

> historically, people have lived for 99 percent of their tenure on earth in conditions very closely approximating a steady state. Economic growth is essentially a phenomenon of the last 200 years, and only in the last 50 years has it become the dominant goal of nations. Growth is an aberration, not the norm.[997]

Even one of the leading utilitarian thinkers in political economy, John Stuart Mill, wrote more than 150 years ago in his *Principles of Political Economy* that a "stationary state" was consistent with the limits of the Earth and could support an evolving and improving society:

> I cannot...regard the stationary state of capital and wealth with the unaffected aversion so generally manifested towards it by political economists of the old school. I am inclined to believe that it would be, on the whole, a very considerable improvement on our present condition.... It is scarcely necessary to remark that a stationary condition of capital and population implies no stationary state of human improvement. There would be as much scope as ever for all kinds of mental culture and moral and social progress; as much room for improving the art of living, and much more likelihood of its being improved.[998]

How would a transition to a steady-state economy be brought about? To be environmentally sustainable, we would first need to set the maximum allowable scale of human economic activities (i.e., the rate of energy and material use that can be sustained as well as the amount of pollution that can be absorbed indefinitely and that does not interfere to any significant extent with the functioning of ecosystems). In many cases, we may have already far exceeded nature's biophysical limits for waste absorption. This is clearly true for the air pollutant carbon dioxide, the increasing concentration of which is driving global climate change. In other cases, as the ecological footprint analysis suggests (see Chapter 6), we may have just reached the maximum scale, the maximum carrying capacity of the environment. The determination of maximum scale is not simple, as it would require many years of environmental research and monitoring. In addition, it should be noted that there is an enormous difference between maximum scale and optimum scale. In most cases, the optimum scale would be much smaller than the maximum scale. Furthermore, the use of the term "optimum" implies a value judgment. Should it be optimal only for humans or should it be more inclusive: optimal for ourselves as well as for all the other species with whom we share the planet? Should we strive for an anthropocentric or biocentric optimum scale?[999] In view of the new paradigm that all of nature, including humans, forms an interconnected whole, it is clear that we must strive toward a biocentric optimum, which is likely to be much smaller than an anthropocentric optimum.

If a society could reach consensus regarding the optimum scale of human economic activities, policies would have to be designed, enacted and enforced to intentionally stabilize, and in many cases reduce, matter-energy use and pollution. Currently, the most common tools for this are regulation and taxation, but it is likely that more innovative approaches will emerge in the future. Regulations have been effectively used in the past to reduce the amount of waste released into the environment as well as to conserve energy. For example, the prohibition of lead in gasoline, PCBs and ozone-depleting chemicals almost completely halted the release of these pollutants into the environment.[1000] Various air and water quality standards assure that waste discharges are strictly limited. Cap and trade schemes are also used to control the total amount of pollutants (such as carbon dioxide) released into the environment while providing some flexibility in meeting overall discharge limits, although the success of some of these programs may be questionable.

It is possible that effective capping schemes could also be applied to limit the total use of renewable and non-renewable minerals and fuels. Policies could be enacted, as has already been done for selected forests and fisheries, to limit the harvest of any renewable resource to its sustainable yield, thereby preventing the degradation of the ecosystems that provide these valuable resources.[1001] Similarly, a maximum extraction rate (i.e., depletion quota) for non-renewable minerals and fossil fuels could be set in order to assure the availability of these limited resources until alternatives have been found or the need for them no longer exists.[1002]

Although the establishment and enforcement of depletion and pollution quotas is probably the best way to protect the environment and ensure sustainability, quotas may be difficult to implement because of widespread political resistance. Instead, it may be politically more feasible to slowly increase the price of mineral and energy resources via taxation to ultimately reduce overall use to the necessary levels. Before implementing various taxation schemes, however, it would be best to first remove "perverse" subsidies and to internalize all resource use and pollution costs that are currently set to be borne by the environment and future generations (see Chapter 8).

If removal of perverse subsidies and internalization of previously externalized costs did not reduce overall resource use, it would be necessary to implement some type of ecological tax to further increase the price of minerals and energy. Because of the strong public resistance to tax increases, innovative approaches would be needed to implement ecological taxes. A revolutionary proposal for ecological tax reform was made by Ernst Ulrich von Weizsacker, who recommended that income taxes be reduced to compensate for new energy taxes, thereby ensuring a fiscally neutral approach that keeps total taxes paid by the average consumer more or less the same. The ultimate goal of von Weizsacker's ecological tax reform is the reduction of resource use and pollution to achieve sustainability while at the same time promoting both full employment via reduction of payroll taxes as well as increasing international competitiveness through (appropriate) technological innovation.[1003]

Placing a limit on the use of renewable and non-renewable resources would signal scarcity and result in commodity price increases. This, in turn, would stimulate technological innovation, including efficiency improvements designed to derive more benefits in the form of consumer goods and services (e.g., per capita GDP) from the now more limited amounts of mineral and energy sources. For example, if there were a worldwide limit to the total amount of crude oil that could be extracted each year, the price of crude would increase, which would then provide a signal to businesses to design and build more efficient technologies (e.g., hybrid cars) to supply the necessary economic goods and services under pressure of decreased resource inputs. However, as we demonstrated in great detail in Chapter 5, there are both thermodynamic and practical limits to the degree to which technological efficiencies can be improved. Thus, while during periods of intense technological innovation, which most likely involve efficiency improvements, dematerialization and pollution control, the total level of material affluence (GDP) would still increase, it must ultimately stabilize at a more or less constant level when technological limits have been reached.[1004] In summary, the total size of the human economy (GDP) will necessarily reach steady-state conditions, which are determined by the limits imposed on matter-energy throughput and pollution to ensure environmental sustainability and by the inherent thermodynamic and practical limits imposed by technology.

What would the final steady-state look like? Most importantly, the total size of the economy, in terms of the monetary value of all goods and services provided (i.e., GDP), would no longer grow but remain more or less constant. Minor fluctuations would still occur. Because GDP is the product of population size (P) times per capita affluence (A=GDP/P) (see the IPAT equation in Chapter 6), society has some flexibility in achieving steady-state conditions. For example, if the population (P) is still growing, the material standard of living (GDP/P) would have to be reduced accordingly to maintain GDP at a constant level. Alternatively, if there is a desire to further increase material affluence (GDP/P), then corresponding reductions in the size of the population (P) would have to be made to keep GDP constant. We cannot have both uncontrolled population growth *and* large increases in material affluence as is currently believed in most of the world. In the end, a steady-state economy would be characterized by a stabilized population living at a constant (non-growing) level of material affluence.[1005]

The transition to a steady-state economy would result in a number of other significant changes. First, because material consumption would be kept at a constant level, the total size of capital and infrastructure needed to produce all goods and services would also be maintained at constant levels (i.e., rates of investment would balance rates of depreciation).[1006] Second, because the size of the human population would remain constant within each nation, there would have to be policies to assure that birth rates equal death rates, and that immigration is balanced by emigration.[1007] Third, while international trade would continue, it would be more limited because it could no longer be used by wealthy nations to circumvent their own sustainability and carrying capacity constraints by importing large amounts of raw materials and energy resources from poor nations while simultaneously exporting waste and pollution to them.[1008] Thus, future economies will necessarily be local rather than global.[1009] Fourth, although a steady-state economy would have quantitative limits to material growth and affluence, there would be no limit to qualitative improvement and social progress. For example, there would be unlimited opportunities to increase the quality of life, to have excellent education and health care accessible to all, improved democratic institutions representing the interests of all people, better community life, more meaningful work and more leisure time that could be used to pursue the many social, cultural, intellectual and spiritual activities known to increase psychological well-being.[1010]

Despite the many positive aspects of life in a steady-state economy, there is likely to be considerable resistance to abandoning the concept of continual economic growth and ever-rising material consumption. As was pointed out in the previous chapter, people are currently pacified by the unrealistic hope of a better future in which everyone will ultimately live in extravagant wealth and affluence. People are unlikely to give up this hope unless it is replaced by the immediate reality of a better quality of life and greater psychological well-being. Furthermore, it is unlikely that the current extreme inequalities in income and wealth, both within and among nations, can be maintained under steady-state conditions. Thus, in order to effect a transition to a sustainable steady-state economy, there will have to be a more equitable distribution of income and wealth. This will likely engender strong resistance by the privileged, who understandably will fear losing some of their material wealth, power and control. As steady-state economist Herman Daly proposes in *Beyond Growth*, we could strive for a more equitable income distribution where the maximum income is no more than ten times the minimum income, as is already generally the case within the US military, the government and universities:

> The minimum income would be some culturally defined amount sufficient for food, clothing, shelter, and basic health and education. The maximum might be ten times the minimum.... There is some evidence that a factor of ten is sufficient to reward real differences and to provide sufficient incentive so that all necessary jobs are filled voluntarily. In the US military, for example, the highest paid general makes ten times the wages of the lowest private. Probably the general has extra fringe benefits that may result in a factor of eleven or twelve. The same range is found in the Civil Service—a GS-18 makes about ten times the salary of a GS-1. In the university a distinguished professor is paid about ten times the salary of a graduate student instructor. In corpora-

tions the range is much greater, but this is largely because the top officers have the privilege of setting their own salaries, unlike generals, senior civil servants, and distinguished professors. I take a factor of ten, then, as a benchmark. The important thing is the change from unlimited to limited inequality.[1011]

In *Agenda for a New Economy*, David Korten suggests changes in policy to limit extreme income inequality:

It is perfectly reasonable that those who carry more responsibility should be compensated accordingly; the ratios of more than 1,000 to 1 between the CEO and the lowest-paid workers, which is common in US corporations, are unconscionable. Public policy should provide incentives to limit this ratio to no more than 15 to 1. If the top jobs in a corporation or another organization are so difficult or distasteful that qualified applicants can be attracted only with an obscene pay package, then perhaps the corporations needs to be broken up into more manageable pieces.[1012]

After the transition to a steady-state economy, science and technology would no longer be employed to promote unrestrained economic growth but rather to assure that all economic activities are carried out within prescribed limits to pollution and matter-energy use in order to assure long-term sustainability. Thus, a paradigm shift will be necessary not only in terms of our view of nature and the economy, but also in terms of how we view and apply science and technology.

A Different View of Science and Technology

As was discussed earlier (Chapters 3 and 11), the primary goal of science and technology in our current economic system is to increase material affluence and to generate profits for the already wealthy by controlling and exploiting both people and the environment. In view of our new paradigm of interconnectedness and interdependence, this arrangement is neither environmentally sustainable nor socially desirable. Clearly, the direction of science and technology must no longer be guided by the outdated values of power, control and exploitation but rather by the values of social and environmental harmony, cooperation and mutual enhancement.

Environmentally and socially appropriate technologies would have the following characteristics: first and most importantly, technological innovation would be employed to maximize the efficiencies of energy and material use, with the objective of maintaining matter-energy throughputs within the biophysical limits needed to assure sustainability. Second, technologies would be designed to treat, minimize and prevent pollution. Third, new technologies would be developed for the large-scale generation of environmentally friendly renewable energy. Finally, there would be efforts toward greater democratic control over the direction of technological development, to ensure that new technologies enhance the quality of work and community life. Various aspects of an environmentally and socially appropriate technology will be discussed in more detail in the next chapter.

Changes need to be made in the way scientific research is conducted. Though most research today is carried out under the presumption of value-neutrality, with scientists busy generating vast quantities of "objective" data, there needs to be increased awareness that both the direction of scientific inquiry (i.e., what questions are asked and what research is funded) and the use of scientific data (for technological applications) are affected by the dominant values of society. Thus, a new "critical science" is desperately needed in order to expose the value dimensions of the scientific enterprise. As a result, it would no longer be possible for scientists to escape responsibility for the consequences of their work, and some scientists might choose to become activists, with the goal of changing the direction of science toward more environmental and humane ends. Topics related to critical science will be discussed further in Chapter 14.

A Different View of Medicine

Currently, the practice of modern medicine is based on an outdated mechanistic view of the body, providing high-tech treatments within the confines of a rigid infrastructure, while ignoring the social and environmental contexts of disease causation and prevention.[1013] Furthermore, as was pointed out in the previous chapter, the primary goal of high-technology medicine is the maximization of profits for the medical-industrial complex. This strategy has led to a tripling of medical care expenditures in the United States from 1960 to 2003 (i.e., from 5.2% to

15.8% GDP[1014]) but unfortunately has not resulted in a corresponding improvement in the overall health of the nation, with the US now ranking number 37 in the world.[1015] Clearly, medicine needs a new vision, a new paradigm acknowledging the need for holistic and preventative approaches, in order to become more effective.

The paradigm of interconnectedness would have profound consequences for the practice of medicine. Rather than devoting the practice of medicine almost exclusively to ineffective and expensive mechanistic treatments for isolated diseases, the new medicine would focus on promoting the overall health of the nation. The concepts of health and disease would be redefined. Health would be seen as a state of balance, in which the individual is in harmony within as well as with the surrounding social and natural environment. Disease would be seen as a disturbance of this equilibrium. It has been observed that disease often occurs when environmental and social changes occur too fast for the individual to adjust.[1016] It has been suggested that most so-called degenerative diseases in industrialized nations are the result of maladaptation, meaning that humans are increasingly unable to adjust to the modern lifestyle because their bodies and minds have evolved for millennia to maximize their chances of survival under rather different environmental and social conditions.[1017]

The first step toward a transition to a holistic and preventative medicine would be to redirect a substantial portion of current medical research toward improving our understanding of the social and environmental factors involved in disease causation. As was discussed in Chapter 7, most of the increases in life-expectancy during the 19th century were not due to medical interventions but rather the result of improved social and environmental conditions, such as better nutrition, hygiene, sanitation, water supply and housing.[1018] Furthermore, there is already significant evidence that lower socio-economic status and poverty are associated with higher incidence of disease and mortality. The reason for this phenomenon is that the poor have less nutritious food, more hazardous jobs, higher stress and lower levels of education and literacy, the latter preventing them from learning about or achieving a healthy lifestyle, such as consuming low-fat diets and exercising while avoiding bad habits such as smoking, drug and alcohol abuse, and high-risk sex.[1019] More recently, it has become apparent that many cancers are caused by dietary and environmental (industrial) carcinogens and that the risk of coronary heart disease is increased by a high-fat animal-based diet.[1020] Overall, there is evidence from epidemiological studies that at least 80 to 90 percent of diseases that significantly contribute to mortality in Western populations are due to non-genetic factors (i.e., environmental, social and lifestyle factors).[1021] In addition, a recent study found that eight of the nine leading causes of death in the United States were preventable and, if prevented, would reduce national health care costs by approximately 70 percent.[1022]

With an understanding of the environmental, social and lifestyle factors involved in disease causation, large-scale community disease-prevention and health-promotion programs could be implemented. As has long been recognized from epidemiological studies, it is not necessary to know the molecular mechanisms of disease causation in order to establish effective prevention measures.[1023] A basic understanding of general cause-and-effect relationships is sufficient. For example, knowledge of the fact that people who smoke or eat a high-fat meat-based diet have a higher incidence of cancer and heart disease is sufficient to warn against these practices, even if the exact molecular and cellular mechanisms of these diseases are not known. Since the prevention of disease is much more cost-effective than treatment, a large reduction in health care costs would result.[1024] This, in turn, would make universal health care coverage more affordable and implementable. Despite these positive aspects, a shift from treatment to prevention will undoubtedly be resisted by the profit-oriented medical-industrial complex (see Chapter 11).

Despite large-scale disease-prevention and health-promotion programs, illness will still occur, albeit at a much reduced rate, and medical intervention will still be required. Instead of treating symptoms in isolation from their environmental, social and lifestyle contexts, medical practice would involve removing causative factors and other obstacles to recovery and promoting the patients' innate healing abilities.[1025] (See discussion of the placebo effect in Chapter 7.) As Norman Ford comments in *Lifestyle for Longevity,*

> A major principle of holistic healing is that when the cause of disease is removed, the body becomes a self-healing entity. When

we stop punishing our bodies with self-destructive habits and re-place them with healthful [ones], most degenerative diseases will gradually disappear.[1026]

Similarly, Andrew Weil, MD, writes in *Health and Healing*,

The body has innate healing abilities: Healing comes from in-side, not outside. It is simply the body's natural attempt to re-store equilibrium when equilibrium is lost. Healing cannot be prevented from occurring (though it can be obstructed in its expression), nor can it be obtained from anyone or anything ex-ternal.... Medicines and medicine men can sometimes catalyze a healing response or remove obstructions to it, but they never give you what you do not already have.[1027]

Finally, we must recognize the obvious fact that death cannot be pre-vented but only postponed. When death finally comes, it should be accepted as natural and not fought with ineffective and expensive high-technology medical interventions.[1028] Clinging to a life that will inevitably soon end causes both the patient and family extreme suffer-ing. Would it not be better to die with dignity and caring, and pass on one's wealth to one's family or charitable organizations instead of en-riching the medical-industrial complex through ineffective, expensive and agonizing treatments? In conclusion, implementation of this model of preventative and holistic medicine would significantly improve the overall health of the nation and at a much lower cost than the current system. Only the medical-industrial complex and those they manipu-late through fear and fanaticism would oppose such a respectful and humane change for the better.

The Need for Increased Awareness

A paradigm shift to a view of reality based on interconnectedness rather than separateness will not come about unless there is a signifi-cant maturation of consciousness. Before discussing how this change may be encouraged, we need first to understand that there are different levels of human consciousness. Dr. David Korten, who reviewed the per-tinent literature on this subject in his excellent book, *The Great Turning*, defined five levels, or orders, of consciousness, which to some degree parallel the stages of moral development identified by Piaget and others, stages through which the normal child passes during development toward adulthood. The first order, magical consciousness, corresponds to the awareness of a young child of two to six years of age. At this stage, the lines between fantasy and reality are blurred (e.g., belief in Santa Claus, Easter Bunny); there is only a rudimentary ability to recognize causal relationships; and therefore one is unable to recognize the consequences of one's actions or take responsibility for them. At age six or seven, most children make the transition to the second order, imperial consciousness, which allows them to distinguish between real and imagined events and to recognize that actions have consequences. There is an understanding of the concept of reciprocity but the idea of justice is limited to a primitive and personally enforced "eye for an eye and a tooth for a tooth" retributive justice.[1029] Furthermore, there is a recognition that conforming to the expectations of authority figures generally results in rewards.

At age 11 or 12, children reach the third order of socialized consciousness, in which cultural norms of a larger reference group (peer pressure) become internalized. A person at this level of consciousness is capable of empathy and is willing to subordinate his or her own needs and desires to those of the reference group. As Korten notes,

> The Socialized Consciousness brings a growing appreciation of the need for rules, laws, and properly constituted political and religious authority to maintain essential social and institutional order, and it internalizes a play-by-the-rules, law-and-order morality.... The Socialized Consciousness constructs its identity through its primary reference groups, as defined by gender, age, race, ethnicity, religion, nationality, class, political party, occupation, employer, and perhaps a favored sports team.... It is the consciousness of the Good Citizens, who have a "Small World" view of reality defined by their immediate reference group, play by the existing rules, and expect a decent life in return for themselves, their families, and their communities.... On the downside, it is susceptible to manipulation by advertisers, propagandists, and political demagogues, and it is prone to demand rights for

the members of its own identity group that it is willing to deny others.[1030]

According to Korten, most adults do not progress beyond the level of socialized consciousness.[1031] This is very unfortunate, because almost all major problems we face, including those discussed in this book, are a direct consequence of the limited worldview of people trapped at this level. The principal shortcoming of socialized consciousness is that it is based on an artificially constructed "in group–out group" duality, where only people in the "in group" are treated with respect and moral consideration, while those in the "out group" may be freely subjected to mistreatment, exploitation and even murder.

Many atrocities have been committed by people functioning at the level of socialized consciousness who, unaware of their limited view of reality, believe that their actions are morally justified. For example, while patriotism encourages people to transcend their own selfish interests for the larger good of the nation, it becomes dangerous if it is used to incite the populace to engage in intrinsically immoral and unjust wars of conquest and exploitation. Even a casual study of history demonstrates that rulers and politicians have manipulated citizens who function at the limited level of socialized consciousness to carry out horrendous acts in war with other nations that would be considered criminal and immoral if they were done in one's own country. Similarly, many atrocities have been committed by people who honestly believe that their own race or religion is superior. Many civil wars, religious wars, wars of conquest, slavery and the Holocaust have to a large degree been caused and carried out by people trapped at the level of socialized consciousness. Furthermore, even the exploitation of the poor and middle classes by the wealthy occurs at the level of restricted socialized consciousness among the latter. As long as the wealthy consider themselves intrinsically different and separate from working people, they will create and enforce an economic system that enables them to appropriate obscene profits without experiencing any moral qualms whatever. Moreover, as long as people feel no kinship with animals and consciously exclude them from their reference group, they will continue to mistreat, exploit, murder and eat them without the slightest moral hesitation. Finally, as long as people exclude the environment from their circle of concern, they will feel morally justified to exploit and pollute it.

Clearly, if we want to avoid these types of atrocities in the future, it will be necessary for large numbers of people to progress to higher levels of consciousness at which the overlying principles of unity and interconnectedness are applied to something larger than one's primitively selected reference group. According to Korten, the next level, or fourth order, of consciousness is cultural consciousness, which is concerned with equal justice for all people and universal respect for human rights. The transition from socialized to cultural consciousness is actively discouraged by most social organizations such as corporations, political parties, churches, labor unions and even educational institutions because each of these reference groups has its own belief system that demands loyalty. Individuals who mature past the level of socialized consciousness will often face rejection by their previous affinity group.[1032] Finally, a few people progress to the highest level, or fifth order, of spiritual consciousness. According to Korten,

> The Spiritual Consciousness, the highest expression of what it means to be human, manifests the awakening to Creation as a complex, multi-dimensional, interconnected, continuously unfolding whole... [It] supports an examined morality grounded in the universal principles of justice, love, and compassion common to the teachings of the most revered religious prophets.... The Spiritual Consciousness joins the Cultural Consciousness in seeking to change unjust laws. It recognizes, however, that at times it must engage in acts of principled nonviolent civil disobedience both to avoid being complicit in the injustice and to call the injustice to public attention.[1033]

It is obvious that if everyone were to function at the level of spiritual consciousness, seeing themselves as an inseparable part of the whole of creation, all motivation for control, exploitation, mistreatment and murder would instantly disappear.[1034] It is, of course, unrealistic to expect a general population-wide leap to the level of cultural or spiritual consciousness, particularly in the short time frame needed to solve the most serious social and environmental problems. However, as a first step, it would be most advantageous if the leaders of society—politicians, business people, teachers, professors, scientists and engineers—were to function at the level of cultural or spiritual consciousness. Those in the most influential positions need to have a highly developed sense of moral obligation and responsibility. (See Chapter 14 on the social responsibility of scientists.)

The crucial question is how to accelerate the expansion of moral awareness. There are a number of possible ways. First, formal education may help, but only if it is intentionally focused on topics that foster a deeper understanding of our joint humanity and interconnectedness to each other and nature. Courses in anthropology, comparative religions, philosophy, ethics, psychology, biological evolution, environmental sciences and even quantum physics and astronomy may serve this purpose. Unfortunately, education alone is not necessarily sufficient to raise people to a higher level of consciousness, as history has repeatedly shown. Some of the most criminal and morally repugnant acts have been committed by very intelligent and highly educated people. Second, being exposed to the views and practices of other cultures, such as participating in cultural exchanges or serving in the Peace Corps, may facilitate the transition to a broader level of consciousness and concern. Third, crises of various kinds may force people to reevaluate, analyze and identify the root causes of a problem and to entertain novel and more constructive approaches. For example, the environmental crisis has taught us that our view that we are separate from nature is wrong.

A traditional and effective way to increase one's awareness of the interconnectedness of life is to practice spiritual disciplines and acts of charity. Unfortunately, our modern, industrialized, consumer culture only strengthens selfishness. Transportation, computer and mass media technologies have accelerated the pace of life to the point that it is difficult to slow down and find sufficient peace to simply stop and think. Furthermore, the electronic mass media provide a stream of distractions that make it difficult to entertain profound thoughts.[1035] Aggressive advertising campaigns encourage materialism and a truly unwholesome level of individualism. Immersion in the lifestyle of our modern industrialized society is not conducive to the development of a higher level of awareness through which social and environmental problems can be perceived and solved.

We began this chapter with a quotation from Albert Einstein about the fact that no problem can be solved at the same level of consciousness at which it was created. Thus, it is appropriate to close this chapter with another quotation from this great philosopher-scientist:

A human being is a part of the whole called by us Universe, a part limited in time and space. He experiences himself, his thoughts and feelings as something separated from the rest, a kind of optical delusion of his consciousness. This delusion is a kind of prison for us, restricting us to our personal desires and to affection for a few persons nearest to us. Our task must be to free ourselves from this prison by widening our circle of compassion to embrace all living creatures and the whole of nature in its beauty.[1036]

CHAPTER 13

The Design of Environmentally Sustainable and Socially Appropriate Technologies

A S HAS BEEN MENTIONED repeatedly throughout this book, the primary goal of technology in our current economic system is to increase material affluence and to generate profits for the wealthy by controlling and exploiting both people and the environment. In view of the reality of interconnectedness, this is neither environmentally sustainable nor socially desirable. In this chapter we discuss how to design technologies that reflect the values of environmental sustainability and social appropriateness. We also emphasize the importance of heeding the precautionary principle in order to prevent unintended consequences, as well as the need for participatory design to ensure greater democratic control of technology. Finally, as a specific example of an environmentally sustainable and socially appropriate technology, we discuss the positive contribution of local, organic, small-scale agriculture.

Design Criteria for Environmental Sustainability

In the previous chapter, we discussed the need for a paradigm shift to a worldview that is based on the reality of interconnectedness rather than the illusion of separateness. This new orientation would result in profound changes in the way technology is designed and applied. Technology would no longer be used to control, exploit and pollute but would, instead, be designed and applied to guarantee long-term environmental sustainability. We outlined the three principal sustainability conditions in Chapter 6, and will repeat them here:

1. Sustainable Energy Generation Condition
All energy used for industrial and economic activities must be supplied from renewable sources at rates that do not cause any disruptive environmental side effects. If renewable energy is obtained from biomass, its harvest must not exceed the regenerative capacity of the respective ecosystem (sustainable yield criterion).

2. Sustainable Materials Use Condition
a. All raw materials used in industrial processes must be supplied from renewable resources at rates that do not exceed the regenerative capacity of the respective ecosystem (sustainable yield criterion) and do not cause any environmental disruptions.
b. The dissipative use of non-renewable materials must be discontinued, or at least greatly minimized, by either finding renewable substitutes or by recycling them to the greatest extent possible.

3. Sustainable Waste Discharge Condition
Wastes can be released into the environment only at a rate compatible with the assimilation capacity of the respective ecosystems, and without negative impact on biodiversity.

We will discuss these three sustainability conditions in more detail, particularly in terms of the design and application of future technologies. Regarding the generation of energy, only renewable energy is environmentally sustainable over the long run. Though obvious, that is worth repeating because the supply of fossil energy sources is limited and nuclear power plants generate unmanageable radioactive wastes; neither fossil nor nuclear energy are long-term options and therefore should not be given serious consideration when planning for a sustainable energy future.

In terms of renewable energy generation, the primary options are direct solar heating, solar thermal, photovoltaics, wind power, hydroelectric and biomass energy. All of these renewable energy technologies have the advantage of capturing diffuse solar energy in a decentralized fashion for local use, which makes them, in principle, more amenable to local control.[1037] (See below.) Furthermore, with few exceptions such as hydroelectric power, energy generation by these various renewable technologies can, in theory, be greatly expanded. However, as explained in detail in Chapter 6, the generation of renewable energy on the enormous scale required to meet current and future energy needs is likely to have significant undesirable environmental and social impacts and is expected to result, and already has resulted in, increasing public opposition.[1038] Therefore, before embarking upon a massive campaign to generate very large quantities of renewable energy, it is extremely important to assess all potential hidden costs by carefully performing environmental impact and life-cycle analyses (see also Chapter 8).

If the boundaries for these analyses are chosen to include the effects on the totality of the environment and future generations, the findings will most likely indicate that renewable energy generation can be expanded only very moderately. This will be particularly the case if society were to demand that renewable energy generation be limited to a scale that is compatible with a biocentric, rather than an anthropocentric, optimum (see Chapter 12). Consequently, given that renewable energy generation will have to be limited, economic growth must slow and eventually stop, leading to a steady-state economy described in the previous chapter. The transition to renewable energy can be facilitated by removing all subsidies for fossil and nuclear energy, by instituting carbon taxes and by providing temporary subsidies for the development and deployment of (localized) renewable energy technologies.

Regarding the sustainable materials use conditions, a choice is possible between renewable and non-renewable resources. For renewable resources, which most likely will be derived from biomass, it must be assured that harvests occur at rates not exceeding the sustainable yield. As in the case of renewable energy, the supply of renewable materials derived from biomass is likely to be limited, given its already disruptive competition with agriculture and forestry. The use of non-renewable resources such as metals, minerals and products derived from petroleum can, in principle, be continued as long as they can be completely recycled. However, as discussed in Chapter 6, only certain materials such as scrap metal, glass, paper and limited kinds of plastics have been successfully recycled. A large fraction of non-renewable materials becomes dispersed in the environment, either during their use or as a result of their disposal, making it virtually impossible to recover them for reuse. This dissipative use is not sustainable because the stock of non-renewables will become depleted with time. In addition, many dispersed materials, particularly heavy metals and chlorinated organic chemicals, cause significant environmental disruptions. Consequently, to satisfy the sustainable materials use condition, future technologies will have to be designed with the objective of assuring the complete recycling of non-renewables and avoiding their dissipative use. To promote the development of sustainable technologies, it may be necessary to provide economic incentives to increase recycling and to completely prohibit the dissipative use of environmentally harmful substances.[1039]

Regarding the "sustainable pollution" condition, wastes must be discharged into the environment only at rates compatible with the assimilation capacity of the receiving ecosystems. Ideally, all wastes should be completely biodegradable, with their breakdown products being recycled in the environment and assimilated naturally by the surrounding biota. For example, human and animal excreta are completely biodegraded by microorganisms, and the resulting breakdown products (carbon dioxide, nitrogen compounds, etc.) are readily taken up by plants and used for their growth. Similarly, many products derived from plant and animal biomass, such as wood, fiber and even biopolymers, are in principle biodegradable, but only if they are disposed of in environments that provide optimum conditions for microbial biodegradation. It will be of no use to discard biodegradable waste materials into hermetically sealed landfills that do not provide enough oxygen for microorganisms to aerobically biodegrade these compounds. Thus, a transition to completely biodegradable materials must be accompanied by changes in the way the resulting wastes are managed; that is, composting facilities rather than landfills will be needed. Finally, even if wastes are biodegradable and their breakdown products are readily taken up by biota, they can be discharged only at rates compatible with the assimilative capacity of the environment. For example, agricultural runoff rich in nitrogen nutrients has caused algal blooms and associated oxygen dead zones in lakes, rivers and certain coastal areas, because the receiving bodies of water are overwhelmed by the high rate of fertilizer input. Similarly, although carbon dioxide (CO_2) from the burning of fossil fuels is readily assimilated by plants via photosynthesis, the rate of CO_2 generation is greater than the rate at which plants can absorb it, resulting in rising atmospheric CO_2 concentrations and global climate change.

If a waste is not biodegradable, then it would have to be completely biologically inert (i.e., not assimilable) and non-toxic to be acceptable for disposal into receiving ecosystems. However, even inert and non-toxic materials can cause damage if discharged at excessively high rates. For example, the dispersal of inert sand and silt from mining and forestry operations can result in siltation of nearby rivers and lakes, thereby causing harm to aquatic species living in the affected habitats.

Although "sustainable pollution" conditions are very simple and straightforward, it must be recognized that the majority of wastes that are currently generated in industrialized nations do not meet the criteria of biodegradability or biological inertness. For example, a large fraction of the more than 100 million tons of synthetic organic chemicals manufactured each year, representing over 65,000 different compounds in regular commercial use,[1040] is likely to be resistant to biodegradation and has the potential to bioaccumulate in the food chain and negatively affect the exposed biota. Synthetic herbicides and pesticides that are sprayed onto agricultural land are anything but biologically inert and cause serious harm to both plant and animal life. In addition, almost all fossil-fuel derived plastics, including synthetic fibers used in clothing, are not biodegradable, and microscopic plastic particles have been observed to bioaccumulate in the tissues of marine animals in all of the world's oceans, potentially causing harmful effects yet unknown.[1041] Finally, consumer products containing heavy metals, such as paints, batteries, electronics, fluorescent lights and many others, have negative impacts if dispersed in the environment. Therefore, to avoid problems associated with the disposal of these wastes, their production would have to be either completely prohibited ("If you don't want it in the environment, don't make it") or there would have to be strict regulations ensuring that all wastes are collected and then either recycled or treated to become harmless. The dissipative use of harmful products cannot be permitted.

Sustainability conditions could be achieved faster and easier through efficiency improvements. It must be noted that efficiency improvements are very important for achieving sustainability but only if they are specifically used for that objective, and not for promoting further economic growth, as is currently the case (see Chapter 5). Increasing energy efficiency could reduce the demand for energy, thereby easing environmental pressures associated with large-scale renewable energy generation. Increasing materials-use efficiency (i.e., dematerialization, or using fewer materials in consumer products) and making consumer goods more durable could reduce the demand for both renewable and non-renewable materials, thereby making it easier to meet sustainable yield targets and easier to recycle and treat wastes. Improving the efficiency of waste treatment could reduce the amount of waste discharged into the environment.

Although the three sustainability conditions are conceptually simple, they are not easy to implement because it would require a complete restructuring of most major industries. In fact, many would have to go out of business, such as oil and petrochemical companies, nuclear power plants, mining companies, pesticide and herbicide factories, and any manufacturing operation that violates the above-mentioned sustainability criteria. Because conforming to sustainability conditions would threaten not only profits but the very survival of many corporations, very strong resistance can be expected against redesigning technologies and industries for sustainability. Thus, in the final analysis, achieving sustainability is not primarily a scientific or technical problem but rather an economic and political challenge.

Design Criteria for Social Appropriateness

Future technologies should be designed not only for environmental sustainability but also for social appropriateness, meaning that their primary function should be to meet genuine human needs and improve social welfare rather than to maximize profits. To assure that future technologies are socially appropriate,[1042] their design must be guided by the following three criteria: first, they should be local and decentralized; second, they should be human-scale and simple to operate; and third, they should provide an environment for humane, satisfying and meaningful work.

Technologies should be compatible with the local culture and should encourage local production for local consumption, employing local labor and local financial resources. Ideally, work and living locations should be in close proximity, and cities should be redesigned so that most destinations can be reached by foot or bicycle.[1043] The need for long-distance transport should be minimized by providing both excellent work and leisure opportunities locally.[1044] The use of local and decentralized technologies would have the following advantages. First, they could be easily managed by the people involved, encouraging the control of technology by local democratic consensus rather than control by distant bureaucracies.[1045] According to the principle of subsidiarity, decisions, including those related to technologies, should always move closer to the people affected by them. (See the section below on the democratic control of technology.)[1046] Second, local production for local consumption increases self-sufficiency and reduces dependence on imported and scarce resources such as fossil fuels. Finally, the local and decentralized use of technology is likely to result in a better balance of local benefits and local costs; that is, it would be more difficult to externalize costs to distant locations, thereby reducing the possibility of exploitation. For example, if people live in close proximity to their work, they will be less likely to ignore potential negative effects that their work may have. As Wendell Berry observes,

> When people do not live where they work, they do not feel the effects of what they do. The people who make wars do not fight them. The people responsible for strip-mining, clear-cutting of forests, and other ruinations do not live where their senses will be offended or their homes or livelihoods or lives immediately threatened by the consequences. The people responsible for the various depredations of "agribusiness" do not live on farms.[1047]

Another way to design more socially appropriate technologies is to assure that they are of human scale and simple (i.e., easy to understand, maintain and repair). As E. F. Schumacher once said, "Any fool can make things complicated, but it requires a genius to make things simple."[1048] Therefore, great emphasis should be placed on designing technologies that are relatively small scale and simple, thereby reducing the dependence on large financial and physical capital, as well as on technical experts. As John Cavanagh and Jerry Mander write,

> Megatechnologies require megainvestments and megamanagement. They are invariably biased toward absentee, usually global,

ownership and operation, run by principles that are anathema
to communities and democratic governance. If manufacture is
to be more community based and designed to be smaller, then all
technological systems and infrastructures need to be appropriately designed and scaled. Local infrastructures for reuse, recycling, and recovery of materials are mandatory, as well as a strong
emphasis on local energy systems, including solar, small-scale
electrical generation, wind, and local hydro, operating on a scale
that fits the needs of small, flexible, manufacturing processes.[1049]

There are a number of obvious advantages to having human-scale and
simple technologies: first, simple, small-scale technologies are more
amenable to personal or community control. For example, craft technologies can be handled easily by a single person or a few people and
do not require complex mega-management. Second, the employment
of simple technologies that are understandable by their users increases
self-reliance. For example, growing food in community gardens using
simple implements makes it possible to increase self-sufficiency. Third,
small and simple technologies are more readily affordable by all, thereby
avoiding the exclusion of poorer people from the benefits of available
technologies.[1050] For example, the use of bicycles, affordable by all, is
much more egalitarian than the use of automobiles, which are too expensive for many people worldwide to own and operate. Finally, simple
technologies are likely to encourage a return to a more simple life. As E.F.
Schumacher comments,

> Simplicity, from a Christian point of view, is a value in itself. Mak-
> ing a living should not absorb all or most of a man's attention,
> energy, or time, as if it were the primary purpose of his existence
> on Earth. Complexity forces people to become so highly special-
> ized that it is virtually impossible for them to attain to wisdom or
> higher understanding.... A life-style full of complexity and spe-
> cialization, while conducive to the acquisition of knowledge of
> the lower things, normally involves such agitation and constant
> strain that it tends to act as a complete barrier against the acquisi-
> tion of any higher knowledge.[1051]

Finally, technologies should be designed to assure that work is both satisfying and meaningful. As was discussed in Chapter 9, most work in industrialized societies does not promote happiness because it is repetitive, boring, alienating and meaningless. In the final analysis, the problem is that technological efficiency, labor productivity and corporate profits are considered to be more important than the physical and psychological well-being of the people doing the work. If we strive toward a sustainable society with less material consumption, we need to change the way we work so that most people feel fulfilled with their occupations and thus have less need for compensatory satisfactions in the form of consumer products and other detrimental distractions. In short, as E. F. Schumacher points out, we need to develop a humane technology:

> If one thing stares us in the face, it is that insane work cannot produce a sane society.... Mindless work is as intolerable in a society that wishes to be sane and civilized as filthy or stinking water, nay, it is even more intolerable. Why can't we set new tasks to our scientists and engineers, our chemists and technologists, many of whom are becoming increasingly doubtful about the human relevance of their own work? Has the affluent society nothing to spare for anything really new? Is "bigger, faster, richer" still the only line of development we can conceive, when we know that it entails the perversion of human work so that, as one of the Popes put it, "from the factory dead matter goes out improved, whereas men there are corrupted and degraded?"... Could we not devote at least a fraction of our research and development efforts to create what might be called technology with a human face?[1052]

Technology should be designed to allow maximum autonomy and self-directed work, minimum supervision, relaxed interactions with co-workers and opportunities to develop one's talents and skills, and it should provide meaningful and socially constructive products and services.[1053] In addition, technologies should also be designed with the goal of providing full employment instead of eliminating jobs in order to maximize corporate profits. David Dickson writes in *The Politics of Alternative Technology* that technology should be non-alienating, in the sense of remaining directly under the individual's control, and linking him with, rather than separating him from, fellow members of the community and the natural environment.... This approach would closely resemble that of the traditional craft techniques. It would attempt to ensure that the machine remained an appendage of man, rather than the other way round.[1054]

In summary, future technologies could easily be designed to be socially appropriate by becoming local, decentralized, small-scale and simple, and by allowing for satisfying and meaningful work. Most likely, socially appropriate technologies would not be as efficient as current industrial mass-production in terms of providing large amounts of consumer goods and services at low prices. But in most cases, this would become acceptable and even preferred because, with the support of socially appropriate technologies, people would derive their sense of well-being from creative and meaningful work and enjoyable family and community life rather than from mindless materialistic consumption and distractions. This change would also promote environmental sustainability.

The Prevention of Unintended Consequences

In Chapter 1 we discussed how scientific reductionism generates incomplete knowledge of complex systems by focusing on only a small number of selected cause-effect relationships while ignoring all others. If this scientific "half knowledge" is then exploited for technological applications, it will inevitably lead to many unintended consequences as described in Chapter 2. The critical question therefore becomes, How can the negative side effects of technology be avoided? Before addressing this question, it must be recognized that it is impossible for technology to create solely positive effects (Chapter 1). Negative side effects are inherently unavoidable. The best we can do is to minimize them.

In an effort to prevent the environmental problems associated with human activities and various technological applications, environmentalists have urged the application of the precautionary principle in all policy decisions. The precautionary principle recognizes the inherent limitations of scientific knowledge and therefore favors the most conservative, low risk approach.[1055] As Stephen Dovers and John Handmer point out in their article "Ignorance, The Precautionary Principle, and Sustainability,"

> (a) Uncertainty is unavoidable in sustainability issues; (b) uncertainty as to the severity of the environmental impacts resulting from a development decision or an ongoing human activity should not be an excuse to avoid or delay environmental protection measures; (c) the principle recommends an anticipatory or preventative approach, rather than a defensive one which simply reacts to environmental damage when it becomes apparent; and (d) the onus of proof shifts away from the environment or those advocating its protection, towards those proposing an action that might harm it.[1056]

The precautionary principle is, of course, nothing new. Societies that are guided by tradition automatically view anything that deviates too much from customary values and practices with suspicion, actively discouraging change and risk-taking. Even in medicine, the guiding principle for thousands of years has been "First, do no harm."[1057] The precautionary principle should be applied not only to medical practice but to all new technologies because they may very well have extremely negative effects on individuals, society or the environment.[1058] In fact, John Cavanagh and Jerry Mander suggest that technologies should be allowed only if insurance companies are willing to cover their potential risks:

> And the precautionary principle…should apply to every technological intervention. All technologies should be assumed potentially guilty of harms until proven innocent, a reversal of present standards. Insurance liability coverage is a crucial indicator for precautionary standards: if the insurance industry refuses coverage, that is a clear signal of danger and unsustainability. At present, the risks endemic to nuclear power, genetic engineering, and large-scale chemical production, for example, have all been externalized, with governments and taxpayers assuming all the risk for insurance companies that refuse to cover losses beyond a certain amount. The nuclear and genetic engineering industries are specifically protected by law from the full risk of coverage of their inventions…. If the risk is beyond the means of the perpetrators, then it should not be permitted.[1059]

If we were to follow this entirely reasonable suggestion, many technologies would have to be abandoned immediately and many new technological innovations would not be allowed to enter the market. For example, it is unlikely that the oil and automobile industries could obtain insurance coverage for the potential risks associated with global climate change, which, after all, are increased substantially by the use of their products. Similarly, there would be no insurance companies willing to cover the risks associated with pesticides, herbicides, toxic and persistent organic chemicals, genetically engineered plants, nano-materials, nuclear power or radioactive wastes.

In summary, if we want to avoid the negative consequences associated with new technologies, we must heed the precautionary principle. As Robert Costanza suggests, "We should tentatively assume the worst and then allow ourselves to be pleasantly surprised if we are wrong, rather than assume the best and face disaster if we are wrong."[1060]

The Democratic Control of Technology

In Chapter 10, we observed that there is little public debate on topics related to the development and adoption of new technologies and that, instead, entrepreneurs and government bureaucracies, with the help of scientists and engineers, make most of the decisions regarding the technological fate of society. In most cases, the decisions do not reflect the values of average people. And, as was discussed in Chapter 11, the sole interest of companies and corporations is the maximization of profits for owners or shareholders. Thus, the control of technology in industrialized societies is inherently undemocratic as corporations and bureaucracies, rather than the public, are the key decision makers in matters of technological innovation.[1061]

The development of socially appropriate technologies is possible only if their design is guided by the values and interests of the people who will be using them rather than corporate or government entities. A first step toward the democratic control of technology is to question and reform all those technologies that are inherently incompatible with democracy. As Jerry Mander writes,

> If we believe in democratic processes, then we must also believe
> in resisting whatever subverts democracy. In the case of technol-

ogy, we might wish to seek a line beyond which democratic control is not possible and then say that any technology which goes beyond this line is taboo. Although it might be difficult to define this line precisely, it might not be so difficult to know when some technologies are clearly over it. Any technology which by its nature encourages autocracy would surely be over such a line. Any technology that benefits only a small number of people to the physical, emotional, political, and psychological detriment of larger numbers of other people would also certainly be over that line. In fact, one could make the argument that any technology whose operations and results are too complex for the majority of people to understand would also be beyond this line of democratic control.... Either we believe in democratic control or we do not. If we do, then anything which is beyond such control is certainly anathema to democracy.[1062]

Consider, for example, how difficult it is to exert true democratic control over large, complex, centralized and entrenched technologies and technological systems such as nuclear power plants, oil refineries, chemical factories, industrialized agriculture, the automobile industry and associated transportation infrastructure, weapons manufacturing or even television. To avoid technologies that are difficult or impossible to control by democratic consensus and to create technologies that are more compatible with democracy and reflect the will of the people, citizens will need to become involved in the so-called participatory design of future technologies.

The concept of participatory design is, of course, nothing new because research and development (R&D) activities are already directed by many people who are not technical experts, such as entrepreneurs, bureaucrats, managers, planning and marketing professionals, bankers, lawyers, accountants, policy analysts and congressional staff.[1063] Thus, no great conceptual leap is required to imagine that average citizens could be included in the R&D planning process. The participation of citizen representatives would ensure that a more diverse range of social needs, concerns and experiences are reflected in the design process. Furthermore, broadened popular involvement in design would provide opportunities to publicize new technological developments at an early stage when there is still the greatest potential to change course (i.e., before they become entrenched as a result of institutional commitments, enduring material embodiment or social dependence).[1064]

Because universities and government research facilities receive public funding and therefore must in principle serve the public interest, they are the most obvious places to initiate the democratic involvement of citizen representatives in research and development planning activities.[1065] One possibility is to have a three-stage decision process, in which research proposals and design concepts are first subjected to a rigorous peer review by technical experts to judge the scientific and technical merits, followed by a second evaluation by business experts to determine the business potential and economic promise, which is then followed by a third assessment by a panel of citizen representatives who appraise potential social and environmental issues.[1066] Participation by all employees in research and development decisions of corporations offers an additional possibility for democratizing technological design.[1067]

Already a number of successful cases demonstrate the possibility and benefits of greater democratic control of science and technology.[1068] In Sweden, national research priorities are directed by the Council for Planning and Coordination of Research, whose members consist of six leading scientists, five members of parliament, four representatives of labor unions and three at-large members.[1069] Similarly, in Denmark, research and development proposals are evaluated not only by technical authorities but also by experts in the technologies' social dimensions and effects and by representatives of organized interest groups. The final judgment is then made by representative citizens.[1070] In the United States, there are efforts to include representatives of organized stakeholder groups such as business people, workers and environmentalists in the decision-making process, thereby increasing the chances that economic, workplace and ecological concerns are addressed.[1071] However, it would also be beneficial to include citizen representatives in the research and development planning process, as is already done in Scandinavia. In the United States, there are at least two major examples in which large numbers of laypeople have been involved in evaluating innovative technologies and even participating directly in technological research and design: one is the participation of people with physical disabilities in barrier-free design, and the other is the environmental movement.[1072]

As these success stories demonstrate, it is certainly possible to guide technological innovation in a more democratic fashion than is done currently. However, as Richard Sclove points out in *Democracy and Technology*, there is likely to be strong resistance against efforts to democratize technological development:

> Many participatory design exercises have encountered outright opposition from powerful institutions—opposition engendered not because the exercises were failing, but because they were succeeding. The absence of more examples of participatory R&D would seem to have much less to do with issues of layperson competency than it does with economic, political, and cultural resistance.... It follows that if participatory design of democratic technologies is ever to become normal social practice, then political pressure must be developed for diverting economic resources to it or else for incorporating it into the core social institutions that today undertake most R&D. Otherwise control of the technological order will remain substantially in elite hands.... Support for technological democratization may be strongest on the progressive side of the political spectrum. But democracy is a motherhood issue (few are willing to admit publicly against it) and strong democracy's respect for cultural tradition, democratic community, and political decentralization could appeal to many conservatives, too. But one could anticipate opposition among the elite circle of business, military, and academic leaders who have exercised hegemony over science and technology decision making since World War II.[1073]

Sclove suggests that the funding structure for research and development must be revised to include strong citizen participation and influence.

Local Organic Agriculture: A Model of Environmentally Sustainable and Socially Appropriate Technology

Given the conditions for environmental sustainability and social appropriateness, it may appear to some readers that it is virtually impossible to design technologies that meet these criteria. While it may indeed be difficult to replace the numerous technologies in modern industrialized societies, it is demonstrably possible. Consider, for example, that almost all technologies prior to the Industrial Revolution were environmentally sustainable and socially appropriate. Even today, local organic agriculture, sometimes also called ecoagriculture or organic farming, serves as a model of how food can be produced in a manner that is optimal both for the environment and for people.[1074]

Organic agriculture generally requires less fossil fuel because it is more labor intensive. Ideally, to become completely sustainable, from an energy perspective, organic agriculture would have to use only biofuels and other renewable energy sources, including human and animal power. Gardening is the perfect example of producing high-quality food without the need of fossil-fuel-powered machines.

Ecoagriculture also satisfies to a large degree both the sustainable materials use and the sustainable waste discharge conditions. Completely biodegradable organic animal and plant wastes are recycled back onto the land, thereby preventing soil erosion and increasing soil fertility without the need for nitrogen fertilizers derived from petroleum. Crop rotation and biological pest control are used to reduce problems caused by weeds and insects, thereby minimizing or eliminating the use of harmful herbicides and pesticides.[1075] Ideally, only food that could be grown under local climatic conditions (e.g., dry land farming) would be produced, thereby avoiding the unsustainable use of irrigation water obtained from rivers or aquifers.

Local organic farming also meets the criteria of social appropriateness. It addresses the needs of the community for healthy and nutritious food by growing food locally for local consumption, thereby providing produce that is fresh while avoiding fossil-fuel-intensive long-distance transport. By buying only food that is in season, people feel more connected to the land and climate. Furthermore, social and community life is improved by buying organic food directly from the farmers at local farmers markets. The direct interaction between food producers and consumers also makes local democratic control of organic farming possible. For example, people could express directly to farmers their preferences for certain types of foods prior to the growing season. Even the financing for seeds and supplies at the beginning of each year could be obtained directly from the customers instead of commercial banks.

This practice currently exists in the form of community supported agriculture (CSA) in many parts of the United States. Community supported agriculture should be expanded and ultimately replace corporate, industrialized agriculture.

Local organic farming is generally small scale and employs only relatively simple machinery, thereby providing conditions for more satisfying work. "Old-fashioned" gardening, which is an excellent example of small-scale, simple and organic agriculture, is so satisfying that many people actually choose it as a leisure activity. Finally, ecoagriculture also heeds the precautionary principle by shunning the use of potentially dangerous herbicides, pesticides and genetically engineered crops, while successfully relying on more natural approaches.

In conclusion, as the example of small-scale, local organic agriculture demonstrates, it is possible to design technologies that are both environmentally sustainable and socially appropriate. In the next chapter, we will discuss how the conduct of science must change in order to better support the needs of both people and the environment.

CHAPTER 14

Critical Science
and Social Responsibility

The Myth of Value-Neutrality

In Chapter 10, we exposed the myth of the value-neutrality of technology by demonstrating that the dominant social and cultural values of control, exploitation and violence guide the design of many modern technologies. As a result, these values become permanently embodied in technological structures and products, making it extremely difficult to apply those technologies to uses that reflect different values. While most technological accomplishments, including the many unintended consequences discussed in Chapter 2, could not have been realized without the help of modern science, many people, including most scientists, believe that science, because it merely provides objective knowledge of natural phenomena, is entirely value neutral. This belief is described by Steven and Hilary Rose in *The Myth of the Neutrality of Science*:

> The activities of science are morally and socially value-free. Science is the pursuit of natural laws, laws which are irrespective of the nation, race, politics, religion, or class position of their discoverer. Although science proceeds by a series of approximations to a never attained truth, the laws and facts of science have an immutable quality.... Because this is the case, although the uses to which society may put science may be good or evil, the scientist carries no special responsibility for those uses.[1076]

In *Towards a Science for the People,* William Zimmerman and colleagues show how scientists cite value-neutrality in order to escape responsibility:

> Others still argue that science should be an unbridled search for truth, not subject to political and moral critique. J. Robert Oppenheimer, the man in charge of the Los Alamos project which built and tested the first atomic bomb, said in 1967 that, "our work has changed the conditions in which men live, but the use made of these changes is the problem of governments, not of scientists."... The attitude of Oppenheimer and others, justified by the slogan of truth for truth's sake, is fostered in our society and has prevailed.[1077]

In Europe, beginning in the 17th century, it came to be understood that for rational inquiry to flourish, facts derived from observable phenomena must take precedence over dogma.[1078] Science, if it were to exist at all, would have to become objective and unencumbered by the extraneous requirements of the church. This was a necessary and positive development.

However, many scientists today are likely to confuse objectivity with value-neutrality.[1079] Most scientists are taught to generate experimental data in a maximally objective manner and to avoid subjective biases whenever possible. When subjectivity does enter, it is generally minimized or eliminated via the peer-review process and through consensus building within the scientific community. The result is objective and scientifically valid knowledge largely devoid of subjectivity. However, the absence of subjectivity does not imply value-neutrality. It is quite possible to accumulate objective, valid facts which reflect a value orientation. As a result of this confusion between objectivity and value-neutrality, most scientists and engineers are not aware of the value dimension of their work; that is, they may be said to be value-blind.[1080]

Furthermore, not recognizing the difference between scientific knowledge and the act of performing science may also contribute to the myth of value-neutrality. As William Lowrance writes in *Modern Science and Human Values,*

Although scientific knowledge, once attained, may be considered ambi-potent for good and evil, the work of pursuing new science and development technologies is by no means value-neutral.... No major creative activities in society...should be allowed to claim valuative or moral immunity. Jacob Bronowski's distinction is key: "Those who think that science is ethically neutral confuse the findings of science, which are, with the activities of science, which are not.".... Surely it is wrong to view not-yet accomplished research, which cannot be undertaken without commitment of will and resources, as being anything other than value-laden.[1081]

Scientific endeavors are ultimately based on human choice operating within a social context and associated paradigms. Choices of research topic, even the design of experiments, are not culture free.[1082] All choices necessarily reflect a distinct culture and value orientation. All scientific activities are, by definition, value laden.[1083]

Consider, for example, the choices individual scientists must make when performing research. The topic and hypotheses to be tested must be selected, and experiments, or other data gathering techniques, must be designed and employed to test those hypotheses. In addition, decisions must be made as to which data will or will not be included in the analyses and how they will be interpreted.[1084] Not only are these choices subjective, they necessarily reflect a certain value-orientation— what is considered important and what is not. Given the large number of choices that must be made, it is not surprising that scientists often differ in their approach to and interpretation of observed phenomena. Conflicting views are not necessarily the result of faulty science but rather may reflect the differences in choices and values of individual scientists.[1085]

Currently most scientific research requires major financial support, which is generally provided either by governments or large corporations. As a result, the topics addressed by pure or applied research are those that support the goals and values of the funding agents.[1086] As Steven and Hilary Rose observe, making funding decisions inevitably involves value judgments:

> If we recognize that Big Science is State financed, and that there
> is always more possible science than actual science, more ideas
> about what to do than men or money to do them, the debate is,
> in a sense, short-circuited. Science policy means making choices
> about what science [ought] to do. Whoever makes these choices,
> by definition they cannot be ideology- or value-free; they imply
> an acceptance of certain directions for science and not others;
> opening certain routes means closing others.... The science they
> generate cannot be neutral.[1087]

The value orientation of those who provide the funds for research is
often adopted, knowingly or unknowingly, by the scientists carrying out
the research. In most cases, scientists have little choice in the matter
because they are financially dependent upon the source of funding.[1088]
The differences in goals and values of the entities providing the research
funding are often responsible for scientific discrepancies and controversies. For example, a survey of scientific studies on four of the most
common chemicals used in the United States found that only 14 percent
of the industry-funded studies made unfavorable comments compared
to 60 percent of the non-industry funded studies.[1089] Clearly, the value
orientation of those who fund science has a major effect on the outcome
and interpretation of research.[1090] This has led some to claim that "science is politics by other means."[1091]

Finally, general worldviews and widely shared paradigms, and the
socio-cultural values they inform, have profound effects on the conduct
of science.[1092] As discussed in Chapter 12, people are generally not consciously aware of the worldview and paradigms that shape their opinions and guide their behavior. Similarly, many scientists are unlikely to
recognize that scientific knowledge is socially constructed and therefore
biased by contemporary worldviews. For example, the approach of modern science, which attempts to understand nature in terms of underlying mechanisms, is a direct result of the socio-economic and cultural climate of 15th- and 16th-century Europe. The mechanistic view of nature,
together with its emphasis on objective inquiry, emerged victorious over
previously competing alternative magic worldviews and continued to
be supported for centuries by the dominant utilitarian goals and values
of industrialized societies.[1093] Even today, entire scientific disciplines

are shaped by larger social priorities; for example, physics is influenced by the need for military defense and offense, chemistry by the need of industry to develop new materials and products for mass consumption, biology by the need of the medical industrial complex for new treatments.[1094]

It must also be recognized that a number of scientific theories, particularly those related to the processes of nature, are heavily biased by the socio-economic and cultural context in which they developed. The main reason for this, according to anthropologist Mary Douglas and political scientist Aaron Wildavsky, is that "nature is a mirror reflecting whatever version of reality the looker wishes to see in it.... It has enough diversity for everyone."[1095] And, as Jeremy Rifkin writes in *The Biotech Century*, the prevailing socio-economic order shapes our concept of nature which then, in turn, is used to justify that order:

> Every major economic and social revolution in history has been accompanied by a new explanation of the creation of life and the workings of nature.... In each instance, the new cosmology serves to justify the rightness and inevitability of the new way human beings are organizing their world by suggesting that nature itself is organized along similar lines. Thus, every society can feel comfortable that the way it is conducting its activities is compatible with the natural order of things, and, therefore, a legitimate reflection of nature's grand design.[1096]

Let us consider two interesting historical examples of how the concept of nature was biased by the prevailing socio-economic order. Prior to the Scientific Revolution, St. Thomas of Aquinas believed that nature is populated with "many creatures different among themselves in gradation of intellect, in form, and in species"[1097] and that this "diversity and inequality" guarantees the orderly functioning of the system as a whole. St. Thomas's characterization of nature as a hierarchy of species strikingly mirrors the feudal hierarchy of Medieval Europe.[1098]

During the height of the Industrial Revolution, Charles Darwin published *On the Origin of Species*, which postulated a theory of biological evolution based on the concepts of ruthless competition, struggle for existence and survival of the fittest. It is no surprise that Darwin's theory reflects a view of nature that is remarkably similar to the socio-economic order of Victorian England at that time, when men, women and children were struggling for their livelihood by working 12 to 14 hours per day under appalling conditions. [1099] The theory of biological evolution was then used by the ruling classes of industrialized societies to justify competition, greed and the exploitation of workers and the environment by claiming that such behavior conforms with the laws of nature. [1100] In this context, as was noted previously, as working conditions in industrialized societies became more humane, with management and labor often negotiating cooperative agreements, scientists began to observe that cooperation, rather than ruthless competition, plays an important role in the survival of many species. [1101]

As science is obviously not value neutral, why has the myth of value-neutrality remained so common and persistent? The concept of "purity" and value-neutrality of science seems to have first emerged in 19th-century Germany when the natural sciences struggled to preserve their autonomy and freedom. [1102] As Robert Proctor comments in *Value-Free Science*,

> Value-freedom is an ideology of science under siege—a defensive reaction to threats to the autonomy of science from political tyrants, religious zealots, secular moralists, government bureaucrats, methodological imperialists, or industrial pragmatists asking that science be servile or righteous or politically correct or practical or profitable.... This may help to explain why it is that the ideal of pure science arises, curiously enough, at the very time that science is perceived as vital for the realization of political, military, and industrial objectives. [1103]

Even today, the desire for scientific freedom and autonomy remains one of the motives behind the persisting myth of value-free science. The ideology of value-neutrality offers many advantages both to those who fund and to those who conduct scientific research. First, individual scientists are, in principle, free to carry out any research they like, being motivated primarily by their desire for intellectual pleasure, money, status or agreeable working conditions. [1104] Second, scientists often feel they need not be concerned about questions of responsibility, because they are, after all, only generating objective knowledge. How the scientific knowledge is likely to be used is not their concern. Designing nuclear weapons is perceived to be as respectable and morally justifiable as finding a cure for cancer.[1105] Third, the ideology of value-neutrality also stifles social criticism—after all, science is value neutral and therefore, scientists do not make value judgments. How can such practice objectively be faulted? Even more ominously, as Robert Proctor summarizes in *Value-Free Science*,

> Value-freedom also had come to mean a willingness to sell one's skills to the highest bidder. Value-freedom allowed one to believe it was no worse to work to spread disease than to cure it; value-freedom meant that there was nothing wrong with doing market research for tobacco companies to help them sell cigarettes. The cash value of neutrality was to stifle social criticism—value-freedom translated into the commandment: "Thou shalt not commit a critical or negative value judgement—especially of one's own society."[1106]

Thus, the veil of value-neutrality serves the purpose of isolating science from critique by non-scientist citizens, thereby ensuring that questionable research is not prohibited and continues to receive adequate funding. Fourth, by confusing value-freedom with objectivity, the myth of value-neutrality legitimizes science as the only reliable source of value-free, objective, unbiased and valid knowledge, casting aside all other forms of knowledge as value laden, subjective, biased and invalid. Finally, and probably most importantly, value-neutrality serves to camouflage the interests of those who fund science, allowing them to pursue their agendas under the cloak of scientific neutrality and objectivity. For example, although most basic research in particle-, nuclear-, and even astro-physics is funded and carried out because of its military promise, it is presented to the public only as the generation of exciting new knowledge about the fundamental workings of the universe. As Proctor concludes in *Value-Free Science*,

> The prospect of a value-laden science is, for many, the prospect of a science whose results are continually in contestation. For others, it is the more frightening prospect of a science continually at the mercy of dominant interests, a science that, under the guise of neutrality, helps create a world to serve those interests....

> Science must not be let off the hook—we cannot rest comfortably
> in the view that science needn't become political since it already
> is by definition.[1107]

In conclusion, while scientific knowledge itself can be objective, the acts of performing science and making choices regarding its direction are anything but value neutral. The fact that science is value laden has significant implications not only with respect to questions of moral and social responsibility but also in terms of what kind of science should be conducted in the future.

A New, Critical Science

Many, if not most, of the unintended consequences of modern technologies and the science that made them possible, are directly or indirectly the result of the predominant values of industrialized societies, (i.e., power, control, exploitation and the violence required to implement these). As we argued in Chapter 12, a lasting solution to significant human problems can be found only if there is a paradigm shift, a change from the current worldview of separateness to one of interconnectedness. Similarly, as long as present-day science is directed, consciously or unconsciously, by the above-mentioned dominant values of ethically challenged individuals, corporations, governments or the consumer culture as a whole, it is unlikely that scientific inquiry will be effectively employed to find permanent solutions to our many environmental and social problems.

What is needed is a new kind of science, the goal of which would be to generate not only more knowledge but knowledge for the specific purpose of exposing the limitations as well as the dangers of modern technologies and for designing future technologies that are both environmentally sustainable and socially appropriate. This new activist science, which has been termed "critical science" by Jeremy Ravetz in his book *Scientific Knowledge and Its Social Problems*, has three major goals: to expose the hidden values of mainstream science and technology, to provide a critical analysis of technological innovations as they appear, or preferably before, and to support the interests of people and the environment. As Ravetz points out, critical science is unlikely to appeal to the average scientist who is motivated by the self-centered goals of intellectual curiosity, job security, money or status. It would, however, appeal to those who are seriously concerned about the suffering of humanity and the degraded state of the environment and therefore to those who wish to use their scientific expertise to expose problems associated with modern technologies and to find appropriate, lasting solutions:

> The emergence of a "critical science", as a self-conscious and coherent force, is one of the most significant and hopeful developments of the present period.... Instead of isolated individuals sacrificing their leisure and interrupting their regular research for engagement in practical problems, we now see the emergence of scientific schools of a new sort. In them, collaborative research of the highest quality is done, as part of practical projects involving the discovery, analysis, and criticism of the different sorts of damage inflicted on man and nature by runaway technology, followed by their public exposure and campaigns for their abolition.[1108]

One of the functions of critical science is to create awareness of the underlying values and the political and financial interests that are currently determining the course of science and technology in industrialized society.[1109] This exposure of the value-laden character of science and technology is done with the goal of emancipating both people and the environment from domination and exploitation by powerful interests.[1110] The ultimate objective is to redirect science and technology to support both ordinary people and the environment, instead of causing suffering by enabling oppression and exploitation. Furthermore, by exposing the myth of the value-neutrality of science and technology, critical science attempts to awaken working scientists and engineers to the social, political and ethical implications of their work, making it impossible or, at the very least, uncomfortable for them to ignore the wider context and corresponding responsibilities of their professional activities.[1111]

Another goal of critical science is to promote continuing comprehensive analyses of the effects of science and technology on both society and the environment, placing particular emphasis on identifying root causes, so that lasting solutions, instead of short-term techno-fixes, may be found. Critical science, because it uses the powerful tool of the scientific method, can be a very strong weapon against destructive technologies and any antisocial institutions that support them. Because critical science almost always challenges the status quo, it requires very high standards of scholarship to stand the test of counterattacks. As Ravetz writes,

> Any critical publication is bound to be scrutinized severely by experts on the other side, so high standards of adequacy are required because of the political context of the work. Indeed, a completely solved problem in critical science is more demanding than in either pure science or technology.[1112]

Similarly, T.L. Guidotti observes in "Critical Science and the Critique of Technology,"

> It is an enterprise that is by nature controversial, that requires an interdisciplinary but rigorous approach, and that is often a lonely pursuit, both because its messages are almost always unwelcome by society and because its new insights only rarely seem plausible in light of conventional wisdom.[1113]

It is clear that critical science is not for the faint-hearted. It requires not only technical competence, a skeptical mind and a commitment to high-quality interdisciplinary research but also the ability to handle hostile counterattacks and the willingness to take an activist stance in the interests of the public and the environment. This also requires moral courage, particularly if one's livelihood is at stake. In short, it is not for the person who wants an easy, quiet life.[1114]

In addition to providing an analysis and critique of modern science and technology, critical science also has the goal of redirecting science and technology toward addressing the needs of the oppressed and impoverished, generating socially useful knowledge, improving the human condition, protecting the environment, in short, creating a better world for all.[1115] Because of these attributes, it has also been called "Public Interest Science."[1116] As Bernard Dixon writes in *What is Science For*, it challenges scientists to ask themselves,

> How can my scientific skills be best used to serve the people: to expose and correct the role of science and technology in wreak-

ing genocide in wars, in oppressing individuals and minorities
by acting as an agent of civil war, and in permitting malnutrition
and disease both throughout the world at large, and even in rich
societies?[1117]

Although Enlightenment philosopher and (proto) scientist Francis
Bacon is generally remembered for his view that knowledge is power
and that scientific knowledge, in particular, is useful for the human
control and exploitation of nature, he may have also been the first critical scientist as he urged future researchers to pursue science for higher
ends:

> Lastly, I would address one general admonition to all; that they
> consider what are the true ends of knowledge, and that they seek
> it not either for pleasure of mind, or for contention, or for supe-
> riority to others, or for profit, or fame, or power, or any of these
> inferior things; but for the benefit and use of life; and that they
> perfect and govern it in charity. For it was from lust of power that
> the angels fell, from lust of knowledge that men fell; but of char-
> ity there can be no excess, neither did angel or man ever come in
> danger by it.[1118]

Finally, critical science is incomplete unless its results are translated
into action. This requires that critical scientists communicate their
findings to the public, suggest possible solutions and even become politically active to ensure that problems are adequately addressed and
permanently solved.[1119] This will be a challenge for the average scientist
who still believes that publishing one's research in journals is sufficient
and who does not see the need for social and political action to create a
better world. As Richard Welford writes,

> Researchers have been less willing to engage in a political debate.
> Too often, for example, academics communicate their research
> findings only to other academics and their students and fail
> therefore to grasp the need for change amongst policy-makers
> and decision-makers. They shy away from translating research
> findings into a real agenda for change. The results of research
> should not be judged in terms of publications—that is merely an

indicator of activity. The ultimate measure should be associated with action and change. Unless such research leads to change it is in fact pointless.[1120]

Similarly, Steven and Hilary Rose assert, "It is no longer adequate for the scientists to retire to their ivory-towered laboratories, emerging only to share the mutually congratulatory blandnesses of the scientific establishment. They have to learn to speak out."[1121]

Critical science in the modern age was born after two atomic bombs were dropped by the United States on Japan in the summer of 1945. This catastrophic event prompted a group of physicists who had worked on the Manhattan Project to begin publishing the *Bulletin of the Atomic Scientists*. The primary purpose of the *Bulletin* was to inform the public about nuclear policy debates, specifically in relation to the Cold War, while at the same time advocating international control of nuclear energy. Since then, the *Bulletin* has provided citizens, policy makers, scientists and journalists with non-technical, scientifically sound and policy-relevant information about nuclear weapons and other global security issues.[1122] Unfortunately, there appears to be no other major organization in the United States that critically analyzes ongoing military research and development activities. One reason for the absence of critical science in the area of military R&D is the secrecy under which the development of weapons and defense systems is carried out.[1123]

With the rise of the environmental movement in the 1960s and 1970s, entirely new scientific disciplines came into existence, many of them devoted to the critical analysis of humanity's impact on the environment and the mitigation of that impact. Examples include environmental science and engineering, environmental toxicology and epidemiology, technology and risk assessment, conservation biology, restoration ecology, ecological economics, industrial ecology and environmental ethics.[1124] In response to many serious environmental problems, the Union of Concerned Scientists was founded in 1969 in the United States. The original mission of the Union was to conduct a critical examination of governmental science and technology policy and to redirect research from the present emphasis on military technology toward finding solutions to pressing environmental and social problems.[1125] In the same year, the British Society for Social Responsibility in

Science was established in the United Kingdom. Its mission was to create awareness of scientists' individual and collective responsibilities and to draw the attention of the public to the implications and consequences of scientific and technological developments.[1126]

There are also many opportunities for conducting critical science in medicine. Examples include the critical analysis of social, environmental, economic and lifestyle factors involved in health and disease, the practice of holistic and alternative medicine, and increased attention to the needs of the poor, both in developed and developing nations.[1127] In the United States there are several prominent "critical medicine" organizations, such as the Physicians Committee for Responsible Medicine, whose mission is to promote preventative medicine and encourage higher standards of ethics and effectiveness in research,[1128] as well as several anti-vivisection societies (e.g., the American Anti-Vivisection Society and the National Anti-Vivisection Society), whose goals are the elimination of the use of animals in biomedical research, product testing and education.[1129]

In summary, critical science exposes the hidden values and interests that drive scientific and technological developments, provides an interdisciplinary analysis of the causes of unintended consequences brought about by science and technology, and attempts to support the interests of ordinary people and the environment. Although very few professionals have devoted their lives to critical science, their research and subsequent activism has had a disproportionately large impact on public opinion as well as science and technology policy.[1130] Thus, critical science holds the promise of redirecting science and technology to support the important goals of environmental sustainability, social appropriateness and psychological well-being.

The Question of Responsibility

Before discussing the specific steps needed to transform science as currently practiced into a more aware and critical science, we must first address the issue of responsibility in science: who should be responsible for the choice of research topics, who is accountable for the misuse of scientific knowledge, and who is in a position to predict potential negative effects?[1131] One common view is that scientists are morally responsible for the negative consequences that result from the various applications of their knowledge and inventions.[1132] After all, if scientists take personal pride in the many positive achievements of science, why should they be allowed to escape responsibility for the negative consequences related to the use or abuse of scientific knowledge?[1133] Furthermore, scientists have a collective responsibility for both the choice of research areas and the ethical conduct of science. Committees of scientists are often involved in the planning of governmental and corporate research programs.[1134] Many professional societies and national organizations, such as the National Academy of Science in the United States, have ethical guidelines for the conduct of scientific research.[1135] Clearly, there is recognition that scientists, both individually and collectively, have a special and much greater responsibility than average citizens with respect to the generation and use of scientific knowledge.[1136]

However, scientists may respond with a list of reasons, some of them valid and others not, claiming that the situation is not that simple and that they should not be blamed for all the evils created by the scientific knowledge they generate. First, there is the problem of fragmentation and diffusion of responsibility. Because of the intellectual and physical division of labor, the resulting fragmentation of knowledge, the high degree of specialization, and the complex and hierarchical decision-making process within corporations and government research laboratories, it is extremely difficult for individual scientists to control the applications of their innovations.[1137] This fragmentation of both work and decision making results in fragmented moral accountability, often to the point that "everybody involved was responsible but none could be held responsible."[1138] However, as Ravetz points out in *Scientific Knowledge and Its Social Problems*, while this excuse may have some validity for scientists conducting fundamental research, it is no longer a defense for those working on projects leading to obvious real-world applications:

> The division of labour in large scale technical problems is extremely fine; so that the scientist who publishes a result generally has no more knowledge of its possible functions than does a process-worker assembling a standard component of a device. This does not mean that the position of the agent is one of moral neutrality; rather that it is morally indeterminate. However, to the extent that his research is related to technical applications,

the area of indeterminacy decreases, and the scientist's responsibility becomes defined. Once the scientist is aware of the likely
consequences of his work, his sole disclaimer of responsibility
can be along the lines of "I was only following orders". This is no
longer likely to be acceptable as a defense, in science as anywhere
else.[1139]

Another, and related, excuse is ignorance. Obviously, a scientist performing basic research cannot predict what new knowledge about
nature and the universe will be gained and how this may be misused
for destructive purposes in the near or distant future. While this excuse
appears plausible at first, it loses credibility given that most of Western science during the past 200 years has been guided by the paradigm
that scientific knowledge is good because it allows for greater control
and exploitation of nature (see also Chapters 3 and 12). Therefore, even
basic research is actually conducted for purposes beyond the generation of pure knowledge, and scientists should not be allowed to excuse
themselves because they have not made an effort to study the history
and philosophy of science. The excuse of ignorance is even weaker for
scientists conducting applied research. In most cases, the objectives of
applied research are well known. For example, most corporations conduct research on specific products or services that promise to yield the
greatest profit (see Chapter 11). Similarly, most of the research funded by
governments is mission oriented, such as protecting the environment,
developing new drugs or designing more lethal weapons. In all cases
where the applications of scientific knowledge and innovation are well
known *a priori*, it is impossible for a scientist to escape responsibility
for research that is morally dubious.[1140] In conclusion, as John Forge
writes in *Moral Responsibility and the 'Ignorant Scientist*,' "Ignorance is
not an excuse precisely because scientists can be blamed for being ignorant."[1141]

Finally, another and more valid consideration is that responsibility
also falls on those who provide the funding for the research, which in
most cases are corporations and government agencies. Furthermore, because taxpayers indirectly provide the funds for government-sponsored
research, they and the politicians who represent them (i.e., society at
large) should be held accountable for the uses and abuses of science.[1142]

Compared to earlier times, when scientists could often conduct their research independently, today's experimental research requires expensive laboratories and instrumentation, making scientists completely dependent on those who pay for the studies. As Ravetz writes,

> Research is now capital intensive. Any significant piece of work is almost certain to cost far more than an individual scientist can afford out of his own pocket; it will generally cost much more than his annual income. Hence he is no longer an independent agent, free to investigate whatever problem he thinks best. Nor is he likely to have personal contact with a private patron who will provide for all his needs. Rather, in order to do any research at all, he must first apply to the institutions or agencies that distribute funds for this purpose; and only if one of them considers the project worth the investment can he proceed.[1143]

Thus, in view of the fact that individual scientists have very little control over what research areas are funded, it is understandable that scientists are unwilling to take full responsibility for the misuses and abuses of the scientific knowledge that they have generated. Nevertheless, as we will discuss in the remainder of this chapter, there are a number of ways in which scientists can exert a positive influence and redirect science to the service of people and the environment: enacting and enforcing professional ethics codes and boycotting morally dubious research by withholding their skills and expertise.

The Problem of Professionalism

A professional, according to the original meaning of the Latin-derived word, is someone who professes the truth, who has a calling to make a positive difference. As Ravetz points out, a profession could be defined as "an occupation whose members should be prepared to die for its integrity. If someone would not, then he does not have a calling, and his work is utterly without significance."[1144] Unfortunately, most science and engineering professionals do not have such high standards, and many do not consider themselves as having a special calling. More likely, scientists and engineers are interested instead in the intellectual challenges of their work, the relatively good income and in some cases, status and prestige.[1145]

Pleasure in research and problem solving can easily result in a situation in which science and engineering professionals work on projects that, though intellectually stimulating, may be applied for destructive purposes or may never fundamentally solve or prevent any serious problem. Consequently, many problems will never disappear, which, of course, is beneficial to scientists and engineers because it guarantees their long-term employment. For example, in medical research, working on the development of anti-cancer drugs is likely to provide an intellectually stimulating, handsomely remunerative and secure career but is unlikely to make a significant contribution to the health of society, at least when compared to readily available, simple and highly effective preventative measures (e.g., vegetarian diet, exercise, non-smoking programs, etc.) that can be implemented at much lower cost. Similarly, the engineer working to improve the various efficiencies of industrial processes and consumer products may enjoy doing so but will never solve resource depletion and environmental pollution problems as long as the root causes of overpopulation and overconsumption are not addressed (see Chapter 5). In summary, there are few incentives for science and engineering professionals to permanently solve many of humanity's most serious social and environmental problems.[1146]

One major reason that science and engineering professionals are given few incentives to find permanent solutions to important problems is that those who fund their work are highly motivated to maintain the current industrial and economic system. Any substantial change, which could be provoked by rigorous root-cause analyses, is likely to threaten the profits and social status of the privileged. Consequently, one of the main functions of professionals, including scientists and engineers, is to keep the current industrial, economic and political system functioning smoothly, without asking too many critical questions.[1147] Some have even suggested that professionals are "bribed" by higher-than-average salaries to maintain the status quo,[1148] to behave, in the words of MIT professor Noam Chomsky, like well-trained dogs, never questioning their masters:

> The intellectual class is supposed to be so well trained and so well indoctrinated that they don't need a whip. They just react spontaneously in the ways that will serve external power interests,

without awareness, thinking they're doing honest, dedicated
work. That's a real trained dog.[1149]

As history has repeatedly shown, many professionals have been willing
to "just follow orders," executing the highly questionable missions of
their funding providers.[1150] There are several reasons that professionals
may have problems questioning and confronting the status quo. First, as
mentioned above, science and engineering professionals are generally
completely dependent on government or corporate funding sources for
their livelihood. An overly critical attitude toward those who provide the
funding may result in the abrupt termination of a long career and the
relatively comfortable lifestyle that accompanies it.[1151] Second, because
of widespread authoritarian child-raising practices and the authoritarian character of most major societal institutions such as governments,
corporations and churches, many people, including scientists and
engineers, find it very difficult to question authority.[1152] Even if they
could bring themselves to do so, they would risk being replaced by more
compliant individuals.[1153] Finally, many scientists and engineers may
be conformist because they fear rejection by their peers and the social
and professional isolation that may follow. As studies of whistle-blowers
have repeatedly demonstrated, dissenting professionals are treated with
scorn, dismissed by management and shunned by peers, resulting in a
ruined professional as well as social life.[1154]

In conclusion, professionals may have forgotten the original calling
of their profession and may have become too willing to sell their expertise and knowledge to the highest bidder, causing much of their work
to be applied for destructive, questionable or worthless purposes. This
must change. The professions need to be empowered again to make a
positive difference, to be in agreement with their original missions. The
first step in this direction would be the formulation and enforcement of
strict professional ethical codes.

The Need for Comprehensive Professional Ethics

Modern science and technology have given us unprecedented freedoms and powers. Without strict ethical restraints, these are likely to be
misused or abused. Traditional ethical codes are often inadequate and
outdated because they were developed in earlier, simpler times, when

individuals did not have the opportunity to wield the vast powers that advanced science and technology provide.[1155] Greater power requires greater responsibility, more wisdom and strict ethical restraints if catastrophe is to be avoided.[1156] We can no longer allow science and technology to continue in a moral vacuum. As Bertrand Russell pointed out many years ago, scientific progress without moral progress is likely to end in disaster:[1157]

> Science has not given men more self-control, more kindliness, or more power of discounting their passions in deciding upon a course of action. It has given communities more power to indulge their collective passions.... Men's collective passions are mainly bad; far the strongest of them are hatred and rivalry directed towards other groups. Therefore, at present all that gives men power to indulge their collective passions is bad. That is why science threatens to cause the destruction of our civilization.[1158]

Because science and engineering professionals have greater powers, and thus greater responsibilities, than the average citizen with respect to science and technology, they must unite to empower themselves and their professions by adhering to comprehensive, enforceable codes of ethics.[1159] As quantum physicist Fritjof Capra notes, this first step towards ethical reform is of utmost importance:

> A system of ethics is urgently needed, since most of what scientists are doing today is not life-furthering and life preserving but life-destroying. With physicists designing nuclear weapons that threaten to wipe out all life on the planet, with chemists contaminating our environment, with biologists releasing new and unknown types of microorganisms into the environment without really knowing what the consequences are, with psychologists and other scientists torturing animals in the name of scientific progress, with all these activities occurring, it seems that it is most urgent to introduce ethical standards into modern science.[1160]

Although it is beyond the scope of this book to provide a detailed code of ethics, several suggestions may be offered. First, each profession must develop its own ethical guidelines, which should firmly remind its members of their professional calling or mission. Second, strict adherence to the ethics code should become a requirement for all professionals. Scientists and engineers could be required to take an exam to receive an ethics certificate, similar to a professional license. Ownership of such a certificate could be a prerequisite for engaging in any type of professional work or receiving corporate or government funding. Finally, and probably most importantly, the ethics codes would need to be enforced. Scientists and engineers who violate the code would be at risk of losing their employment or funding. Conversely, scientists and engineers whose employers or funding agencies force them to conduct work that violates their professional code of ethics should have the right to appeal. An official hearing could be scheduled and presided over by an independent third party, with the goal of resolving the ethics dispute and applying sanctions, if necessary. Aggrieved scientists should also be entitled to full legal recourse.[1161]

While the formulation and enforcement of a strict code of ethics initially may appear difficult, it should be kept in mind that enforceable ethics codes would empower scientists and engineers to work only on constructive and worthwhile research and development projects, while allowing them to avoid or resist work that they consider harmful or questionable. Thus, in the final analysis, if professionals can agree upon and adhere to a common code of ethics, their power relative to those who employ them or provide their funding would be significantly increased. This, in turn, should make technology less exploitative and more responsive to the needs of people and the environment.

In this context, an important issue is what kinds of scientific knowledge should and should not be generated. This difficult topic has been almost completely ignored and is rarely discussed among scientists. The standard position is that the generation of scientific knowledge cannot and should not be restrained.[1162] Indeed, any attempt to do so would be, in the words of mathematician John von Neumann, "contrary to the whole ethos of the industrial age."[1163]

Though it is true that it would not be easy to determine what kind of knowledge should be generated, because the ultimate effects of scientific research are difficult to predict,[1164] von Neumann's position nevertheless poses a number of problems. First, it ignores the fact that scientific research is a social activity that ultimately reflects the values of society and those who provide the funding, whether government agencies or corporations. Given that funding is always limited, research topics are generally chosen with great care to reflect the values of the funding agencies. The generation of scientific knowledge is indeed continually constrained by limited budgets so that there is no valid reason why it should not also be constrained in the future by ethical considerations. Second, the position that the acquisition of new knowledge should not be controlled implies that scientific freedom is the highest value of society and thus would have priority over all other values such as human welfare and environmental protection. Clearly, such a position is absurd. Science should always serve the interests of people and the environment rather than damaging them to serve special interests.[1165] Finally, the above "no-restraint" position also ignores the fact that, to a limited extent, there are already areas of scientific inquiry that are considered ethically unacceptable or at least questionable. As Joanna Kempner and colleagues write in a recent *Science* article titled "Forbidden Knowledge,"

> Forbidden knowledge embodies the idea that there are things that we should not know. Knowledge should be forbidden because it can only be obtained through unacceptable means, such as human experiments conducted by the Nazis; knowledge may be considered too dangerous, as with weapons of mass destruction or research on sexual practices that undermine social norms; and knowledge may be prohibited by religious, moral, and secular authority, exemplified by human cloning.[1166]

The most obvious way to exclude certain areas of scientific inquiry is to withhold funding. In addition, access could be denied to instrumentation that is necessary to conduct the ethically questionable research.[1167] However, as Robert Sinsheimer points out, none of these measures is likely to be implemented without the consensus of the scientific community:

> Unless a major portion of the scientific community believes that restraint is necessary, nothing will happen. For this to happen, that community will clearly have to become far more alert to, and aware of, and responsible for the consequences of their

activities. The best discipline is self-discipline. Scientists are keenly sensitive to the evaluations of their peers. The scientific community and the leaders of our scientific and technical institutions will have to develop a collective conscience. They will have to let it be known that certain types of research are looked upon askance, much as biological warfare research is today. It needs to be understood that such research will not be weighed in consideration of tenure and promotions. Societies need to agree not to sponsor symposia on such topics. All of these and similar measures short of law could indeed be very effective.[1168]

In conclusion, it would be much easier to prevent the generation of deleterious scientific knowledge at an early stage than to deal with the abuses and misuses that later result from it.

Toward a Critical Science and Engineering

Clearly, the enactment and enforcement of professional codes of ethics is a necessary first step. However, to successfully reform the science and engineering professions to the point that they truly function in the service of both people and environment, significant work and activism are needed. First, science and engineering education must include instruction in the ethical dimensions and social responsibilities of the profession. As Alan Drengson points out in *The Practice of Technology*, this is almost entirely absent in today's schools, colleges and universities:[1169]

As it is, we train people in specialized technical fields with... scant attention to the ethical requirements such power, skill, and knowledge demand. Current training fails to educate the whole person. As a result, we have people with powerful knowledge and skills who sell their services to the highest bidder, with no concern for their responsibilities to the community, society, or to the natural world. This incomplete training and education also results in the design, development, and use of technology practices that are destructive of most values, such as human health, freedom, community, and culture.[1170]

One relatively easy way to improve this situation would be to design rigorous courses on professional ethics and social responsibilities and to make successful completion of these courses strict requirements for graduation, both at the undergraduate and graduate levels. Such courses should cover all aspects of professional ethics, from adherence to ethical codes to solving ethical dilemmas. These course should also include the social, economic and political implications of science and engineering work; the history of science and engineering, including periods of social activism; an exposé of the myth of value-neutrality; and the urgent need for critical science and engineering.[1171] Ideally, for maximum impact, such courses should be tailored to the specific science or engineering discipline. For example, physics and astronomy majors should be made aware that many of them may later be offered careers in military R&D, and therefore should discuss how they feel about devoting their lives to building ever more lethal weaponry. Chemistry majors should be made aware of the environmental and health consequences of the new chemicals they may be employed to synthesize, develop and manufacture. Biology majors need to become aware of the dangers and abuses of genetic engineering, and engineering majors need to understand how their future inventions may result in unintended environmental, social and economic consequences.[1172]

Following graduation, after entering the workforce, scientists and engineers should stay "ethically active" throughout their careers, ensuring that they follow both the letter and spirit of their professional ethics codes and that their work is socially and environmentally responsible. There are several ways for scientists and engineers to remain engaged. First, recognizing that it is their duty of inform society about the implications of their work,[1173] they must become increasingly involved in public policy discussions and decisions related to science and technology.[1174] As a recent consensus document titled "On Being a Scientist: Responsible Conduct in Research," issued jointly by the United States National Academy of Sciences, National Academy of Engineering, and Institute of Medicine, advocates,

> If scientists do find that their discoveries have implications for some important aspect of public affairs, they have a responsibility to call attention to the public issues involved. They might set up a suitable public forum involving experts with different perspectives on the issue at hand. They could then develop a

consensus of informed judgement that can be disseminated to the public.... The important point is that science and technology have become such integral parts of society that scientists can no longer isolate themselves from societal concerns. Nearly half of the bills that come before Congress have a significant scientific or technological component. Scientists are increasingly called upon to contribute to public policy and to the public understanding of science. ... Concern and involvement with the broader uses of scientific knowledge are essential if scientists are to retain the public's trust.[1175]

Given that almost any scientific finding can be perverted, scientists must remain actively involved in order to ensure that their newly created knowledge is applied only for constructive purposes and not misused.[1176] As Dr. Arthur Galston, the anti-war activist chemist who, while an innocent graduate student, invented the defoliants that years later caused vast destruction in Vietnam, advises,

> In my view, the only recourse for a scientist concerned about the social consequence of his work is to remain involved with it to the end. His responsibility to society does not cease with publication of a definitive scientific paper. Rather, if his discovery is translated into some impact on the world outside the laboratory, he will, in most instances, want to follow through to see that it is used for constructive rather than anti-human purposes.... Science is now too potent in transforming our world to permit random fallout of the social consequences of scientific discoveries.[1177]

One way to prevent potential future problems is for scientists to become more actively involved in redirecting both corporate and governmental research programs to focus on constructive areas of investigation while avoiding participation in research that is morally questionable or is likely to generate knowledge that could easily be misused.[1178] For example, military research could be redirected to focus on the development of technologies that could be used only for defense or that are non-lethal. Similarly, much research in medicine could be redirected toward disease prevention.

When confronted with a situation in which new technological processes or products might present a serious danger to the public, science and engineering professionals should inform management so that corrective actions can be taken. If management is unwilling to take appropriate steps to address and reduce these risks and hazards, it is the duty of scientists and engineers to inform the public directly. The latter procedure is called "whistle-blowing," and, as was mentioned earlier, whistle-blowers are often avoided or rejected by their peers and very frequently fired. Therefore, to encourage greater public protection through whistle-blowing, it is important that effective whistleblower protection laws are passed and vigorously enforced, ideally with the help of professional societies, so that whistle-blowers are rewarded for their bravery rather than punished.[1179]

Finally, it is always possible for scientists and engineers to employ the strategy of passive resistance by withdrawing their support from ethically dubious projects. As Albert Einstein once said, "Organized power can be opposed only by organized power. Much as I regret this, there is no other way."[1180] Thus, science and engineering professionals should organize as a group to resist work that violates their professional ethics codes. As Howard Zinn notes in *A People's History of the United States*, the leaders of society depend on the cooperation of the people, and if the latter fail to cooperate, the leaders are utterly powerless:

> In a highly developed society, the establishment cannot survive without the obedience and loyalty of millions of people who are given small rewards to keep the system going: the soldiers and police, teachers and ministers, administrators and social workers, technicians and production workers, doctors, lawyers, nurses, transport and communication workers, garbagemen and firemen. These people—the employed, the somewhat privileged—are drawn into alliance with the elite. They become the guards of the system, buffers between the upper and lower classes. If they stop obeying, the system fails.[1181]

Thus, if scientists and engineers organized and chose not to work on morally dubious projects, the parties that provide funding would not be able to have these projects completed. Through massive non-cooperation, science and engineering professionals could tip the balance of power in their favor, and could use that power to redirect science and technology toward humane and environmentally constructive purposes.[1182]

Furthermore, scientists must be trained to recognize and resist the psychological tendency toward rationalization, normalization and "group think," often expressed as, "Everybody is doing it," "This is the way things are done around here,"[1183] or "If I don't do it, someone else will." Finally, scientists and engineers must also resist the urge to dissociate their thoughts and actions from their feelings. When asked whether he had ever considered the effects of a nuclear weapons explosion on innocent civilians, one US nuclear scientist responded, "If you start thinking about that too much, you are not going to be able to think... very creatively about new ways of making bombs."[1184]

Many prominent and highly respected scientists have urged other professionals not to work in certain areas of research and development. The famous Renaissance scientist Leonardo da Vinci is known to have invented the first submarine, but deliberately suppressed it "on account of the evil nature of men, who would practice assassination at the bottom of the sea."[1185] As was mentioned earlier, both the *Bulletin of Atomic Scientists* and Union of Concerned Scientists continue to offer forums in the United States to discuss and debate controversial matters related to scientific research and technology development.

In the absence of a binding and enforceable code of professional ethics, there will always be professionals who are willing to offer their services to the highest bidder, no matter how morally dubious the work may be. But even in the absence of an enforceable ethical code, it would be helpful if the best and brightest of scientists and engineers, those who hold advanced degrees and are particularly talented, were to resist the temptation to work in areas of questionable value. In the absence of their assistance, it would be much more difficult to carry out scientifically and technically challenging research and development that is ethically dubious. Their public resistance would encourage many other scientists and engineers to shun questionable or destructive projects.[1186]

For Further Thought

Chapter 1

1. To what extent are the long-term effects of a new technology predictable? Can the disciplines of history, anthropology, psychology and sociology contribute to the prediction of consequences? In what ways?
2. Give examples of the physical, chemical and biological interconnectedness of life on Earth. Select specific linkages and discuss what would happen if they were disrupted.
3. Consider Barry Commoner's third law of ecology, "nature knows best." Give examples of cases in which this principle was violated, in society and in your life.
4. Apply Barry Commoner's fourth law of ecology, "There is no such thing as a free lunch," to five major technologies commonly used in industrialized societies. List both benefits and costs. Are the benefits worth the costs?
5. Make a list of irreversible environmental, social and cultural consequences of modern technologies. Are you concerned about the irreversibility of these effects? If not, why not?
6. Why are "half knowledge" and "ignorance of ignorance" so dangerous?
7. Give examples of human-made scientific ignorance. List at least five major technological innovations and explain how their widespread use has increased ignorance in terms of dealing with their unintended consequences.

Chapter 2

1. What examples can you give of technologies that seem to have no present negative consequences? Can you think of potential future problems they might create?
2. How would you go about evaluating the social and cultural effects of a new technology?
3. Give an example of the second law of thermodynamics—show how a particular case of increase in order (neg-entropy) is associated with an equal or greater increase in disorder (entropy) elsewhere. What implications does the second law of thermodynamics have for sustaining technological development? Is it possible to have highly complex industrial societies without destroying the environment? What solution do you suggest?
4. Suppose you wished to ban the production of harmful synthetic organic chemicals. How would you proceed? What opposition would you expect in response to your proposals, and from whom? How would you respond to this opposition?
5. How would you address the problem of global climate change? What policies would you suggest, and what types of new technologies would you recommend? What opposition would you expect, and from whom?

6. What proposals would you put forward to halt species extinction? What opposition would you expect, and from whom?

7. How would you reform agriculture to make it more environmentally benign and sustainable?

8. Do you believe that the potential benefits of genetic engineering outweigh the risks? Why or why not? Carefully explain your position. Are you concerned that genetically modified organisms (GMOs) are allowed, and without product labeling, in many foods in the United States?

9. Given the many negative impacts of the automobile, would you be willing to give up or significantly reduce car travel? Would that even be possible? Are there alternative modes of transportation that would work for you?

10. If you are a scientist or engineer, would you feel comfortable developing new weapons systems? Carefully explain your position. Would you be in a position to control the way in which they would later be used? Is that of concern to you? Why or why not?

11. How would you go about reducing health care costs (as a percentage of GDP) while at the same time improving the health of the nation?

12. Are you concerned about human overpopulation? If you are concerned, what policies would you suggest to reduce the size of the human population, both worldwide and in the United States? How would population size be reduced if left solely to nature? Which would you prefer?

Chapter 3

1. How has modern technology interacted with the environment? Has the natural world been enhanced or damaged?

2. What is the relationship of human technology to other species? Give both positive and negative examples (e.g., veterinary medicine, habitat destruction and so forth). What are the long-term effects of these?

3. Give examples of how technologies increase the physical or psychological distance among humans and between humans and the natural environment. Does this technology-caused increase in physical (or psychological) distance encourage unethical exploitative behavior? What steps could be taken to minimize this problem?

4. After learning about the conditions on factory farms, do you still consider it ethical to eat meat? If so, why? What steps could you take to reduce the massive suffering of farm animals?

5. Given the extremely large per capita ecological footprint of Americans, do you believe that the goal of an American standard of living for everyone on Earth is even achievable? What solutions would you suggest?

6. Give examples of workers being controlled by the dictates of machine schedules or other technological systems. What are the likely effects of these work arrangements on the psychological well-being of the affected individuals? Can you suggest alternate and more humane ways to organize work?

7. Analyze the effects of television and movies on your behavior. Are your purchasing habits influenced by TV advertising? Do you view news as entertainment or as a call to action? If you were to reduce TV viewing by one hour per day, what would you do instead?

8. The late historian Howard Zinn writes in *The Power of Nonviolence*: "We [need]

new ways of thinking. A $300 billion military budget has not given us security. American military bases all over the world, our warships on every ocean, have not given us security. Land mines and a 'missile defense shield' will not give us security. We need to rethink our position in the world. We need to stop sending weapons to countries that oppress other people or their own people. We need to be resolute in our decision that we will not go to war, whatever reason is conjured up by the politicians or the media, because war in our times is always indiscriminate, a war against innocents, a war against children. War is terrorism, magnified a hundred times."[1187] Do you agree or disagree? What, in your opinion, would increase international security? What specific steps could be taken to prevent violent conflicts?

Chapter 4

1. List a number of counter-technologies (technologies developed in response to the adverse effects of previous technologies). Discuss in what ways these counter-measures may themselves generate problems.

2. Give reasons for the general preference to address symptoms with techno-fixes rather than focusing on root causes to achieve more permanent solutions.

3. Make a list of technologies that have been proposed for the mitigation of global climate change. Rate them in terms of their ability to address the root cause of the problem. Which are likely to produce negative environmental or social side effects?[1188] Based on your analysis, which technological solutions to climate change do you prefer? Are there non-technological solutions you could recommend?

4. Give examples of technologies that attempt to solve social, cultural, economic and political problems. Do these social fixes provide permanent solutions? What solutions would you recommend instead?

5. What motivates arms races? Who benefits and who suffers? What do you believe are the major causes of international conflict? Can you suggest alternative approaches to resolving differences between nations?

6. Give examples of medical treatments or drugs that address diseases caused by poor lifestyle choices. How effective and cost-efficient are these treatments? What alternative, more successful and more economical solutions would you suggest? What policies would you recommend that might motivate individuals to adopt a healthier lifestyle?

Chapter 5

1. Give examples of how technological innovations and efficiency improvements have increased the consumption of mineral and energy resources. Do you think that this trend is sustainable?

2. Consider the following example of the rebound effect: a mandated increase in the fuel efficiency of automobiles leads to more miles being driven, thereby negating some or all of the benefits of mandated fuel conservation. How could such a rebound effect be avoided?

3. Imagine living in an ultra-efficient technological society. Would that be preferable to the current situation? If not, why not?

4. Give examples of how excessive focus on efficiency strengthens materialistic

values and weakens non-materialistic ones. List specific non-materialistic values that are undermined and what effect this may have on society and the psychological well-being of individuals.

5. Consider Herman Daly's warning: "If our ends are perversely ordered, then it is better that we should be inefficient in allocating means to their service."[1189] How should society's ends be re-ordered to make efficiency improvements more constructive?

Chapter 6

1. Does advanced technology make a society less or more vulnerable to collapse? Develop arguments defending each point of view. Which arguments do you find most convincing? Is one point of view true in the short term and the other true in the long term?

2. What are real-world limits to human population growth? Is it safe to approach these limits? What are current indications that some of these limits have already been exceeded?

3. In view of the IPAT equation, consider this statement: "For long-term sustainability, we must not only reduce the size of our ecological footprint but also the number of feet." Do you agree or disagree? Why?

4. What would have the greatest effect in the pursuit of long-term sustainability: (a) recycling, (b) new technologies, (c) population reduction, (d) government regulation, (e) organic agriculture? Defend your choice. Show how it interacts with the others.

5. According to Paul Murtaugh and Michael Schlax, avoiding the birth of a single child in an industrialized nation has a much a greater effect on reducing CO_2 emissions than environmentally conscious consumer behavior. For example, it is estimated that an average citizen participating in personal conservation measures could reduce his/her lifetime carbon dioxide emissions by 486 tons, which is 20 times *less* than the CO_2 emissions avoided by choosing to have one fewer child.[1190] Will this information affect your choice of family size? Why or why not?

6. What are the consequences for individual citizens of economic and social collapse? Describe conditions in "failed states."

7. Give examples of the effects of human population growth on biodiversity.

8. Review the three sustainability conditions. Make a list of your daily activities (cooking, driving to work, shopping, etc.) and a list of your possessions, and determine whether they meet or violate these conditions. What changes would have to be made, at the individual and societal level, to meet these three sustainability criteria?

9. Globally, the number of humans is currently about 7 billion,[1191] and is expected to increase to about 9.2 billion by 2050.[1192] At this rate of growth, the world population will increase by approximately 59 million people per year, which is equivalent to adding a Los Angeles every three weeks or a country the size of the United Kingdom each year.[1193] Following China and India, the United States is the third most populous nation in the world.[1194] In contrast to other industrialized countries whose populations have more or less stabilized (e.g., Japan, European countries), the United States is still growing at a rapid and unsustainable pace. There are currently about 308 million people in the United States,[1195] and this number could increase to 438 million by

2050,[1196] equivalent to adding a city the size of Chicago[1197] each year. Do you feel comfortable with this rapid population growth in the United States? If not, what policies would you recommend that would reduce this unsustainable growth?[1198]

10. Take the online "Ecological Footprint Quiz" at myfootprint.org. Identify personal changes you can make to reduce your footprint. Retake the online quiz to determine how such changes would quantitatively reduce your environmental impact. Are you willing to make such changes? If not, why not?

11. Harvard's socio-biologist Edward O. Wilson warns, "The pattern of human population growth in the 20th century was more bacterial than primate. When *Homo sapiens* passed the six billion mark we had already exceeded by perhaps as much as 100 times the biomass of any large animal species that ever existed on land. We and rest of life cannot afford another 100 years like that."[1199] Do you agree? Why or why not?

12. Eating meat could be as damaging to the climate as driving a car. Consider, for example, that the production of 1 kg (2.2 pounds) of beef does as much damage to the climate as driving an energy-efficient car for 250 km (155 miles).[1200] Furthermore, a recent study by Gidon Eshel and Pamela Martin at the University of Chicago has demonstrated that an individual eating the standard meat-based American diet causes about 1.5 tons *more* greenhouse gas emissions (CO_2 equivalents) per year than a person eating a 100 percent vegetarian (vegan) diet. By comparison, an environmentally conscious consumer who chooses to drive a Toyota Prius (hybrid) instead of a Toyota Camry would reduce annual CO_2 emissions by only about 1 ton.[1201] Adopting a vegan diet has a greater impact on reducing global warming than buying a hybrid car. Why is there more emphasis on better cars than on better diets?

13. E. O. Wilson cautions, "The constraints of the biosphere are fixed. The bottleneck through which we are passing is real. It should be obvious to anyone not in euphoric delirium that whatever humanity does or does not do, Earth's capacity to support our species is approaching the limit.... But long before that ultimate limit was approached, the planet would surely have become a hellish place to exist."[1202] Do you agree? Why or why not?

14. "Frugality is a revolt against some basic values of the Sumptuous Society. For the sake of personal, social, and ecological well-being, frugality rejects the gluttonous indulgence, compulsive acquisitiveness, conforming and competitive consumerism, casual wastefulness, and unconstrained material economic growth promoted by the peddlers of economic 'progress' and embraced to different degrees by all of us who have known the enticements of affluence."[1203] Do you agree or disagree with this statement? Why or why not?

15. According to several estimates, the entire Earth can sustainably support only 1 to 2 billion people at an American standard of living, ensuring sufficient nutrition, health, personal dignity and freedom.[1204] For the United States, the optimum population level has been estimated to range from 40 to 100 million, which is less than one third of the current size.[1205] What do you think is the optimum population size for the United States and the world? Do you prefer an optimum that is biocentric (allowing room for all species to survive and thrive) or anthropocentric (allowing room for humans only)?

16. One common reason many countries, including the United States, have no population policies is concern over potential human rights violations. The

modern Western commitment to individualism, often socially irresponsible individualism, may lead Western observers to identify certain harmless techniques and incentive structures as violations of the rights of individuals. An international consensus must be developed in which the individual is considered not to have the right to endanger the future of his society, his nation, other nations or the Earth through his unwise choice with regard to reproduction. The human rights of future generations to a life free from wars, famine and disease caused by the excessive reproduction of their forebears should be guaranteed. Irresponsible reproduction should not be considered a "right" any more than any other act that endangers the well-being of society. Effective forms of incentive or social pressure, whether appealing or not to the western bias toward individualism, should not be gratuitously faulted.[1206] Do you agree or disagree? Why?

17. Gandhi once said, "Civilization, in the real sense of the term, consists not in the multiplication, but in the deliberate and voluntary reduction of wants."[1207] How do you define civilization?

Chapter 7

1. How do you personally define "progress"? What would "progress" look like in your own life; in your society; in the world?

2. Do you perceive time as linear or cyclical? What aspects of your life and society may have predisposed you to this perspective? Which do you feel more accurately describes reality? Are both true?

3. Are you a techno-optimist or a techno-pessimist? What life experiences and observations have led you to this orientation?

4. Evaluate a current technology from the point of view of (a) an economist and (b) an environmentalist. Why is it often unwise to use current trends as predictors of future trends?

5. What is wrong with this statement: "The costs of raw materials have fallen sharply over the period of recorded history.... These trends are the best basis for predicting the trends of future costs"?[1208] What important factors have been omitted?

6. What are some statements you have noticed in the media (radio, TV, internet) that reflect strong techno-optimism? How would you rewrite these to give a more balanced viewpoint?

7. What changes/reforms of the media might produce more realistic and balanced discussions of new technologies?

8. What are some potential negative consequences of excessive techno-optimism?

9. Give an example of how a particular belief, worldview or philosophy has produced real world consequences.

Chapter 8

1. To what extent should new technologies (including new chemicals) be evaluated before they are widely distributed? What do you feel would constitute adequate testing?

2. Choose three major technologies in common use. Select the following boundaries for analysis: (a) yourself in the present, (b) yourself over your en-

tire lifetime, (c) your nation in the present, (d) humanity into the indefinite future and (e) all species into the indefinite future. For each case, list both costs and benefits. How does the perceived cost-benefit ratio change as the boundaries of the analysis expand? Can you suggest policies for the evaluation and adoption of new technologies?

3. What is your view of the rights of future generations? Is it fair to externalize costs to future generations? Why, compared to certain Native Americans who considered the impacts of their decisions to the "seventh generation," do Western policy makers rarely consider effects on the future of humanity?

4. Listen to and read about the public debates on new or questionable technologies such as genetic engineering, nano-technology or "planetary" engineering. Identify the potential biases (personal, business, institutional, cultural) of the various stakeholders involved in the disputes (see Huesemann (2002)). Which side of the issues do you support? Why?

5. Regarding the monetization of intangible values, consider the conclusion of the National Academy of Sciences: "There is no satisfactory way to summarize all the costs or benefits of regulatory options in dollars or other terms which can be mathematically added, subtracted, or compared."[1209] Why, despite this condemnation, is the monetization of benefits and costs still used to compare alternate technologies or policy options? How does the monetization of intangible values bias the decision-making process, particularly with respect to the development of new technologies? Can you suggest alternatives?

6. Conduct your own "Ivan Illich" analysis of automobile travel speed by dividing your annual miles driven by the total number of hours per year spent driving and working to pay for car transportation-related expenses. How does "travel speed" change with income? Can you suggest alternative transportation solutions?

7. It is believed that in order to control spiraling health care costs, it will be necessary to reimburse health care providers only for effective medical treatments. This would require the establishment of a national center for medical technology assessment, whose mission would be to evaluate all medical treatments in an objective and unbiased fashion in terms of their efficacy and cost-effectiveness. If implemented, numerous expensive and ineffective treatments would likely be discontinued, resulting in significant savings in health care costs. Unfortunately, previous attempts, such as the establishments of the Peer-review Office in the Public Health Service (1970s), the National Center for Health Care Technology (under President Carter) and the Office of Technology Assessment (1972-1995), have all failed.[1210] What entities are likely to oppose the establishment of a national center for medical technology assessment? What solutions do you suggest?

Chapter 9

1. How could engineers use their talents more constructively instead of continuing to create consumer products?

2. Is it ethical for psychologists to use their talents in the advertising industry?

3. What is your opinion of consumerism? How can you turn your opinion into action?

4. Make a list of your needs, classify them according to Maslow's and Max-Neef's proposed categories, and discuss how you satisfy them. Make a list of your wants and desires. Have they changed with time? What is their origin? Have you been successful in satisfying them? Could you do without some of them? If not, why not?

5. List at least five consumer technologies and identify which genuine needs or culturally conditioned wants they attempt to satisfy. Are there any non-technological alternatives?

6. Give examples of technological innovations that originally were non-essential but later became genuine needs. How does advertising create desires?

7. Give examples of rapid obsolescence in common consumer goods. Would you be satisfied with a high-quality product even though it is not the latest model? Why or why not?

8. For one day, count the number of radio and TV commercials you encounter. What values do they reflect? What wants do they stimulate? Do they make you dissatisfied with what you have or with your life? Do you believe you are influenced by these ads? If so, are you concerned about that? Do you believe other people are influenced?

9. Evaluate the type of products and services you purchase. Do any serve to exhibit your social status? Are there other, non-materialistic ways for you to increase your self-esteem?

10. Research has shown that materialistic consumerism is negatively correlated with psychological well-being and happiness. Test this finding by evaluating your short-term and long-term levels of happiness following purchase of specific consumer goods. How long did your immediate positive experience last? Did your overall long-term happiness increase?

11. For one week, keep a log of the hours you watch TV and movies or surf the web for entertainment and news. Do you think this is the best use of your time? If not, what else could you do to improve your sense of well-being?

12. A sense of community has been found to enhance psychological well-being. There are numerous ways to nurture community: buy locally at farmers markets, from food co-ops and from family businesses; support community banks and credit unions; grow food together in urban gardens; organize local festivals and neighborhood picnics; use the local library; take children to the park; organize pot luck dinners; bake extra and share; help lost pets; know your neighbors; greet people; and turn off your television.[1211] Make a list of actions you could personally take to improve community spirit and the quality of relationships in your community.

13. It has been suggested that although the proposal to significantly reduce material consumption may sound draconic, it can easily be done without adversely affecting one's sense of well-being. The first step is to clearly identify what are needs (necessary) and wants (unnecessary). If people focus on the satisfaction of needs and ignore wants, material consumption is likely to decline substantially. It is important to satisfy basic non-material needs, such as the need for acceptance, esteem, self-actualization and spiritual growth, directly instead of attempting this indirectly (and in vain) through material consumption.[1212] A study in the United Kingdom has shown that at least half of all material consumption is currently used in the attempt to satisfy non-

material needs.[1213] Thus, if these psychological, social and spiritual needs were to be satisfied directly, material consumption could be reduced by 50 percent while at the same time increasing a sense of well-being and happiness.[1214] As Donella Meadows and colleagues state,

> People don't need enormous cars; they need admiration and respect. They don't need a constant stream of new clothes; they need to feel that others consider them to be attractive, and they need excitement and variety and beauty. People don't need economic entertainment; they need something interesting to occupy their minds and emotions. And so forth. Trying to fill real but nonmaterial needs—for identity, community, self-esteem, challenge, love, joy—with material things is to set up an unquenchable appetite for false solutions to never satisfied longings. A society that allows itself to admit and articulate its nonmaterial human needs, and to find nonmaterial ways to satisfy them, would require much lower material and energy throughputs and would provide much higher levels of human fulfillment.[1215]

Outline steps to reduce your consumption of products and services while maintaining or even increasing your sense of well-being.

14. Independent of media content, television viewing has been shown to be detrimental to family and community life because it is a solo activity that isolates people from each other (see Chapter 10). Thus, one of the simplest ways to increase happiness and psychological well-being may very well be to turn off the TV and computer to free up time for interacting with family and community. Conduct the following experiment: schedule one TV- and Internet-free day per week, and slowly increase the number of TV-free days.[1216] Record your experiences. Did you feel better or worse? If you are leaning toward activism, you could organize TV-free weekends for whole neighborhoods and communities in an effort to increase people's participation in social events such as picnics, festivals and concerts.[1217] How would you go about planning this? Do you think it will be easy or difficult? Why?

15. One way to increase psychological well-being in society as a whole is to design and implement public policies specifically for this purpose. Currently, the overriding goal of industrialized nations is the promotion of economic growth, an increase in material affluence and the maximization of profits. As industrial societies have become wealthier and rising affluence has failed to increase people's happiness, it becomes necessary to abandon these outdated and ineffective policies and replace them with others that focus specifically on increasing the overall sense of well-being and happiness in society.[1218] A first step would be to devise a new set of indicators for subjective well-being and begin tracking them over time.[1219] It is a well-known management principle that "you get what you measure."[1220] If there were a set of indicators of well-being, it could be measured and reported periodically, as is done with the economy. Rudimentary measures of well-being are already being tracked in Europe (e.g., the German Socioeconomic Panel, a large annual survey of life satisfaction in Germany, and the Eurobarometer, which is conducted regularly in the European Union, include questions related to sense of well-being).[1221] It has been proposed that a national well-being index in the United States could be devised that would systematically assess key variables in

representative samples of the population, such as positive and negative emotions, engagement, perceived life purpose and meaning, optimism and trust, and so forth.[1222]

Suggest survey questions that could be used to track the well-being of an entire nation. What changes in public policy would you recommend to improve a nation's overall "happiness"?

Chapter 10

1. Project: try living one week without your cell phone, without watching TV, without accessing the Internet or using your computer or other electronic device. Record your experiences.
2. Give examples of at least three technologies that embody the values of power, control or exploitation.
3. When watching TV, make a log for one day, listing the values promoted there. Do you agree with those values? If not, what can be done?
4. Give examples of technological megasystems. How do you think they came into being? Is it possible to dismantle them? How?
5. Give an example that you have encountered of the "technological imperative" being expressed to defend the introduction of a new technology. Are any social, environmental or ethical considerations being ignored?
6. Make a list of technologies you depend on. What would happen if a particular technology were no longer available? Are there other alternatives/substitutes?
7. What technologies of those you frequently encounter might not have been adopted in the absence of heavy advertising?
8. List a few technologies that appear to have an inherent social or political bias.
9. What technologies have you encountered that seem to embody the values of cooperation, non-violence and justice? Would you consider these to be "appropriate" technologies?

Chapter 11

1. Select several common technologies or products. Who was responsible for developing them? To what extent was the public engaged? How could the public be more involved in influencing the direction of technological innovation?
2. Consider again Jerry Mander's statement: "Profit is based on paying less than the actual value for workers and resources.... Profit is based on underpayment.... This is called exploitation."[1223] Do you agree or disagree with this statement? How would you design a different economic system?
3. Give five examples of the common conflict between profit maximization and social responsibility from the perspective of a manager or an entire corporation. How would you resolve such conflicts? Who should decide?
4. How would you reform the current system of industrialized agriculture and food manufacturing to better meet the nutritional needs of people, both rich and poor?
5. How would you reform the practice of medicine to improve the health not only of certain individuals but of the entire nation? What policies would you recommend to motivate a change in this direction?
6. Fortunately, individuals have complete control over what they eat. Unfortunately, there is confusion as to what constitutes a healthy diet. Part of this con-

fusion is caused by authors of popular but unproven diet regimens that exploit desperate dieters to make quick book sales. In addition, the food industry promotes many of its highly processed products as essential to health, but the list of what are to be considered healthy ingredients changes constantly. This confusion is of tremendous benefit to agribusinesses and the food industry, as their profits and very existence would be threatened if a majority of people were to decide that a simple but varied, mostly vegetarian diet consisting of organic unprocessed grains, vegetables and fruits would be best in terms of promoting health and preventing animal cruelty.[1224] What are your criteria for deciding what constitutes a healthy diet? What are your sources of information? Can you trust them? Would you be willing to change your current diet if you found reliable evidence that it is unhealthy? Why or why not?

7. Compare the advantages and disadvantages of the current for-profit health care system in the United States with the not-for-profit systems in Europe and Canada. How would you design a national health care system with the explicit goal of improving the nation's health? What policies (i.e., "carrots and sticks") would you suggest for motivating individuals to adopt more healthy habits?

8. President Eisenhower called the billions spent on the preparation for war "a theft" from those who are without food and shelter.[1225] Similarly, Martin Luther King warned that "a nation that continues year after year to spend more money on military defense than on programs of social uplift is approaching spiritual death.[1226] Do you favor a reallocation of funds to better address social and environmental needs? How would you spent such a dividend? (For more information, see Brown (2008), Plan B, Chapter 13, Table 13-2, p. 282; Table 13-3, p. 285, earth-policy.org/index.php?/books/pb4).

Chapter 12

1. Describe the worldview implicit in Western technological society. Is it the same as your own worldview? If not, how does it differ?

2. Give an example of how a particular belief, worldview or philosophy has produced real-world consequences.

3. Describe some paradigm conflicts that you have observed in Western society (e.g., creationism vs. Darwinism, climate change denial vs. climate change awareness).

4. What messages communicated by the mass media would tend to increase one's feelings of alienation and separation?

5. What actions could individuals or groups take to foster social interconnectedness?

6. What entities in society stand to gain from excessive human population increase?

7. What entities in society would (and do) oppose moving to a steady-state economy?

8. How would you define a "biocentric" optimum for the size of the human population?

9. At what level would you place yourself in Korten's "levels of consciousness"? What steps would have you to make in order to move up one level? Two levels? What hurdles would you have to overcome?

10. Make a list of actions you personally could take to improve your own health and decrease the probability of disease.
11. What specific public policies could be implemented to improve the health of the nation?
12. What policies could governments use to reduce material consumption?
13. Clive Hamilton suggests in *Growth Fetish*,

> The transition to a post-growth society would begin by imposing restrictions on the quantity and nature of marketing messages, by first banning advertising and sponsorship from all public spaces and restricting advertising time on television and radio. Tax laws could be changed so that the costs of advertising are no longer a deductible business expense but come out of profits. Second, we should demand legislation requiring truth in advertising, a move that would require nothing more than enshrining in legislation and enforcing the industry's own code of conduct under which advertisers are banned from making misleading claims or ascribing to products properties that they do not possess.[1227]

Do you agree or disagree with this proposal? Why?
14. Given that much of the material consumption in modern industrialized societies serves primarily to display social status, the associated pollution and matter-energy use could be reduced significantly if non-materialistic ways were developed for the expression of status. Here we could learn from the world's militaries, which have for centuries successfully employed very simple, inexpensive symbols such as stars and colored stripes on uniforms to exhibit rank. It has been suggested, for example, that the most considerate, charitable, environmentally conscious and frugal could be accorded the highest social status with certain symbols. Suggest other innovative ways for people to design and exhibit social rank in non-materialistic ways.
15. Audrey Chapman writes in *Consumption, Population, and Sustainability*,

> Most religious traditions have teachings that discourage overconsumption and criticize greed and lack of sharing.... For example, Christian scriptural teachings that denigrate materialism and emphasize spirituality and sharing may have little to do with contemporary religious practice. While Jesus emphasized that it is not possible to serve two masters, God and mammon, many American Christians believe that wealth is a sign of God's favor.[1228]

How would you explain this American interpretation of Christian scripture?
16. It has been suggested that a life of voluntary simplicity and frugality, if adopted by millions of people, could become a serious threat to consumerism.[1229] This could occur without a conflict or confrontation between adherents of frugality and the status quo. If many people were to bypass the current consumer culture by ignoring advertisements and buying only necessary items, ideally from local farmers and merchants, the system of corporate-controlled consumerism would collapse as a result of benign neglect and lack of financial support.[1230] Do you agree or disagree? Why? If you agree, what actions could you take to simplify your life and reduce material consumption?

Chapter 13

1. Explain the "precautionary principle." Do you think it suggests a responsible approach in the absence of important information? Why or why not?

2. In what aspects of life is application of the precautionary principle especially important? What do these aspects have in common?

3. Make a list of what you put into the garbage. What proportion is biodegradable, recyclable or inert? How could you change your shopping habits to reduce your garbage? What changes could manufacturers make? Why has this not been done?

4. List the technologies that are around you and that you are using on a daily basis. Do they meet the criteria of social appropriateness? If not, why not? Are there changes that would make them more socially appropriate? What is your definition of "socially appropriate"?

5. Do you consider your work satisfying and meaningful? What would have to change to make work more fulfilling? Are any changes in technology necessary?

6. Do you agree with John Cavanagh and Jerry Mander's suggestion that no new technologies should be allowed if insurance companies are unwilling to cover them? What technologies, past and present, are unlikely to receive insurance coverage? Why?

7. Make a list of major technologies around you. To what extent did their development reflect the will of the people? Do you think they would they have been developed if a democratic vote had been taken?

8. Consider the potential for growing local organic food. During World War II, Americans planted over 20 million Victory Gardens, and the harvest accounted for nearly a third of all vegetables consumed in the United States in 1943.[1231] A recent United Nations FAO (Food and Agriculture Organization) project has created 8,000 microgardens, one square meter each, for the disadvantaged citizens of Caracas, Venezuela. Each square meter, if continuously cropped and supplied with compost fertilizer, can produce 330 heads of lettuce, 18 kilograms of tomatoes, or 16 kilograms of cabbage per year.[1232] Another option is to create larger urban community gardens, in vacant lots, public parks or even on rooftops.[1233] There is great potential for urban gardening in the United States. For example, a survey has indicated that Chicago and Philadelphia have about 70,000 and 31,000 vacant lots, respectively, many of which could be converted to urban gardens. Nationwide, hundreds of thousands of vacant lots in cities could be used for growing organic food.[1234] As Ted Trainer writes in *The Conserver Society*, gardening has great potential for producing most, if not all, of our food in an environmentally sustainable way:

> It takes about two hectares of crop and rangeland for agribusiness to feed the average American, but one hectare of intense home gardening along permaculture lines can provide all the food (excluding meat) necessary for at least 12 and possibly 25 people, at no cost in non-renewable fertilizer and energy. We would have little difficulty producing all the food and natural materials we need within and close to our settlements, in much more pleasant and ecologically sustainable ways that we now produce them.[1235]

Suggest personal actions and public policies that would encourage an entire nation to grow local organic food.

9. How could citizen involvement in "participatory design" of technological innovations be introduced in Western societies? Discuss the Swedish example.
10. What current technologies in our society lend themselves to undemocratic ends (e.g., data gathering, citizen monitoring)?
11. List five potential or existing technologies that you feel are socially and environmentally appropriate.

Chapter 14

1. Are all disciplines within science and technology equally at risk of creating unfortunate consequences? If not, which ones require more careful evaluation in this regard?
2. Give examples of positive and negative consequences of research and development in five different disciplines.
3. Has the knowledge of human psychology been abused by the advertising industry? If so, what could psychologists do to remediate this situation?
4. Within your own discipline, what positive uses as well as abuses do you recognize?
5. What kind of future would you like to create for yourself and your descendants?
6. What information would help you predict future adverse consequences that could potentially be generated by your own work, profession or discipline?
7. Have you ever had to choose whether or not to become a whistleblower? What did you do? What led you to the choice you made?
8. Consider Hermann Gossen's situation, which he described in 1853:

> I close with the wish that my work would receive rigorous but impartial examination. I am in a strong position to insist on this request because I was forced to fight so many mistaken ideas generally considered correct, ideas that have thus become so much more dear to the heart of many, yes, very many people, because their position in life is partly or wholly dependent upon accepting these ideas as true. Giving up these ideas would put them in the situation in which I now find myself, namely, at a mature age to have to look for a new position.[1236]

How many of your own ideas do you know are convenient but not true?
9. Will the work you are doing, or being trained to do, in all honesty contribute to the well-being of society and the environment? How? Does it have a potential for negative effects? What are these?
10. Examine your own motivation: What percent of your interest is in enhancing yourself (income, prestige, power, entertainment) and what percent of your motivation is driven by the wish to help (other people, other species, society as a whole or the environment)? Are you satisfied with these percentages?
11. Do you have special advantages or abilities that you believe increase your responsibility to society?
12. What values are reflected in different research topics/areas/programs?

13. Is there a standard professional ethics code in your discipline? How could it be improved? Is it enforced?

14. Worldwide, at least half a million scientists are carrying out weapons research, and this accounts for about 50 percent of all research and development expenditures.[1237] What ethical considerations do you believe should be applied? Are scientists able to control the uses to which their research is put? If not, what would you suggest?

15. It has been said that in large-scale, professional enterprises, such as military research and development, "scientists and engineers need to take responsibility for their work, even if they contribute only a small 'piece of the puzzle.'"[1238] How can such responsibility be exercised?

16. "If I don't do the work, they will fire me and hire instead somebody else who will."[1239] How would you respond to this statement?

Bibliography

Abbasi, S.A. et al. 1995. "Environmental Impact of Non-Conventional Energy Sources." *Journal of Scientific and Industrial Research* 54:285-293.

Abraham, M. 2006. "Sustainability—Philosophy vs. Engineering Tools." *Environmental Progress* 25 (2):87-88.

Adriaanse, A., et al. 1997. *Resource Flows—The Material Basis of Industrial Economies.* Washington, DC: World Resources Institute.

Advocates of Peace. 2002. *The Power of Nonviolence—Writings by Advocates of Peace.* Boston, MA: Beacon Press.

Alford, C.F. 2002. *Whistleblowers—Broken Lives and Organizational Power.* Ithaca, NY: Cornell University Press.

Alfredsson, E.C. 2004. "Green Consumption—No Solution for Climate Change." *Energy* 29:513-524.

Allen, S. 2007. "Critics Blast Slow Progress on Cancer." *The Boston Globe*, December 2, 2007.

Allenby, B.R. 1999. *Industrial Ecology—Policy Framework and Implementation.* Englewood Cliffs, NJ: Prentice Hall.

Allenby, B.R., and D.J. Richards, eds. 1994. *The Greening of Industrial Ecosystems.* Washington, DC: National Academcy Press/National Academy of Engineering.

Alloy, L.B., and L.Y. Abramson. 1979. "Judgement of Contingency in Depressed and Nondepressed Students—Sadder but Wiser?" *Journal of Experimental Psychology* 108 (4):441-485.

Altieri, M.A. 1995. *Agroecology—The Science of Sustainable Agriculture.* Boulder, CO: Westview Press.

Alvord, K. 2000. *Divorce Your Car.* Gabriola Island, BC: New Society Publishers.

Anderson, C.A., and B.J. Bushman. 2002. "The Effects of Media Violence on Society." *Science* 295 (March 29):2377-2379.

Anonymous. 1999. *Technology Triumphs, Morality Falters.* Washington, DC: Pew Research Center for the People and the Press.

———. 2000. "Who Wants To Live For Ever?" *The Economist* (December 23):23-24.

Argyle, M. 1987. *The Psychology of Happiness.* New York, NY: Methuen and Co., Ltd.

Armstrong, B., and R. Doll. 1975. "Environmental Factors and Cancer Incidence and Mortality in Different Countries, with Special Reference to Dietary Practices." *International Journal of Cancer* 15:617-631.

Arrow, K., B. Bolin, R. Costanza, P. Dasgupta, C. Folke, C.S. Holling, B.-O. Jansson, S. Levin, K.-G. Maeler, C. Perrings and D. Pimentel. 1995. "Economic Growth, Carrying Capacity, and the Environment." *Science* 268 (April 28):520-521.

Ashkinazy, A. 1972. "Are Engineers Responsible for the Uses and Effects of Technology?" *Professional Engineer* August:46-47.

Atkins, P.W. 1984. *The Second Law.* New York, NY: W.H. Freeman and Company.

Ausubel, J. 1996. "Can Technology Spare the Earth?" *American Scientist* 84 (2): 166-178.

Ausubel, J.H., and A. Gruebler. 1995. "Working Less and Living Longer—Long-Term Trends in Working Time and Time Budgets." *Technological Forecasting and Social Change* 50:113-131.

Ausubel, J.H., and H.E. Sladovich, eds. 1989. *Technology and Environment.* Washington, DC: National Academy Press.

Ayres, R., and K. Martinas. 1995. "Waste Potential Entropy—The Ultimate Eco-toxic?" *Economie Appliquee* 48:95-120.

Ayres, R.U. 1994. "Industrial Metabolism—Theory and Policy" in *The Greening of Industrial Ecosystems,* edited by B.R. Allenby and D.J. Richards. Washington, DC : National Academy Press.

——.1995. "Economic Growth—Politically Necessary but Not Environmentally Friendly." *Ecological Economics* 15:97-99.

——.1998. "Eco-thermodynamics—Economics and the Second Law." *Ecological Economics* 26:189-209.

——.1996a. "Statistical Measures of Unsustainability." *Ecological Economics* 16: 239-255.

——.1996b. "Limits to Growth Paradigm." *Ecological Economics* 19:117-134.

——.1999. "The Second Law, the Fourth Law, Recycling and Limits to Growth" *Ecological Economics* 29 (3):473-483.

Ayres, R.U., and L.W. Ayres. 1996. *Industrial Ecology—Towards Closing the Materials Cycle.* Brookfield, VT: Edward Elgar Publishing Company.

Ayres, R.U., and C.J.M. van den Bergh. 2005. "A Theory of Economic Growth with Material/Energy Resources and Dematerialization—Interaction of Three Growth Mechanisms."*Ecological Economics* 55:96-118.

Ayres, R.U., and I. Nair. 1984. "Thermodynamics and Economics." *Physics Today* (November):62-71.

Ayres, R.U., and U.E. Simonis, eds. 1994. *Industrial Metabolism—Restructuring for Sustainable Development.* New York, NY: United Nations University Press.

Ayres, R.U., and B. Warr. 2005. "Accounting for Growth—The Role of Physical Work." *Structural Change and Economic Dynamics* 16:181-209.

Baark, E., and U. Svedin. 1988. *Man, Nature and Technology—Essays on the Role of Ideological Perceptions.* London, UK: The MacMillan Press LTD.

Bacon, Francis. 1620. *Novum Organum.* Basil Montague, editor and translator, 1854. *The Works of Francis Bacon,* Volume 3. Philadelphia, PA: Parry and MacMillan. http://history.hanover.edu/texts/Bacon/novorg.html, accessed 3/8/2007.

Balzhiser, R.E., M.R. Samuels and J.D. Eliassen. 1972. *Chemical Engineering Thermodynamics—The Study of Energy, Entropy, and Equilibrium.* Englewood Cliffs, NJ: Prentice Hall.

Barbour, I.A. 1980. *Technology, Environment, and Human Values.* New York, NY: Praeger.

Barnaby, W. 2000. "Science, Technology, and Social Responsibility." *Interdisciplinary Science Reviews* 25 (1):20-23.

Barnard, N.D., A. Nicholson and J.L. Howard. 1995. "The Medical Costs Attributable to Meat Consumption." *Preventative Medicine* 24:646-655.

Barsam, A.P. 2008. *Reverence for Life—Albert Schweitzer's Great Contribution to Ethical Thought.* New York, NY: Oxford University Press.

Barsamian, D., and N. Chomsky. 2001. *Propaganda and the Public Mind—Conversations with Noam Chomsky.* Cambridge, MA: South End Press.

Baye, M.R. 2006. *Managerial Economics and Business Strategy.* New York, NY: McGraw-Hill Irwin.

Beaudreau, B.C. 2005. "Engineering and Economic Growth." *Structural Change and Economic Dynamics* 16:211-220.

Beckerman, W. 1996. *Through Green-Colored Glasses—Environmentalism Reconsidered.* Washington, DC: Cato Institute.

Beckwith, J. 1972. "Science for the People." *Annals of the New York Academy of Sciences* 196 (4):236-240.

Beckwith, J., and F. Huang. 2005. "Should We Make a Fuss? A Case for Social Responsibility in Science." *Nature Biotechnology* 23 (12):1479-1480.

Bedient, P.B., H.S. Rifai and C.J. Newell. 1994. *Ground Water Contamination—Transport and Remediation.* Englewood Cliffs, NJ: Prentice Hall.

Beecher, H.K. 1955. "The Powerful Placebo." *Journal of the American Medical Association* 159 (17):1602-1606.

———. 1961. "Surgery as Placebo—A Quantitative Study of Bias." *Journal of the American Medical Association* 176 (13):1102-1107.

Bender, W.H. 1994. "An End Use Analysis of Global Food Requirements." *Food Policy* 19 (4):381-395.

Bentzen, J. 2004. "Estimating the Rebound Effect in US Manufacturing Energy Consumption." *Energy Economics* 26:123-134.

Bereano, P.L., ed. 1976. *Technology as a Social and Political Phenomenon.* New York, NY: John Wiley and Sons.

Bergh, J.C.J.M. van den, and J. van der Straaten, eds. 1994. *Toward Sustainable Development—Concepts, Methods, and Policy.* Washington, DC: Island Press.

Berry, W. 1977. *The Unsettling of America—Culture and Agriculture.* San Francisco: Sierra Club Books.

Bezdek, R.H. 1993. "The Environmental, Health, and Safety Implications of Solar Energy in Central Station Power Production." *Energy* 8 (6):681-685.

Bezdek, R.H., R. Wendling, G.E. Bennington and H.R. Chew. 1982. "National Goals for Solar Energy—Economic and Social Implications." *National Resources Journal* 22:337-360.

Bianciardi, C., E. Tiezzi and S. Ulgiati. 1993. "Complete Recycling of Matter in the Frameworks of Physics, Biology, and Ecological Economics." *Ecological Economics* 8:1-5.

Binswanger, M. 2001. "Technological Progress and Sustainable Development—What About the Rebound Effect? "*Ecological Economics* 36:119-132.

Birdsall, N. 1994. *Another Look at Population and Global Warming.* Paper read at United Nations Expert Group Meeting on Population, Environment, and Development, at United Nations Headquarter, New York City, NY.

Blaisdell, B., ed. 2003. *The Communist Manifesto and Other Revolutionary Writings.* Mineola, NY: Dover Publications, Inc.

Blum, W. 2000. *Rogue State—A Guide to the World's Only Superpower.* Monroe, ME: Common Courage Press.

Board on Agriculture and National Research Council. 1989. *Alternative Agriculture—Committe on the Role of Alternative Farming Methods in Modern Production Agriculture.* Washington, DC: National Academy Press.

Boardman, A.E., D.H. Greenberg, A.R. Vining and D.L. Weimer. 1996. *Cost-Benefit Analysis—Concepts and Practice.* Upper Saddle River, NJ: Prentice Hall.

Borgmann, A. 1984. *Technology and the Character of Contemporary Life—A Philosophical Inquiry.* Chicago, IL: The University of Chicago Press.

Bosch, R. van den. 1978. *The Pesticide Conspiracy.* Garden City, NY: Doubleday & Company.

Boserup, A., and A. Mack. 1975. *War Without Weapons—Non-Violence in National Defense.* New York, NY: Schocken Books.

Boudon, R. 2001. *The Origin of Values—Sociology and Philosophy of Beliefs.* New Brunswick, NJ: Transaction Publishers.

Boyd, E.S., and M. Konner. 1985. "Paleolithic Nutrition—A Consideration of its Nature and Current Implications." *New England Journal of Medicine* 312 (5): 283-289.

Boyden, S. 1973. "Evolution and Health." *The Ecologist* 3:304-309.

Boyle, G., ed. 1996. *Renewable Energy—Power for a Sustainable Future.* Oxford, UK: Oxford University Press.

Bradshaw, G.A., and M. Bekoff. 2001. "Ecology and Social Responsibility—The Re-Embodiment of Science." *Trends in Ecology and Evolution* 16 (8):460-465.

Braine, D., and H. Lesser. 1988. *Ethics, Technology, and Medicine.* Aldershot, UK: Gower Publishing Company.

Braun, E. 1995. *Futile Progress—Technology's Empty Promise.* London, UK: Earthscan Publications Ltd.

Brooks, D. 2003. "The Triumph of Hope Over Self-Interest." *The New York Times,* January 12, 2003, 15.

Brower, M. 1992. *Cool Energy—Renewable Solutions to Environmental Problems.* Cambridge, MA: MIT Press.

Brower, S. 1988. *Sharing the Pie.* Carlisle, PA: Big Picture Books.

Brown, L.R. 2008. *Plan B 3.0—Mobilizing to Save Civilization.* Washington DC: Earth Policy Institute.

Brown, L.R., C. Flavin, S. Postel and L. Starke, eds. 1990. *State of the World 1990—A Worldwatch Institute Report on Progress Toward a Sustainable Society.* New York, NY: W.W. Norton and Company.

Brown, S. 1987. *The Causes and Prevention of War.* New York, NY: St. Martin's Press.

Bruni, L., and P.L. Porta, eds. 2005. *Economics and Happiness—Framing the Analysis.* Oxford, UK: Oxford University Press.

Burhop, E.H.S. 1971. "The British Society for Social Responsibility in Science." *Physics Education* 6:140.

Burns, D.D. 1999. *Feeling Good—The New Mood Therapy.* New York, NY: Harper.

Burroughs, J.E., and A. Rindfleisch. 2002. "Materialism and Well-Being—A Conflicting Values Perspective." *Journal of Consumer Research* 29:348-370.

Calaprice, A. 1996. *The Quotable Einstein.* Princeton, NJ: Princeton University Press.

Callahan, D. 1990. *What Kind of Life—The Limits of Medical Progress.* New York, NY: Simon and Schuster.

———. 1999. *False Hopes—Overcoming the Obstacles to a Sustainable, Affordable Medicine.* Piscataway, NJ: Rutgers University Press.

Callicott, J.B. 1982. "Traditional American Indian and Western European Attitudes Toward Nature—An Overview." *Environmental Ethics* 4 (Winter):293-318.

Campbell, C.J., and J.H. Laherrere. 1998. "The End of Cheap Oil." *Scientific American* (March):78-83.

Carlson, R. 2006. *You Can Be Happy No Matter What—Five Principles for Keeping Life in Perspective.* 15th ed. Novato, CA: New World Library.

Carlsson-Kanyama, A. 1998. "Climate Change and Dietary Choices—How Can Emissions of Greenhouse Gases from Food Consumption Be Reduced?" *Food Policy* 23 (3/4):277-293.

Carpenter, D.O. 1998. "Human Health Effects of Environmental Pollutants—New Insights." *Environmental Monitoring and Assessment* 53:245-258.

Carson, R. 2005 *Silent Spring.* New York, NY: Houghton Mifflin.

Catton, W.R., and R.E. Dunlap. 1980. "A New Ecological Paradigm for Post-Exuberant Society." *American Behavioral Scientist* 24 (1):15-47.

Cauthen, K. 1975. *The Ethics of Enjoyment—The Christian's Pursuit of Happiness.* Atlanta, GA: John Knox Press.

Cavanagh, J., and J. Mander, eds. 2004. *Alternatives to Economic Globalization.* San Francisco, CA: Berrett-Koehler Publishers, Inc.

Chalkley, A.M., E. Billett, and D. Harrison. 2001. "An Investigation of the Possible Extent of the Re-Spending Rebound Effect in the Sphere of Consumer Products." *The Journal of Sustainable Product Design* 1:163-170.

Chang, E.C., ed. 2001. *Optimism and Pessimism—Implications for Theory, Research, and Practice.* Washington, DC: American Psychological Association.

Chapman, A.R., R.L. Petersen and B. Smith-Moran, eds. 2000. *Consumption, Population, and Sustainability—Perspectives from Science and Religion.* Washington, DC: Island Press.

Chertow, M.R. 2001. "The IPAT Equation and Its Variants." *Journal of Industrial Ecology* 4 (4):13-29.

Clark, C.W. 1997. "Renewable Resources and Economic Growth." *Ecological Economics* 22:275-276.

Clark, R. 1972. *The Science of War and Peace.* New York, NY: McGraw-Hill Book Company.

Clarke, A. 1994. "Comparing the Impacts of Renewables." *International Journal of Ambient Energy* 5 (2):59-72.

Clemings, R. 1996. *Mirage—The False Promise of Desert Agriculture.* San Francisco, CA: Sierra Club Books.

Cleveland, C.J., and M. Ruth. 1997. "When, Where and by How Much Does Thermodynamics Constrain Economic Processes? A Survey of Nicholas Georgescu-Roegen's Contribution to Ecological Economics." *Ecological Economics* 22:203-223.

———. 1999. "Indicators of Dematerialization and the Materials Intensity of Use." *Journal of Industrial Ecology* 2:15-50.

Cobb, C., and T. Halstead. 1996. "The Need for New Measurements of Progress." In *The Case Against the Global Economy—and For a Turn Toward the Local,* edited by J. Mander and E. Goldsmith. San Francisco, CA: Sierra Club Books.

Cohen, J.E. 1996. *How Many People Can the Earth Support?* W.W. Norton & Co.

Colborn, T., and C. Clement, eds. 1992. *Chemically Induced Alterations in Sexual and Functional Development—the Wildlife/Human Connection.* Princeton, NJ: Princeton Scientific Publishing.

Colborn, T., D. Dumanoski and J. Peterson Myers. 1996. *Our Stolen Future—Are*

We Threatening Our Fertility, Intelligence, and Survival? New York, NY: Penguin Books USA, Inc.

Colen, B.D. 1986. *Hard Choices—Mixed Blessings of Modern Medical Technology.* New York, NY: G.P. Putnam's Sons.

Collins, F. 1972. "Social Ethics and the Conduct of Science—Specialization and the Fragmentation of Responsibility." *Annals of the New York Academy of Sciences* 196 (4):213-222.

Colwell, T.B. 1969. "The Balance of Nature—A Ground for Human Values." *Main Currents in Modern Thought* 26 (1):46-52.

Commoner, B. 1971. *The Closing Circle—Nature, Man, and Technology.* New York, NY: Alfred A. Knopf.

Connelly, L., and C.P. Koshland. 1997. "Two Aspects of Consumption—Using an Exergy-Based Measure of Degradation to Advance the Theory and Implementation of Industrial Ecology." *Resources, Conservation and Recycling* 19:199-217.

Costanza, R., and H.E. Daly. 1992. "National Capital and Sustainable Development." *Conservation Biology* 6:37-46.

Cooper, T. 2005. "Slower Consumption—Reflections on Product Life Spans and the Throwaway Society." *Journal of Industrial Ecology* 9 (1-2):51-67.

Cotgrove, S. 1982. *Catastrophe or Cornucopia.* New York, NY: John Wiley and Sons.

Cottrell, W.F. 1972. *Technology, Man, and Progress.* Columbus, OH: Charles E. Merrill Publishing Company.

Cowan, R.S. 1983. *More Work for Mother.* New York, NY: Basic Books.

Crow, J.F., ed. 1973. "Population Perspective." *Ethical Issues in Human Genetics,* D.C.B. Hilton, M. Harris, P. Condliffe and B. Berkley, Eds. New York, NY: Plenum Publishing.

———. 1997. "The High Spontaneous Mutation Rate: Is It a Health Risk?" *Proceedings of the National Academy of Science* 94 (August):8380-8386.

———. 1998. "Overdominance—A Half-Century Later." *Evolutionary Biology* 30:1-13.

———. 1999. "The Odds of Losing at Genetic Roulette." *Nature* 397 (28. January): 293-294.

Csikszentmihalyi, M. 1991. *Flow—The Psychology of Optimal Experience.* New York, NY: Harper & Row.

Cummings, C. 1991. *Eco-Spirituality—Toward a Reverent Life.* Mahwah, NY: Paulist Press.

Daily, G.C., S. Alexander, P.R. Ehrlich, L. Goulder, J. Lubchenco, P.A. Matson, H.A. Mooney, S. Postel, S.H. Schneider, D. Tilman and G.M. Woodwell. 1997. "Ecosystem Services—Benefits Supplied to Human Societies by Natural Ecosystems." *Issues in Ecology* 2 (Spring):2-16.

Daily, G.C., A.H. Ehrlich and P.R. Ehrlich. 1994. "Optimum Human Population Size." *Population and Environment* 15 (6):469-475.

Daly, H. 1994. "Steady-State Economics" in *Ecology—Key Concepts in Critical Theory,* edited by C. Merchant. Atlantic Highlands, NJ: Humanities Press.

———. 2005. "Economics in a Full World." *Scientific American* (September):100-107.

Daly, H.E., ed. 1980. *Economics, Ecology, Ethics—Essays Toward a Steady-State Economy.* San Francisco, CA: W.H. Freeman and Company.

———. 1996. *Beyond Growth—The Economics of Sustainable Development.* Boston, MA: Beacon Press.

Daly, H.E., and J. Cobb. 1989. *For the Common Good—Redirecting the Economy To-*

wards Community, the Environment, and Sustainable Development. London, UK: Green Print.

Deevey, E.S. 1960. "The Human Population." Scientific American 203:195–204.

Deikman, A.J. 1982. The Observing Self—Mysticism and Psychotherapy. Boston, MA: Beacon Press.

Delucchi, M. 1996. A Total Cost of Motor-Vehicle Use. Davis, CA: Institute of Transportation Studies, University of California.

Demir, M., M. Ozdemir, and L.A. Weitekamp. 2006. "Looking to Happy Tomorrows with Friends—Best and Close Friendships as They Predict Happiness." Journal of Happiness Studies 8:243–271.

DeNavas-Walt, C., R. Cleveland and B.H. Webster. 2003. Income in the United States: 2002, edited by U.S.C. Bureau: US Government Printing Office.

Denison, E.F. 1985. Trends in American Economic Growth, 1929–1982. Washington, DC: The Brookings Institution.

DeSimone, L.D., and F. Popoff. 1997. Eco-Efficiency—The Business Link to Sustainable Development. Cambridge, MA: MIT Press.

Des Jardins, J.R. 1993. Environmental Ethics—An Introduction to Environmental Philosophy. Belmont, CA: Wadsworth Publishing Company.

Devall, B., and G. Sessions. 1985. Deep Ecology—Living as if Nature Mattered. Salt Lake City, UT: Gibbs Smith (Peregrine Smith Books).

Deyo, R.A., and D.L. Patrick. 2005. Hope or Hype—The Obsession with Medical Advances and the High Cost of False Promises. New York, NY: American Management Association.

Diamond, J. 2005. Collapse—How Societies Choose to Fail or Succeed. New York, NY: Viking Penguin.

Dickson, D. 1975. The Politics of Alternative Technology. New York, NY: Universe Books.

Diener, E. 2000. "Subjective Well-Being—The Science of Happiness and a Proposal for a National Index." American Psychologist (January):34–43.

———. 2006. "Guidelines for National Indicators of Subjective Well-Being and Ill-Being." Journal of Happiness Studies 7:397–404.

Diener, E., M. Diener and C. Diener. 1995. "Factors Predicting the Subjective Well-Being of Nations." Journal of Personality and Social Psychology 69 (5):851–864.

Diener, E., and S. Oishi. 2000. Money and Happiness—Income and Subjective Well-Being Across Nations" in Culture and Subjective Well-Being, edited by E. Diener and E.M. Suh. Cambridge, MA: MIT Press.

Diener, E., and M.E.P. Seligman. 2004. "Beyond Money—Toward an Economy of Well-Being." Psychological Science in the Public Interest 5 (1):1–31.

Dietz, T., P. Stern and R. Rycroft. 1989. "Definitions of Conflict and the Legitimation of Resources—The Case of Environmental Risk." Sociological Forum 4 (1): 47–70.

Dixon, B. 1973. What is Science For? London, UK: Collins.

———. 1978. Beyond the Magic Bullet. New York, NY: Harper and Row.

Dohmen, F., and F. Hornig. 2004. "Der Windmuehlen Wahn—Die Grosse Luftnummer" (The Windmill Craze - The Great Air Number). Der Spiegel 14:80–97.

Doll, R., and R. Peto. 1981. The Causes of Cancer—Quantitative Estimates of Avoidable Risks of Cancer in the United States Today. Oxford, UK: Oxford University Press.

Domhoff, G.W. 1998. Who Rules America? Power and Politics in the Year 2000. 3 ed. Mountain View, CA: Mayfield Publishing Company.

Douglas, M., and A. Wildavsky. 1982. *Risk and Culture—A Essay on the Selection of Technical and Environmental Dangers*. Berkeley, CA: University of California Press.

Dovers, S.R., and J.W. Handmer. 1995. "Ignorance, the Precautionary Principle, and Sustainability." *Ambio* 24 (2):92-97.

Dower, R. 1997. *Frontiers of Sustainability—Environmentally Sound Agriculture, Forestry, Transportation, and Power Production*. Washington, DC: Island Press.

Dower, R., D. Ditz, P. Faeth, N. Johnson, K. Kozloff and J.J. MacKenzie. 1997. *Frontiers of Sustainability—Environmentally Sound Agriculture, Forestry, Transportation, and Power Production*. Washington, DC: Island Press.

Drengson, A. 1995. *The Practice of Technology—Exploring Technology, Ecophilosophy, and Spiritual Disciplines for Vital Links*. Albany, NY: State University of New York Press.

Dresner, S. 2002. *The Principles of Sustainability*. London, UK: Earthscan Publications.

Dubos, R. 1959. *Mirage of Health—Utopias, Progress, and Biological Change*. New York, NY: Harper and Brothers Publishers.

Duchin, F., and G.-M. Lange. 1994. *The Future of the Environment—Ecological Economics and Technological Change*. Oxford, UK: Oxford University Press.

Dukes, J.S. 2003. "Burning Buried Sunshine—Human Consumption of Ancient Solar Energy." *Climatic Change* 61:31-44.

Dunn, P.D. 1978. *Appropriate Technology—Technology with a Human Face*. New York, NY: Schocken Books.

Durning, A.T. 1992. *How Much Is Enough—The Consumer Society and the Future of the Earth*. New York, NY: W.W. Norton and Company.

Eagleton, T. 1991. *Ideology—An Introduction*. New York, NY: Verso.

Eckersley, R. 2000. "The Mixed Blessings of Material Progress—Diminishing Returns in the Pursuit of Happiness." *Journal of Happiness Studies* 1:267-292.

Edsall, J.T. 1975. "Scientific Freedom and Responsibility." *Science* 188 (4189):687-693.

———. 1981. "Two Aspects of Scientific Responsibility." *Science* 212 (4490):11-14.

Ehrenfeld, D. 1981. *The Arrogance of Humanism*. Oxford, UK: Oxford University Press.

Ehrlich, P.R. 1989. "The Limits to Substitution—Meta-Resource Depletion and a New Economic-Ecological Paradigm." *Ecological Economics* 1:9-16.

Ehrlich, P.R., and A.H. Ehrlich. 1981. *Extinction—The Causes and Consequences of the Disappearance of Species*. New York, NY: Random House.

———. 1991. *The Population Explosion*. New York, NY: Touchstone Books, Simon and Schuster, Inc.

———. 1996. *Betrayal of Science and Reason—How Anti-Environmental Rhetoric Threatens Our Future*. Washington, DC: Island Press.

———. 2005. *One with Nineveh—Politics, Consumption, and the Human Future*. Washington, DC: Island Press.

———. 2009. *The Dominant Animal: Human Evolution and the Environment*. Washington, DC: Island Press.

Ehrlich, P.R., and J. Holdren. 1971. "Impact of Population Growth." *Science* 171: 1212-1217.

Ehrlich, P.R., and D. Kennedy. 2005. "Millennium Assessment of Human Behavior." *Science* 309 (July 22):562-563.

Ehrlich, P.R., G. Wolff, G.C. Daily, J.B. Hughes, S. Daily, M. Dalton and L. Goulder. 1999. "Knowledge and the Environment." *Ecological Economics* 30:267-284.

Ehrman, B.D. 1996. *The Orthodox Corruption of Scripture—The Effect of Early Christological Controversies on the Text of the New Testament*. Oxford, UK: Oxford University Press.

———. 2005. *Lost Christianities—The Battles for Scripture and the Faiths We Never Knew*. Oxford, UK: Oxford University Press.

Elgin, D. 1993. *Voluntary Simplicity—Toward a Way of Life that is Outwardly Simple, Inwardly Rich*. New York, NY: William Morrow.

Elliott, D. 1997. *Energy, Society, and Environment—Technology for a Sustainable Future*. New York, NY: Routledge.

Elliott, D.L., L.L. Wendell and G.L. Gower. 1991. "Wind Energy Potential in the United States Considering Environmental and Land-Use Exclusions." Paper read at the Solar Wind Congress, in Denver, Colorado.

———. 1992. "Wind Energy Potential in the United States Considering Environmental and Land-Use Exclusions" in *Proceedings of the Biennial Congress of the International Solar Energy Society—Solar World Congress in Denver, Colorado*, edited by M.E. Ardan, S.M.Á. Burley and M. Coleman. Oxford, UK: Pergamon.

Ellul, J. 1976. *The Technological Society*. New York, NY: Alfred A. Knopf.

Engelman, R. 2008. *More—Population, Nature, and What Women Want*. Washington, DC: Island Press.

Entman, R. 1993. Framing—"Toward Clarification of a Fractured Paradigm." *Journal of Communication* 43 (4):51-58.

Epstein, S.S. 1978. *The Politics of Cancer*. San Francisco, CA: Sierra Club Books.

Eshel, G., and P.A. Martin. 2006. "Diet, Energy, and Global Warming." *Earth Interactions* 10 (9):1-17.

Esselstyn, C.B. 2008. *Prevent and Reverse Heart Disease—The Revolutionary, Scientifically Proven, Nutrition-Based Cure*. New York, NY: Avery.

Evans, R.G., M.L. Barer, and T.R. Marmor, eds. 1994. *Why Are Some People Healthy and Others Not? The Determinants of Health in Populations*. New York, NY: Aldine de Gruyter.

Faber, M., N. Niemes and G. Stephan. 1995. *Entropy, Environment, and Resources*. Berlin, Germany: Springer Verlag.

Fairhurst, G.T., and R.A. Sarr. 1996. *The Art of Framing—Managing the Language of Leadership*. San Francisco, CA: Jossey-Bass Publisher.

Farinelli, U., and P. Valant. 1990. "Energy as a Source of Potential Conflicts." *International Journal of Global Energy Issues* 2 (1):31-40.

Farrell, A.E., R.J. Plevin, B.T. Turner, A.D. Jones, M. O'Hare and D.M. Kammen. 2006. "Ethanol Can Contribute to Energy and Environmental Goals." *Science* 311:506-508.

Feenberg, A. 1999. *Questioning Technology*. London, UK: Routledge.

Ferre, F. 1988. *Philosophy of Technology*. Englewood Cliffs, NJ: Prentice Hall.

Ferris, K.R, and J.S. Wallace. 2009. *Financial Accounting for Executives*. Cambridge, UK: Cambridge Business Publishers.

Flavin, C. 1990. "Slowing Global Warming" in *State of the World 1990—A Worldwatch Institute Report on Progress Toward a Sustainable Society*, edited by L.R. Brown. New York, NY: W.W. Norton and Company.

Flink, J.J. 1975. *The Car Culture*. Cambridge, MA: MIT Press.

Ford, N. 1984. *Lifestyle for Longevity*. Gloucester, MA: Para Research, Inc.

Forge, J. 2000. "Moral Responsibility and the 'Ignorant Scientist.'" *Science and Engineering Ethics* 6 (3):341-349.

———. 2008. *The Responsible Scientist*. Pittsburgh, PA: University of Pittsburgh Press.

Fox, M.W. 1997. *Eating with Conscience—The Bioethics of Food*. Troutdale, OR: NewSage Press.

Frank, R.H. 1999. *Luxury Fever—Why Money Fails to Satisfy in an Era of Excess*. New York, NY: The Free Press.

Frey, B.S., and A. Stutzer. 2000. "Happiness, Economy, and Institutions." *Economics Journal* 110 (446):918-938.

———. 2002. "What Can Economists Learn from Happiness Research?" *Journal of Economic Literature* 40:402-435.

Fries, J.F., and L.M. Crapo. 1981. *Vitality and Aging—Implications of the Rectangular Curve*. San Francisco, CA: W.H. Freeman and Company.

Fries, J.F., C.E. Koop, C.E. Beadle, P.P. Cooper, M.J. England, R.F. Greaves, J.J. Sokolov and D. Wright. 1993. "Reducing Health Care Costs by Reducing the Need and Demand for Medical Services." *The New England Journal of Medicine* 329 (5): 321-325.

Fromm, E. 1967. *The Revolution of Hope—Toward a Humanized Technology*. New York, NY: Harper Collins.

Fuhrman, E. 1979. "The Normative Structure of Critical Theory." *Human Studies* 2:209-228.

Gaffin, S.R. 1998. "World Population Projections for Greenhouse Gas Emission Scenarios." *Mitigation and Adaptation Strategies for Global Change* 3:133-170.

Gaffin, S.R., and B.C. O'Neill. 1997. "Population and Global Warming With and Without CO_2 Targets." *Population and Environment* 18 (4):289-412.

———. 1998. "Combat Climate Change by Reducing Fertility." *Nature* 396:307.

Gallagher, R., and T. Appenzeller. 1999. "Beyond Reductionism." *Science* 284 (5411):79.

Galston, A.W. 1972. "Science and Social Responsibility—A Case History." *Annals of the New York Academy of Sciences* 196 (4):223-235.

Garnett, T. 2009. "Livestock-Related Greenhouse Gas Emissions—Impacts and Options for Policy Makers." *Environmental Science and Policy* 12:491-503.

Gentzler, Y.S. 1999. "What is Critical Theory and Critical Science?" in *Family and Consumer Sciences Curriculum—Toward a Critical Science Approach*, edited by J. Johnson and C.G. Fedje. Peoria, IL: Glencoe-McGraw-Hill.

Georgescu-Roegen, N. 1971. *The Entropy Law and the Economic Process*. Cambridge, MA: Harvard Universtiy Press.

———. 1977. "The Steady-State and Ecological Salvation—A Thermodynamic Analysis." *BioScience* 27 (4):266-270.

Gerbens-Leenes, P.W., and S. Nonhebel. 2002. "Consumption Patterns and their Effects on Land Required for Food." *Ecological Economics* 42:185-199.

Giampietro, M., S. Ulgiati and D. Pimentel. 1997. "Feasibility of Large-Scale Biofuel Production." *BioScience* 47 (9):587-600.

Ginzberg, E. 1990. "High-tech Medicine and Rising Health Care Costs." *Journal of the American Medical Association* 263 (13):1820-1822.

Gitlin, T. 1972. "Sixteen Notes on Television and the Movement." *Triquarterly* (Winter-Spring):356.

Glasby, G.P. 1988. "Entropy, Pollution and Environmental Degradation." *Ambio* 17 (5):330-335.

Glazer, M.P. 1991. *Whistleblowers*. New York, NY: Basic Books.

Glennon, R. 2009. *Unquenchable—America's Water Crisis and What To Do About It*. Washington, DC: Island Press.

Glickman, T.S., and M. Gough, eds. 1990. *Readings in Risk*. Washington, DC: Resources for the Future.

Glynn, S., ed. 1988. *Objectivity and Alienation—Towards a Hermeneutic of Science and Technology*. Edited by D. Braine and H. Lesser, Ethics, Technology, and Medicine. Brookfield, VT: Gower Publishing Company.

Goodell, J. 2010. *How to Cool the Planet—Geoengineering and the Audacious Quest to Fix Earth's Climate*. New York, NY: Houghton Mifflin Harcourt.

Goodland, R. 1997. "Environmental Sustainability in Agriculture—Diet Matters." *Ecological Economics* 23:189-200.

Goodland, R., H.E. Daly, and S.E. Serafy, eds. 1992. *Population, Technology, and Lifestyle*. Washington, DC: Island Press.

Goodpaster, K.E., and K.M. Sayre, eds. 1979. *Ethics and Problems of the 21st Century*. Notre Dame, IN: University of Notre Dame Press.

Gottlieb, R.S. 2006. *The Oxford Handbook of Religion and Ecology*. New York, NY: Oxford University Press.

Graedel, T.E., and B.R. Allenby. 1995. *Industrial Ecology*. Englewood Cliffs, NJ: Prentice Hall.

———. 1998. *Industrial Ecology and the Automobile*. Upper Saddle River, NJ: Prentice Hall.

Grant, L., ed. 1992. *Elephants in the Volkswagen—Facing the Tough Questions about our Overcrowded Country*. New York, NY: W.H. Freeman and Company.

———. 1996. *Juggernaut—Growth on a Finite Planet*. Santa Ana, CA: Seven Locks Press.

Greene, D.L. 1992. "Vehicle Use and Fuel Economy—How Big is the Rebound Effect?" *The Energy Journal* 13 (1):117-143.

Greene, R. 1998. *The 48 Laws of Power*. New York City, NY: Viking Adult Press.

Greenhalgh, G. 1990. "Energy Conservation Policies." *Energy Policy* (April):293-299.

Greening, L.A., D.L. Greene and C. Difiglio. 2000. "Energy Efficiency and Consumption—The Rebound Effect—A Survey." *Energy Policy* 28:389-401.

Gruebler, A. 1994. "Industrialization as a Historical Phenomenon" in *Industrial Ecology and Global Change*, Robert Socolow et al., eds. Cambridge, England: Cambridge University Press.

Guidotti, T.L. 1994. "Critical Science and the Critique of Technology." *Public Health Reviews* 22:235-250.

Gura, T. 2003. "Obesity Drug Pipeline Not So Fat." *Science* 299:849-852.

Gutes, M.C. 1996. "The Concept of Weak Sustainability." *Ecological Economics* 17:147-156.

Haas, R., and P. Biermayr. 2000. "The Rebound Effect for Space Heating—Empirical Evidence from Austria." *Energy Policy* 28:403-410.

Haefele, W. 1981. *Energy in a Finite World—A Global Systems Analysis*. Cambridge, MA: Ballinger Publishing Company.

Hall, C.A., R.G. Pontius, L. Coleman and J. Ko. 1994. "The Environmental Consequences of Having a Baby in the United States." *Population and Environment* 15 (6):505-524.

Hall, C.A.S., D. Lindenberger, R. Kuemmel, T. Kroeger and W. Eichhorn. 2001. "The

Need to Reintegrate the Natural Sciences with Economics." *BioScience* 51:663–673.

Hamilton, C. 2003. *Growth Fetish.* Crows Nest, New South Wales, Australia: Allen and Unwin.

Hannon, B. 1975. "Energy Conservation and the Consumer." *Science* 189 (4197): 95–102.

Hardin, G. 1968. "The Tragedy of the Commons." *Science* 162:1243–1248.

Harris, M., and E. B. Ross. 1987. *Death, Sex, and Fertility—Population Regulation in Pre-industrial and Developing Societies.* New York, NY: Columbia University Press.

Harrison, E. 2000. *Cosmology: The Science of the Universe.* 2nd edition. Cambridge, UK: Cambridge University Press.

Harte, J. 2007. "Human Population as a Dynamic Factor in Environmental Degradation." *Population and Environment* 28:223–236.

Hartung, W. D. 1995. *And Weapons for All.* HarperCollins, New York, NY.

Hayes, D. 1977. *Rays of Hope—The Transition to a Post-Petroleum World.* New York, NY: W. W. Norton.

Heinberg, R. 2003. *The Party's Over—Oil, War and the Fate of Industrial Societies.* Gabriola Island, BC: New Society Publishers.

Herman, E. S. 1992. *Beyond Hypocrisy—Decoding the News in an Age of Propaganda.* Boston, MA: South End Press.

Herman, E. S., and N. Chomsky. 1988. *Manufacturing Consent—The Political Economy of the Mass Media.* New York, NY: Pantheon Books.

Herring, H. 1999. "Does Energy Efficiency Save Energy? The Debate and Its Consequences." *Applied Energy* 63:209–226.

Hertwich, E. G. 2005. "Consumption and the Rebound Effect—An Industrial Ecology Perspective." *Journal of Industrial Ecology* 9 (1–2):85–98.

Hess, K. 1979. *Community Technology.* New York, NY: Harper and Row Publishers.

Hill, J., E. Nelson, D. Tilman, S. Polasky and D. Tiffany. 2006. "Environmental, Economic, and Energetic Costs and Benefits of Biodiesel and Ethanol Biofuels." *Proceedings of the National Academy of Science* 103 (30):11206–11210.

Ho, M.-W. 1998. *Genetic Engineering—Dream or Nightmare?* Bath, UK: Gateway Books.

Hoffert, M. I., and et al. 2002. "Advanced Technology Paths to Global Climate Stability—Energy for a Greenhouse Planet." *Science* 298:981–987.

Hoffren, J. 2006. "Reconsidering Quantification of Eco-Efficiency—Application to a National Economy." *Progress in Industrial Ecology—An International Journal* 3 (6):538–558.

Hoffren, J., and J. Korhonen. 2007. "Eco-efficiency Is Important When It Is Strategic." *Progress in Industrial Ecology—An International Journal* 4 (1/2):1–18.

Holden, C. 2002. "The Quest to Reverse Time's Toll." *Science* 295 (February 8): 1032–1033.

Holdren, J. P. 1990. "Energy in Transition." *Scientific American* (September):157–163.

———. 1991. "Population and the Energy Problem." *Population and Environment: A Journal of Interdisciplinary Studies* 12 (3):231–255.

Holdren, J. P., and P. R. Ehrlich. 1972. "One-Dimensional Ecology Revisited—A Rejoinder." *Bulletin of the Atomic Scientists* (June):42–45.

Holdren, J. P., G. Morris and I. Mintzer. 1980. "Environmental Aspects of Renewable Energy Sources." *Annual Review Energy* 5 (241–291).

Holmes, R.L. 1989. *On War and Morality*. Princeton, NJ: Princeton University Press.

Holtz-Kay, J. 1997. *Asphalt Nation—How the Automobile Took Over America, and How We Can Take It Back*. New York, NY: Crown Publishers, Inc.

Homer-Dixon, T.F. 1994. "Environmental Scarcity and Violent Conflict." *International Security* 19 (1):5-40.

Homer-Dixon, T.F., J.H. Boutwell and G.W. Rathjens. 1993. "Environmental Change and Violent Conflict." *Scientific American* (February):38-45.

Hopper, D.H. 1991. *Technology, Theology, and the Idea of Progress*. Louisville, KY: John Knox Press.

Houghton, J.T. 1997. *Global Warming—The Complete Briefing*. 2 ed. Cambridge, UK: Cambridge University Press.

Houston, J. 1996. *In Search of Happiness—The Quest for Personal Fulfillment*. Colorado Springs, CO: NavPress.

Huesemann, M.H. 2001. "Can Pollution Problems Be Effectively Solved by Environmental Science and Technology? An Analysis of Critical Limitations." *Ecological Economics* 37:271-287.

———. 2002. "The Inherent Biases in Environmental Research and Their Effects on Public Policy." *Futures* 34:621-633.

———. 2003. "The Limits of Technological Solutions to Sustainable Development." *Clean Technologies and Environmental Policy* 5:21-34.

———. 2006. "Can Advances in Science and Technology Prevent Global Warming? A Critical Review of Limitations and Challenges." *Mitigation and Adaptation Strategies for Global Change* 11:539-577.

Huesemann, M.H., and J.A. Huesemann. 2008. "Will Progress in Science and Technology Avert or Accelerate Global Collapse? A Critical Analysis and Policy Recommendations." *Environment, Development and Sustainability* 10:787-825.

Huesemann, M.H., A.D. Skillman and E.A. Crecelius. 2002. "The Inhibition of Marine Nitrification by Ocean Disposal of Carbon Dioxide." *Marine Pollution Bulletin* 44:142-148.

Hueting, R. 1996. "Three Persistent Myths in the Environmental Debate." *Ecological Economics* 18:81-88.

Hueting, R., and L. Reijnders. 1998. "Sustainability Is an Objective Concept." *Ecological Economics* 27:139-147.

Hughes, T. Parke, ed. 1975. *Changing Attitudes Toward American Technology*. New York, NY: Harper & Row Publishers.

Ihde, D. 1993. *Philosophy of Technology—An Introduction*. New York, NY: Paragon House.

Illich, I. 1975. *Medical Nemesis—The Expropriation of Health*. London, UK: Calder and Boyars, Ltd.

Jackson, T. 2005. "Live Better by Consuming Less? Is There a Double Dividend in Sustainable Consumption?" *Journal of Industrial Ecology* 9 (1-2):19-36.

Jackson, T., and N. Marks. 1999. "Consumption, Sustainable Welfare and Human Needs—with Reference to UK Expenditure Patterns between 1954 and 1994." *Ecological Economics* 28:421-441.

Jackson, W. 1987. *Altars of Unhewn Stone—Science and the Earth*. San Francisco, CA: North Point Press.

Jacobson, J.L. 1990. "Holding Back the Sea." *State of the World 1990—A Worldwatch Institute Report on Progress Toward a Sustainable Society*, L.R. Brown, ed. New York, NY: W.W. Norton and Company.

Jacobson, M.F. 2006. *Six Arguments for a Greener Diet—How a More Plant-Based Diet Could Save Your Health and the Environment.* Washington, DC: Center for Science in the Public Interest.

Jennett, B. 1986. *High Technology Medicine—Benefits and Burdens.* Oxford, UK: Oxford University Press.

Jensen, D. 1999. "War on Truth—The Secret Battle for the American Mind." *The Sun*, 7–15.

Jevons, W.S. 1865. *The Coal Question—Can Britain Survive?* London, UK: Republished by Macmillan, 1906.

Jobling, M.A., M.E. Hurles and C. Tyler-Smith. 2003. *Human Evolutionary Genetics—Origin, Peoples and Disease.* New York, NY: Garland Science/Francis and Taylor Group.

Jochem, E. 1991. "Long-Term Potentials of Rational Energy Use—The Unknown Possibilities of Reducing Greenhouse Gas Emissions." *Energy and Environment* 2 (1):31–44.

——. 2000. "Energy and End-Use Efficiency. In *World Energy Assessment—Energy and the Challenge of Sustainability*, edited by J. Goldemberg. New York, NY: United Nations Development Programme, United Nations Department of Economic and Social Affairs, World Energy Council.

Johnson, B.B., and V.T. Covello, eds. 1987. *The Social and Cultural Construction of Risk—Essays on Risk Selection and Perception.* Boston, MA: Reidel.

Johnson, C. 2000. *Blowback—The Costs and Consequences of American Empire.* New York, NY: Henry Holt and Company.

Johnson, J.T. 1984. *Can Modern War Be Just?* Westford, MA: Murray Printing Company.

Jonas, H. 1973. "Technology and Responsibility—Reflections on the New Tasks of Ethics." *Social Research* 40 (Spring):31–54.

Joseph, A., R.B. Spicer, and J. Chesky. 1949. *The Desert People: a Study of the Pagago Indians.* Chicago, IL: University of Chicago Press.

Jowit, J. 2007. "One in Four Mammals Under Threat." *The Observer*, Sunday, September 9, 2007.

Juenger, F.G. 1956. *The Failure of Technology.* Chicago, IL: Henry Regnery Company.

Kahneman, D., A.B. Krueger, D. Schkade, N. Schwarz and A.A. Stone. 2006. "Would You Be Happier if You Were Richer? A Focusing Illusion." *Science* 312 (June 30):1908–1910.

Kasser, T. 2002. *The High Price of Materialism.* Cambridge, MA: MIT Press.

Kempner, J., C.S. Perlis and J.F. Merz. 2005. "Forbidden Knowledge." *Science* 307 (11 February):854.

Kerr, R. 2010. "Do We Have the Energy for the Next Transition?" *Science* (329): 780–781.

Ketcham, B., and C. Komanoff. 1992 (draft). *Win-Win Transportation—A No-Losers Approach to Financing Transport in New York City and the Region.* New York, NY: KEA.

Kheshgi, H.S., R.C. Prince and G. Marland. 2000. "The Potential of Biomass Fuels in the Context of Global Climate Change—Focus on Transportation Fuels." *Annual Review Energy Environment* 25:199–244.

Kimbrell, A. 1993. *The Human Body Shop—the Engineering and Marketing of Life.* New York, NY: HarperCollins.

Kimbrell, A., J. Mendelson, M. Briscoe, E. Harrje, B. Ethridge, A. Bricker, K. Kallio, J. Beck and J. Dixon-Streeter. 1998. *The Real Price of Gasoline—Report No. 3, An Analysis of the Hidden External Costs Consumers Pay to Fuel their Automobiles.* Washington, DC: International Center for Technology Assessment.

King, C.W., and M.E. Webber. 2008. "Water Intensity in Transportation." *Environmental Science and Technology* 42 (21):7866-7872.

King, J., and M. Slesser. 1995. "Can the World Make the Transition to a Sustainable Economy Driven by Solar Energy?" *International Journal Environment and Pollution* 5 (1):14-29.

Kintisch, E. 2010a. *Hack the Planet: Science's Best Hope—or Worst Nightmare—for Averting Climate Catastrophe.* Hoboken, NJ: John Wiley & Sons.

———. 2010b. "Out of Site." *Science* 329:788-789.

Kivel, P. 2004. *You Call This a Democracy? Who Benefits, Who Pays and Who Really Decides.* New York, NY: The Apex Press.

Klare, M.T. 2001. *Resource Wars—The New Landscape of Global Conflict.* New York, NY: Henry Holt and Company.

Klee, R.J., and T.E. Graedel. 2004. "Elemental Cycles—A Status Report on Human or Natural Dominance." *Annual Review Environmental Resources* 29:69-107.

Kleinman, D.L., ed. 2000. *Science, Technology, and Democracy.* Albany, NY: State University of New York Press.

Kohn, L.T., J.M. Corrigan and M.S. Donaldson. 2001. *To Err Is Human—Building a Safer Health System.* Washington, DC : National Academy Press.

Konner, M. 1993. *The Trouble with Medicine.* London, UK: BBC Books.

Korten, D. 1995. *When Corporations Rule the World.* West Hartford, Connecticut and San Francisco, CA: Kumarian Press and Berrett-Koehler Publishers.

———. 2006. *The Great Turning—From Empire to Earth Community.* San Francisco, CA: Berrett-Koehler Publishers, Inc. and Kumarian Press.

———. 2009. *Agenda for a New Economy—From Phantom Wealth to Real Wealth.* San Francisco, CA: Berrett-Koehler Publishers, Inc.

Kosterlitz, J. 1993. "Paying for Miracles." *National Journal* 25 (32):19671971.

Kramer, K.J., H.C. Moll, S. Nonhebel, and H.C. Wilting. 1999. "Greenhouse Gas Emissions Related to Dutch Food Consumption." *Energy Policy* 27:203-216.

Kranczer, S. 1999. "Continued United States Longevity Increases." *Statistical Bulletin of the Metropolitan Life Foundation* 80 (4):20-27.

Krebs, A.V. 1992. *The Corporate Reapers—The Book of Agribusiness.* Washington, DC: Essential Books.

Krimsky, S. 2004. *Science in the Private Interest—Has the Lure of Profits Corrupted Biomedical Research?* Lanham, MD: Roman and Littlefield Publishers, Inc.

Kubey, R., and M. Csikszentmihalyi. 2002. "Television Addiction is No Mere Metaphor." *Scientific American* (February):74-80.

Kuemmel, R. 1989. "Energy as a Factor of Production and Entropy as a Pollution Indicator in Macroeconomic Modeling." *Ecological Economics* 1:161-180.

Kuhn, T.S. 1996. *The Structure of Scientific Revolutions.* 3 ed. Chicago, IL: The University of Chicago Press.

Lane, R.E. 2001. *The Loss of Happiness in Market Democracies.* New Haven, CT: Yale University Press.

LaPorte, T.R., and D. Metlay. 1975. "Technology Observed—Attitudes of a Wary Public." *Science* 188 (April 11):121-127.

Lappe, F. Moore. 1984. *Diet for a Small Planet*. New York, NY: Ballentine Books.

Lappe, F.M., J. Collins, and P. Rosset. 1998. *World Hunger—12 Myths*. New York, NY: Grove Press.

Lappe, M. 1994. *Evolutionary Medicine—Rethinking the Origins of Disease*. San Francisco: Sierra Club Books.

Lappe, M., and B. Bailey. 1998. *Against the Grain—Biotechnology and the Corporate Takeover of Your Food*. Monroe, MN: Common Courage Press.

Lau, C., and L. Kramer. 2005. *Die Relativitaetstheorie des Gluecks. Ueber das Leben von Lottomillionaeren (The Relativity of Luck. About the Life of Lottery Millionaires)*. Herbolzheim, Germany: Centaurus.

Leitenberg, M. 1971. "Social Responsibility (II)—The Classical Scientific Ethic and Strategic-Weapons Development." *Impact of Science on Society* 21 (2):123–136.

LeVay, S. 2010. *Gay, Straight, and the Reason Why—The Science of Sexual Orientation*. New York, NY: Oxford University Press.

Levy, S.B. 1992. *The Antibiotic Paradox—How Miracle Drugs Are Destroying the Miracle*. New York, NY: Plenum Press.

Lewinsohn, P.M., W. Mischel, W. Chaplin and R. Barton. 1980. "Social Competence and Depression—The Role of Illusory Self-Perceptions." *Journal of Abnormal Psychology* 89 (2):203–212.

Lifton, R.J., and E. Markusen. 1990. *The Genocidal Mentality—Nazi Holocaust and Nuclear Threat*. New York, NY: Basic Books, Inc.

Litman, T. 1992. *Transportation Cost Survey*. Victoria, BC: Victoria Transport Policy Institute.

Longino, H.E. 1990. *Science as Social Knowledge—Values and Objectivity in Scientific Inquiry*. Princeton, NJ: Princeton University Press.

Lovins, A.B. 1977. *Soft Energy Paths—Toward a Durable Peace*. Cambridge, MA: Ballinger Publishing Company.

Lovins, A.B., and L. Hunter-Lovins. 1995. "Reinventing the Wheels." *The Atlantic Monthly*. http://www.theatlantic.com/past/docs/issues/96apr/oil/wheels.htm, accessed 4/25/11.

Lovins, A.B., L.H. Lovins, and E. von Weizsacker. 1997. *Factor Four: Doubling Wealth-Halving Resource Use*. London, UK: Earthscan Publications Ltd.

Lowrance, W.W. 1985. *Modern Science and Human Values*. Oxford, UK: Oxford University Press.

Lutter, M. 2007. "Book Review—Winning Lottery Brings No Happiness." *Journal of Happiness Studies* Vol 8, pp. 155–160.

Lyman, H.F. 1998. *Mad Cowboy—Plain Truth from the Cattle Rancher Who Won't Eat Meat*. New York, NY: Scribner.

Lynn, R. 1996. *Dysgenics—Genetic Deterioration in Modern Populations*. Westport, CT: Praeger.

MacKay, D.J.C. 2009. *Sustainable Energy—Without the Hot Air*. Cambridge, UK: UIT Cambridge Ltd.

MacKellar, F.L. 1996. "On Human Carrying Capacity—A Review Essay on Joel Cohen's 'How Many People Can the Earth Support?'" *Population and Development Review* 22 (1):145–156.

MacKenzie, J., R. Dower and D. Chen. 1992. *The Going Rate—What It Really Costs to Drive*. Washington DC 2006: World Resources Institute.

Maddison, A. 1991. *Dynamic Forces in Capitalist Development—A Long-Run Comparative View*. Oxford, UK: Oxford University Press.

Maddison, D., D. Pearce, O. Johansson, E. Calthrop and E. Verhoef. 1996. *The True Costs of Road Transport*. London, UK: Earthscan Publications Ltd.

Mander, J. 1978. *Four Arguments for the Elimination of Television*. New York, NY: HarperCollins, (Reprinted as a Harper Perennial 2002.)

———. 1991. *In the Absence of the Sacred*. San Francisco, CA: Sierra Club Books.

Marks, L.V. 2001. *Sexual Chemistry—A History of the Contraceptive Pill*. New Haven, CT: Yale University Press.

Marshall, E. 1979. "Public Attitudes to Technological Progress." *Science* 205 (July 20):281-285.

Martensen, R. 2008. *A Life Worth Living—A Doctor's Reflections on Illness in a High-Tech Era*. New York, NY: Ashgate Publishing Co.

Martin, M.W. 2007. *Albert Schweitzer's Reverence for Life*. Burlington, VT: Ashgate Publishing Co.

Martin, P. 2005. *Making People Happy—The Nature of Happiness and Its Origins in Childhood*. London, UK: Fourth Estate, Harper Collins Publishers.

Maslow, A. 1954. *Motivation and Personality*. New York, NY: Harper and Row.

Matson, P.A., W.J. Parton, A.G. Power and M.J. Swift. 1997. "Agricultural Intensification and Ecosystem Properties." *Science* 277 (5325):504-509.

Max-Neef, M. 1991. *Human-Scale Development—Conception, Application, and Further Reflection*. London, UK: Apex Press.

May, R.M. 1988. "How Many Species Are There on Earth?" *Science* 241 (September 16):1441-1449.

Mayr, E. 2002. *What Evolution Is*. New York, NY: Basic Books.

McCuen, R.H. 1979. "The Ethical Dimension of Professionalism." *Issues in Engineering* April:89-105.

McDonough, W., and M. Braungart. 2002. *Cradle to Cradle—Remaking the Way We Make Things*. New York, NY: North Point Press.

McGregor, S. 2003. *Critical Science Approach—A Primer*. NA.

McKeown, T. 1979. *The Role of Medicine—Dream, Mirage or Nemesis*. Princeton, NJ: Princeton University Press.

———. 1988. *The Origins of Human Disease*. Oxford, UK: Basil Blackwell Ltd.

McKinlay, J.B. 1977. "The Business of Good Doctoring or Doctoring as Good Business—Reflections on Freidson's View of the Medical Game." *International Journal of Health Services* 7 (3):459-483.

———. 1981. "From 'Promising Report' to 'Standard Procedure'—Seven Stages in the Career of a Medical Innovation." *Milbank Memorial Fund Quarterly, Health and Society* 59 (3):374-411.

McKinlay, J.B., and S.M. McKinlay. 1977. "The Questionable Contribution of Medical Measures to the Decline of Mortality in the United States in the Twentieth Century." *Milbank Memorial Fund Quarterly, Health and Society* 55:405-428.

McNeill, W.H. 1982. *The Pursuit of Power—Technology, Armed Force, and Society Since A.D. 1000*. Chicago, IL: The University of Chicago Press.

Meadows, D., J. Randers and D. Meadows. 2004. *Limits to Growth—The 30 Year Update*. White River Junction, VT: Chelsea Green Publishing Company.

Meadows, D.H. 1972. *Limits to Growth*. New York, NY: Penguin Group.

Melman, S. 1970. *Pentagon Capitalism—The Political Economy of War*. New York, NY: McGraw-Hill.

———. 1983. *Profits without Production*. New York, NY: Alfred A. Knopf.

———. 1984. *The Permanent War Economy*. New York, NY: Simon and Schuster, Inc.

Merchant, C., ed. 1994. *Ecology—Key Concepts in Critical Theory*. Atlantic Highlands, NJ: Humanities Press International, Inc.

Mesthene, E. G. 1970. *Technological Change—Its Impact on Man and Society*. Cambridge, MA: Harvard University Press.

Mill, J. S. 1848. *Principles of Political Economy*. London, UK: John W. Parker, West Strand.

Mitcham, C., and J. Grote, eds. 1984. *Theology and Technology—Essays in Christian Analysis and Exegesis*. New York, NY: University Press of America.

Moffet, J., and P. Miller. 1993. *The Price of Mobility*. San Francisco, CA: Natural Resources Defense Council.

Moss, R. W. 1980. *The Cancer Industry—The Classic Exposé on the Cancer Establishment*. New York, NY: Paragon House.

Mumford, L. 1934. *Technics and Civilization*. San Diego, CA: Harcourt Brace and Company.

Murtaugh, P. A., and M. G. Schlax. 2009. "Reproduction and the Carbon Legacies of Individuals." *Global Environmental Change* 19:14-20.

Myers, D. G., and E. Diener. 1996. "The Pursuit of Happiness." *Scientific American* 274 (5 (May)):70-72.

Myers, N. 1986. "The Environmental Dimension to Security Issues." *The Environmentalist* 6 (4):251-257.

———. 2000. "Sustainable Consumption." *Science* 287:2419.

Myers, N., and J. Kent. 2001. *Perverse Subsidies—How Tax Dollars Can Undercut the Environment and the Economy*. Washington, DC: Island Press.

Myers., N. 2003. "Consumption—Challenge to Sustainable Development." *Science* 276:53-55.

Naess, A., and D. Rothenberg. 1989. *Ecology, Community and Lifestyle—Outline of an Ecosophy*. Cambridge, UK: Cambridge University Press.

Nakicenovic, N. 1996. "Freeing Energy from Carbon." *Daedalus* 125 (3):95-112.

Nakicenovic, N., and A. Gruebler. 1993. "Energy Conversion, Conservation, and Efficiency." *Energy* 18 (5):421-435.

Nakicenovic, N., N. Victor and T. Morita. 1998a. "Emission scenario database and review of scenarios." *Mitigation and Adaptation Strategies for Global Climate Change* 3:95-120.

Nakicenovic, N., A. Gruebler and A. McDonald. 1998b. *Global Energy Perspectives*. Cambridge, UK: Cambridge University Press.

Nardin, T., ed. 1996. *The Ethics of War and Peace—Religious and Secular Perspectives*. Princeton, NJ: Princeton University Press.

National Resource Council. 1992. *Automotive Fuel Economy—How Far Can We Go?* Washington, DC: The National Academies Press.

National Science Board. 2004. *Science and Engineering Indicators 2004—National Science Foundation (NSF)*.

Nelson, R. R. 1996. *The Sources of Economic Growth*. Cambridge, MA: Harvard University Press.

Nesse, R. N., and G. C. Williams. 1994. *Why We Get Sick—The New Science of Darwinian Medicine*. New York, NY: Times Books.

Nestle, M. 2002. *Food Politics—How the Food Industry Influences Nutrition and Health*. Berkeley, CA: University of California Press.

Nickerson, C., N. Schwarz and E. Diener. 2007. "Financial Aspirations, Financial

Success, and Overall Life Satisfaction—Who? and How?" *Journal of Happiness Studies* 8:467-515.

Nicolis, G., and I. Prigogine. 1997. *Self-Organization in Non-Equilibrium Systems—From Dissipative Structures to Order through Fluctuations.* New York, NY: John Wiley and Sons.

Nisbet, M.C., and C. Mooney. 2007. "Framing Science." *Science* 316:56.

Norman, R.J. 1995. *Ethics, Killing and War.* Cambridge, UK: Cambridge University Press.

O'Connor, M. 1994. "Entropy, Liberty and Catastrophe—The Physics and Metaphysics of Waste Disposal" in *Economics and Thermodynamics—New Perspectives on Economic Analysis*, P. Burley and J. Foster, eds. Boston, MA: Kluwer Academic Publishers.

O'Riordan, T. 1993. "The Politics of Sustainability" in *Sustainable Environmental Economics and Management*, edited by R.K. Turner. New York, NY: Belhaven Press.

Oeppen, J., and J.W. Vaupel. 2002. "Broken Limits to Life Expectancy." *Science* 296 (May 10):1029-1031.

Olsen, F.A., ed. 1973. *Technology—A Reign of Benevolence and Destruction.* New York, NY: MSS Information Corporation.

O'Neill, B.C., L.F. MacKellar and W. Lutz. 2001. *Population and Climate Change.* New York, NY: Cambridge University Press.

O'Neill, B.C., and L. Wexler. 2000. "The Greenhouse Externality to Childbearing—A Sensitivity Analysis." *Climatic Change* 47:283-324.

Organization for Economic Cooperation and Development (OECD). 1998. *Eco-Efficiency.*

Organization for Economic Cooperation and Development (OECD) and International Energy Agency (IEA). 2004. *30 Years of Energy Use in IEA Countries.* Paris, France: International Energy Agency (IEA).

Oswald, A.J. 1997. "Happiness and Economic Performance." *Economics Journal* 107 (445):1815-1831.

Otake, K., S. Shimai, J. Tanaka-Matsumi, K. Otsui, and B.L. Fredrickson. 2006. "Happy People Become Happier through Kindness—A Counting Kindness Intervention." *Journal of Happiness Studies* 7:361-375.

Parenti, M. 1988. *The Sword and the Dollar—Imperialism, Revolution, and the Arms Race.* New York, NY: St. Martin's Press.

Peale, N.V. 2007. *The Power of Positive Thinking.* New York, NY: Simon & Schuster.

Pearce, D.W., and G.D. Atkinson. 1993. "Capital Theory and the Measurement of Sustainable Development—An Indicator of Weak Sustainability." *Ecological Economics* 8:103-108.

Pearce, F. 2006. *When the Rivers Run Dry: Water—the Defining Crisis of the Twenty-First Century.* Boston, MA: Beacon Press.

Peet, J. 1992. *Energy and the Ecological Economics of Sustainability.* Washington, DC: Island Press.

Perelman, M. 1977. *Farming for Profit in a Hungry World—Capital and the Crisis in Agriculture.* Montclair, NJ: Allanheld, Osmun & Co. Publishers, Inc.

Perrow, C. 1999. *Normal Accidents: Living with High-Risk Technologies.* Princeton, NJ: Princeton University Press.

Peterson, C., N. Park and M.E.P. Seligman. 2005. "Orientations to Happiness and

Life Satisfaction—The Full Life Versus the Empty Life." *Journal of Happiness Studies* 6:25-41.

Pimentel, D. 2003. "Ethanol Fuels—Energy Balance, Economics, and Environmental Impacts Are Negative." *Natural Resources Research* 12 (2):127-134.

——. 2008. *Biofuels, Solar and Wind as Renewable Energy Systems—Benefits and Risks.* New York, NY: Springer.

Pimentel, D., C. Fried, L. Olson, S. Schmidt, K. Wagner-Johnson, A. Westman, A. Whelan, K. Foglia, P. Poole, T. Klein, R. Sobin and A. Bochner. 1984a. "Environmental and Social Costs of Biomass Energy." *BioScience* 34 (2):89-94.

Pimentel, D., J. Gardner, A. Bonnifield, X. Garcia, J. Grufferman, C. Horan, J. Schlenker and E. Walling. 2009. "Energy Efficiency and Conservation for Individual Americans." *Environment Development and Sustainability* 11 (3):523-546.

Pimentel, D., and M. Giampietro. 1994. "Implications of the Limited Potential of Technology to Increase the Carrying Capacity of our Planet." *Human Ecology Review* Summer/Autumn (1):248-250.

Pimentel, D., C. Harvey, P. Resosudarmo, K. Sinclair, D. Kurz, M. McNair, S. Crist, L. Shpritz, L. Fitton, R. Saffouri and R. Blair. 1995. "Environmental and Economic Costs of Soil Erosion and Conservation Benefits." *Science* 267:1117-1123.

Pimentel, D., M. Herdendorf, S. Eisenfeld, L. Olander, M. Carroquino, C. Corson, J. McDade, Y. Chung, W. Cannon, J. Roberts, L. Bluman and J. Gregg. 1994a. "Achieving a Secure Energy Future—Environmental and Economic Issues." *Ecological Economics* 9:201-219.

Pimentel, D., J. Krummel, D. Gallahan, J. Hough, A. Merrill, I. Schreiner, P. Vittum, E. Back, D. Yen and S. Fiance. 1979. "Benefits and Costs of Pesticide Use in U.S. Food Production." *BioScience* 28 (12):772-784.

Pimentel, D., and H. Lehman, eds. 1993. *The Pesticide Question—Environment, Economics, and Ethics.* New York, NY: Chapman and Hall.

Pimentel, D., L. Levitan, J. Heinze, M. Loehr, W. Naegeli, J. Bakker, J. Eder, B. Modelski and M. Morrow. 1984b. "Solar Energy, Land and Biota." *SunWorld* 8 (3):70-95.

Pimentel, D., and T.W. Patzek. 2005. "Ethanol Production Using Corn, Switchgrass, and Wood; Biodiesel Production Using Soybean and Sunflower." *Natural Resources Research* 14 (1):65-76.

Pimentel, D. and M. Pimentel. 1979. *Food, Energy, and Society.* London, UK: Edward Arnold, Ltd.

——. 1996. *Food, Energy, and Society.* Boulder, CO: University Press of Colorado.

Pimentel, D., G. Rodrigues, T. Wang, R. Abrams, K. Goldberg, H. Staecker, E. Ma, L. Brueckner, L. Trovato, C. Chow, U. Govindarajulu and S. Boerke. 1994b. "Renewable Energy—Economic and Environmental Issues." *BioScience* 44:536-547.

Pimm, S.L., G.J. Russell, J.L. Gittleman, and T.M. Brooks. 1995. "The Future of Biodiversity." *Science* 269 (July 21):347-350.

Platt, J. 1969. "What We Must Do—A Large-Scale Mobilization of Scientists May Be the Only Way to Solve Our Crisis Problems." *Science* 166 (November 28): 1115-1121.

Podrid, P.J. 1989. "Resuscitation in the Elderly—A Blessing or a Curse?" *Annals of Internal Medicine* 111 (3):193-195.

Polak, E.L., and M.E. McCullough. 2006. "Is Gratitude an Alternative to Materialism?" *Journal of Happiness Studies* 7:343-360.

Pollan, M. 2007. *The Omnivore's Dilemma—A Natural History of Four Meals.* New York, NY: The Penguin Group.

———. 2009. *In Defense of Food—An Eater's Manifesto.* New York, NY: The Penguin Group.

Porte, T. La, and D. Metlay. 1975. "Public Attitudes Toward Present and Future Technologies—Satisfactions and Apprehensions." *Social Studies of Science* 5: 373-398.

Postel, S. L. 1999. *Pillar of Sand—Can the Irrigation Miracle Last?* New York, NY: W. W. Norton and Company.

Postel, S. L., G. C. Daily and P. R. Ehrlich. 1996. "Human Appropriation of Renewable Fresh Water." *Science* 271:785-788.

Postman, N. 1992. *Technopoly—The Surrender of Culture to Technology.* New York, NY: Alfred A. Knopf.

Powles, J. 1973. "On the Limitations of Modern Medicine." *Science, Medicine, and Man* 1:1-30.

President's Council on Sustainable Development (PCSD). 1996. *Sustainable America—A New Consensus for the Prosperity, Opportunity and a Healthy Environment for the Future.* Available from clinton2.nara.gov/PCSD/Publications/TF_Reports/amer-top.html, accessed 4/25/2011.

Prigogine, I. 1961. *Thermodynamics of Irreversible Processes.* New York, NY: John Wiley and Sons.

Prigogine, I. 1989. "What is Entropy?" *Naturwissenschaften* 76:1-8.

Princen, T. 2005. *The Logic of Sufficiency.* Cambridge, MA: The MIT Press.

Pritchett, L. H. 1994. "Desired Fertility and the Impact of Population Policies." *Population and Development Review* 20 (1):1-55.

Proctor, R. N. 1991. *Value-Free Science? Purity and Power in Modern Knowledge.* Cambridge, MA: Harvard University Press.

———. 1995. *Cancer Wars—How Politics Shapes What We Know and Don't Know About Cancer.* New York, NY: Basic Books.

Ravetz, J. R. 1990. *The Merger of Knowledge with Power—Essays in Critical Science.* London, UK: Mansell Publishing Limited.

———. 1996. *Scientific Knowledge and Its Social Problems.* New Brunswick, NJ: Transaction Publishers.

Rebane, K. K. 1995. "Energy, Entropy, Environment—Why Is Protection of the Environment Objectively Difficult?" *Ecological Economics* 13:89-92.

Reid, D.. 1995. *Sustainable Development—An Introductory Guide.* London, UK: Earthscan Publications Ltd.

Reijnders, L. 1998. "The Factor X Debate—Setting Targets for Eco-Efficiency." *Journal of Industrial Ecology* 2 (1):13-21.

Reisner, M. 1986. *Cadillac Desert—The American West and Its Dissappearing Water.* New York, NY: Viking Penguin Inc.

Relman, A. S. 1980. "The New Medical-Industrial Complex." *New England Journal of Medicine* 303 (17):963-970.

Renning, K., and H. Wiggering. 1997. "Steps Towards Indicators of Sustainable Development—Linking Economic and Ecological Concepts." *Ecological Economics* 20:25-36.

Rescher, N. 1980. *Unpopular Essays on Technological Progress.* Pittsburgh, PA: University of Pittsburgh Press.

Rhodes, R., ed. 1999. *Visions of Technology.* New York, NY: Simon and Schuster.

Rifkin, J. 1980. *Entropy—A New World View.* New York, NY: The Viking Press.

———, ed. 1990. *The Green Lifestyle Handbook.* New York, NY: Henry Holt & Company.

———. 1992. *Beyond Beef—The Rise and Fall of the Cattle Culture*. New York, NY: Penguin Books USA Inc.

———. 1998. *The Biotech Century—Harnessing the Gene and Remaking the World*. New York, NY: Jeremy P. Tarcher/Putnam.

Rissler, J., and M. Mellon. 1996. *The Ecological Risks of Engineered Crops*. Cambridge, MA: MIT Press.

Robbins, J. 1987. *Diet for a New America*. Walpole, NH: Stillpoint Publishing.

———. 2010. *The Food Revolution: How Your Diet Can Save Your Life and Our World*. Newburyport, MA: Conari Press.

Rockstrom, J., W. Steffen and K. Noone et al. 2009. "Planetary Boundaries—Exploring the Safe Operating Space for Humanity." *Ecology and Society* 14 (2), Article 32.

Rogers, E.M. 2003. *Diffusion of Innovations*. 5 ed. New York, NY: The Free Press.

Rojstaczer, S., S.M. Sterling and N.J. Moore. 2001. "Human Appropriation of Photosynthesis Products." *Science* 294:2549-2552.

Romm, J.J., and C.B. Curtis. 1996. "Mideast Oil Forever?" *Atlantic Monthly* 227 (4):57.

Ropke, I. 1999. "The Dynamics of Willingness to Consume." *Ecological Economics* 28 (3):399-420.

Rose, S., and H. Rose. 1971. "Social Responsibility (III)—The Myth of the Neutrality of Science." *Impact of Science on Society* 21 (2):137-149.

Rosegrant, M.W., S. Msangi, T. Sulser and R. Valmonte-Santos, 2006. "Biofuels and the Global Food Balance," edited by P. Hazell and R.K. Pachauri. *Bioenergy and Agriculture—Promises and Challenges*, Brief #3 vols. Vol. 2020 Focus 14 Report. Washington, DC: International Food and Policy Research Institute.

Rosner, L. 2004. *The Technological Fix—How People Use Technology to Create and Solve Problems*. New York, NY: Routledge.

Roszak, T., M.E. Gomes and A.D. Kanner, eds. 1995. *Ecopsychology—Restoring the Earth, Healing the Mind*. San Francisco, CA: Sierra Club Books.

Rothschild, E. 1973. *Paradise Lost—The Decline of the Auto-Industrial Age*. New York, NY: Random House.

Routley, R., and V. Routley. 1978. "Nuclear Energy and Obligations to the Future." *Inquiry* 21:133-179.

Runge, C.F., B. Senauer, P.G. Pardey and M.W. Rosegrant. 2003. *Ending Hunger in our Lifetime—Food Security and Globalization*. Baltimore, MD: The Johns Hopkins University Press.

Ruskin, J. 1967. *Unto this Last—Four Essays on the First Principles of Political Economy*. Lincoln, NE: University of Nebraska Press.

Russell, B. 2000. *Unpopular Essays*. London, UK: Routledge.

Ruth, M. 1993. *Integrating Economics, Ecology and Thermodynamics*. Dortrecht, The Netherlands: Kluwer Academic Publishers.

———. 1995. "Information, Order and Knowledge in Economic and Ecological Systems—Implications for Material and Energy Use." *Ecological Economics* 13:99-114.

Sagan, L.A. 1987. *The Health of Nations*. New York, NY: Basic Books.

Sagoff, M. 1995. "Carrying Capacity and Ecological Economics." *BioScience* 45 (9):610-620.

Sahlins, M. 2005. *Stone Age Economics*. Hawthorne, NY: Aldine de Gruyter.

Sailor, W.C., D. Bodansky, C. Braun, S. Fetter and B. van der Zwaan. 2000. "A Nuclear Solution to Climate Change?" *Science* 288 (May):1177-1178.

Samuelson, P.A., and W.D. Nordhaus. 1989. *Economics*. 13 ed. New York, NY: McGraw-Hill.

Santillo, D., R.L. Stringer, P.A. Johnson and J. Tickner. 1998. "The Precautionary Principle—Protecting Against Failures of Scientific Method and Risk Assessment." *Marine Pollution Bulletin* 36 (2):939-950.

Sarewitz, D. 1996. *Frontiers of Illusion—Science, Technology, and the Politics of Progress*. Philadelphia, PA: Temple University Press.

———. 2004. "How Science Makes Environmental Controversies Worse." *Environmental Science and Policy* 7:385-403.

Saunders, H.D. 1992. "The Khazzoom-Brookes Postulate and Neoclassical Growth." *Energy Journal* 13 (4):131-148.

———. 2000. "A View from the Macro Side—Rebound, Backfire, and Khazzoom-Brookes." *Energy Policy* 28:439-449.

Scharlemann, J.P.W., and W.F. Laurance. 2008. "How Green Are Biofuels?" *Science* 319:43-44.

Schmidheiny, S. 1992. *Changing Course—A Global Business Perspective on Development and Environment*. Cambridge, MA: MIT Press.

Schnaiberg, A., and K.A. Gould. 1994. *Environment and Society—The Enduring Conflict*. New York, NY: St. Martins Press.

Schor, J. 2000. *Do Americans Shop Too Much?* Boston, MA: Beacon Press.

Schor, J.B. 1992. *The Overworked American—The Unexpected Decline of Leisure*. New York, NY: Basic Books.

———. 2005. "Sustainable Consumption and Worktime Reduction." *Journal of Industrial Ecology* 9 (1-2):37-50.

Schumacher, E.F. 1973. *Small Is Beautiful—Economics as if People Mattered*. New York, NY: Harper and Row Publishers.

———. 1977. *A Guide for the Perplexed*. New York, NY: Harper and Row Publishers.

———. 1979. *Good Work*. New York, NY: Harper and Row Publishers.

———. 1997. *This I Believe and Other Essays*. Dartington, UK: Green Books Ltd.

Schumpeter, J. 1934. *The Theory of Economic Development*. Cambridge, MA: Harvard University Press.

Schwartz, E. 1971. *Overskill—The Decline of Technology in Modern Civilization*. Chicago, IL: Quadrangle Books.

Sclove, R.E. 1995. *Democracy and Technology*. New York, NY: The Guilford Press.

———. 1998. "Better Approaches to Science Policy." *Science* 279 (5355):1283.

Searchinger, T., R. Heimlich, R.A. Houghton, F. Dong, A. Elobeid, J. Fabriosa, S. Tokgoz, D. Hayes and R.H. Yu. 2008. "Use of U.S. Croplands for Biofuels Increases Greenhouse Gases Through Emissions From Land-Use Change." *Science* 319:1238-1240.

Segal, P.H. 1994. *Future Imperfect—The Mixed Blessings of Technology in America*. Amherst, MA: The University of Massachusetts Press.

Seligman, M.E.P. 2002. *Authentic Happiness*. New York, NY: Free Press (Simon and Schuster, Inc.).

———. 2006. *Learned Optimism—How To Change Your Mind and Your Life*. New York, NY: Vintage Books, Random House.

Senauer, B., and M. Sur. 2001. "Ending Global Hunger in the 21st Century—Projections of the Number of Food Insecure People." *Review of Agricultural Economics* 23 (1):68-81.

Sessions, G. 1995. *Deep Ecology for the Twenty-First Century*. Boston, MA: Shambhala.

Sharp, G. 1970. *National Security through Civilian-Based Defense*. Omaha, NE: Association for Transarmament Studies.

———. 1973. *The Politics of Nonviolent Action. Part I (Power and Struggle); Part II (The Methods of Nonviolent Action); Part III (The Dynamics of Nonviolent Action)*. Boston, MA: Porter Sargent Publishers.

Sheldon, K.M., and T.H. Hoon. 2006. "The Multiple Determination of Well-Being—Independent Effects of Positive Traits, Needs, Goals, Selves, Social Supports, and Cultural Contexts." *Journal of Happiness Studies* 8:565-592.

Sheldon, K.M., and S. Lyubomirsky. 2006. "Achieving Sustainable Gains in Happiness—Change Your Actions, Not Your Circumstances." *Journal of Happiness Studies* 7:55-86.

Sherrow, V. 1996. *Bioethics and High-Tech Medicine*. New York, NY: Henry Holt and Company.

Shi, D.E. 1985. *The Simple Life—Plain Living and High Thinking in American Culture*. Oxford, UK: Oxford University Press.

Shiva, V. 1993. *Monocultures of the Mind—Perspectives on Biodiversity and Biotechnology*. London, UK: Zed Books Ltd.

———. 2000. *Stolen Harvest—The Hijacking of the Global Food Supply*. Cambridge, MA: South End Press.

Shrader-Frechette, K.S., and E.D. McCoy. 1993. *Methods in Ecology—Strategies for Conservation*. Cambridge, UK: Cambridge University Press.

Sieferle, R.P. 2004. "Sustainability in a World History Perspective" in *Exploitation and Overexploitation in Societies Past and Present*, B. Benzing and B. Hermann, eds. Muenster, Germany: Lit Verlag.

Silverman, M., and P.R. Lee. 1974. *Pills, Profits, and Politics*. Berkeley, CA: University of California Press.

Simon, J.L. 1996. *The Ultimate Resource II*. Princeton, NJ: Princeton University Press.

Skinner, B.J. 1987. "Supplies of Geochemically Scarce Metals" in *Resources and World Development*, edited by D.J. McLaren and B.J. Skinner. New York, NY: John Wiley.

Skolimowsky, H. 1978. "Eco-Philosophy Versus the Scientific World View." *Ecologist Quarterly* (Autumn 1978):227-248.

Socolow, R., C. Andrews, F. Berkhout and V. Thomas, eds. 1994. *Industrial Ecology and Global Change*. Cambridge, UK: Cambridge University Press.

Solow, R.M. 1957. "Technical Change and the Aggregate Production Function." *Review of Economics and Statistics* 39 (August):312-320.

Speth, J.G. 2009. *The Bridge at the Edge of the World—Capitalism, the Environment, and Crossing the Crisis to Sustainability*. New Haven, CT: Yale University Press.

Stanford, P.K. 2006. *Exceeding Our Grasp—Science, History, and the Problem of Unconceived Alternatives*. New York, NY: Oxford University Press.

Starfield, B. 2000. "Is US Health Really the Best in the World"? *Journal of the American Medical Association* 284 (4):483-485.

Steinhart, J.S., and C.E. Steinhart. 1974. "Energy Use in the U.S. Food System." *Science* 184:307-316.

Stevens, R.W., ed. 1991. *Appropriate Technology—A Focus for the Nineties*. New York, NY: Intermediate Technology Development Group of North America.

Stewart, E.C., and M.J. Bennett. 1991. *American Cultural Patterns—A Cross-Cultural Perspective*. Yarmouth, ME: Intercultural Press.

Stokey, E., and R. Zeckhauser. 1978. *A Primer for Policy Analysis*. New York, NY: W.W. Norton.

Stokstad, E. 2000. "Drug-Resistant TB on the Rise."*Science* 287 (March 31).

Stone, C.D. 1996. *Should Trees Have Standing? And Other Essays on Law, Morals and the Environment*. Oxford, UK: Oxford University Press.

Stork, N.E. 2010. "Re-Assessing Current Extinction Rates." *Biodiversity and Conservation* 19:357-371.

Stunkel, K.R., and S. Sarsar. 1994. *Ideological Values and Technology in Political Life*. Lanham, MD: University Press of America, Inc.

Susiarjo, G., S.N. Sreenath and A.M. Vali. 2006. "Optimum Supportable Global Population—Water Accounting and Dietary Consideration." *Environment, Development and Sustainability* 8:313-349.

Susskind, C. 1973. *Understanding Technology*. Baltimore, MD: The Johns Hopkins University Press.

Tainter, J.A. 1988. *The Collapse of Complex Societies*. Cambridge, UK: Cambridge University Press.

———. 1995. "Sustainability of Complex Societies." *Futures* 27 (4):397-407.

Tatzel, M. 2003. "The Art of Buying—Coming to Terms with Money and Materialism." *Journal of Happiness Studies* (4):405-435.

Taviss, I. 1972. "A Survey of Popular Attitudes Toward Technology." *Technology and Culture* 13 (4):606-621.

Taylor, R. 1979. *Medicine Out of Control—The Anatomy of a Malignent Technology*. Melbourne, Australia: Sun Books.

Teich, A.H., ed. 1993. *Technology and the Future*. New York, NY: St. Martin's Press.

Teller, E., L. Wood and R. Hyde. 1997. "Global Warming and the Ice Ages—I. Prospects for Physics-Based Modulation of Global Change" in *22nd International Seminar on Planetary Emergencies*. Erice (Sicily), Italy: Lawrence Livermore National Laboratory.

Temple, N.J., and D.P. Burkitt. 1994. *Western Diseases—Their Dietary Prevention and Reversibility*. Totowa, NJ: Humana Press.

Tenner, E. 1996. *Why Things Bite Back—Technology and the Revenge of Unintended Consequences*. New York, NY: Alfred A. Knopf.

Trainer, T. 1995b. "Can Renewable Energy Sources Sustain Affluent Society?" *Energy Policy* 23 (12):1009-1026.

———. 1995a. *The Conserver Society—Alternatives for Sustainability*. London, UK: Zed Books.

———. 2010. *The Transition to a Sustainable and Just World*. Canterbury, Australia: Envirobook.

Tudge, C. 1999. *Neanderthals, Bandits, and Farmers—How Agriculture Really Began*. New Haven, CT: Yale University Press.

Turner, G. 2008. "A Comparison of the Limits to Growth with 30 Years of Reality." *Global Environmental Change* 18:397-411.

Turner, J.A. 1999. "Realizable Renewable Energy Future." *Science* 285:687-689.

US Congress, Office of Technology Assessment. 1994. *Saving Energy in U.S. Transportation*. Washington DC: U.S. Congress.

Ullman, R.H. 1983. "Redefining Security." *International Security* 8 (1):129-153.

Vanek, J. 1974. "Time Spent in Housework." *Scientific American* 231 (November): 116-120.

Vitousek, P.M., P.R. Ehrlich, A.H. Ehrlich and P.A. Matson. 1986. "Human Appropriations of the Products of Photosynthesis. "*BioScience* 36 (6):368-373.

Vitousek, P.M., H.A. Mooney, J. Lubchenco and J.M. Melillo. 1997. "Human Domination of Earth's Ecosystems." *Science* 277:494-499.

Vohs, K.D., N.L. Mead and M.R. Goods. 2006. "The Psychological Consequences of Money." *Science* 314 (November 17):1154-1156.

Wachtel, P.L. 1989. *The Poverty of Affluence—A Psychological Portrait of the American Way of Life*. Philadelphia, PA: New Society Publishers.

Wackernagel, M., and W.E. Rees. 1996. *Our Ecological Footprint—Reducing Human Impact on the Earth*. Gabriola Island, BC: New Society Publishers.

Wade, N. 1975. "Karl Hess—Technology with a Human Face." *Science* 187 (January 31):332-334.

Waltner-Toews, D. 2000. "The End of Medicine—The Beginning of Health." *Futures* 32:655-667.

Wang, M. 2005. *Updated Energy and Greenhouse Gas Emission Results of Fuel Ethanol*. Paper read at 15th International Symposium on Alcohol Biofuels, 26-28 September, at San Diego, CA.

Watson, R.T., et al. 2001. *Climate change 2001: Synthesis Report—Summary for Policy Makers*. Intergovernmental Panel on Climate Change. Cambridge, UK: Cambridge University Press

Weed, D.L., and R.E. McKeown. 2003. "Science and Social Responsibility in Public Health." *Environmental Health Perspectives* 111 (14):1804-1808.

Weil, A. 1998. *Health and Healing*. New York, NY: Mariner Books.

Weisman, A. 2007. *The World Without Us*. New York, NY: St. Martin's Press.

Weizsacker, E. von, and J. Jesinghous. 1992. *Ecological Tax Reform: a Policy Proposal for Sustainable Development*. London, UK: Zed Books.

Welford, R. 1997. *Hijacking Environmentalism—Corporate Responses to Sustainable Development*. London, UK: Earthscan.

Westing, A.H., ed. 1986. *Global resources and International Conflict—Environmental Factors in Strategic Policy and Action*. Oxford, UK: Oxford University Press.

Whitaker, J.M. 1985. *Reversing Heart Disease*. New York, NY: Warner Books, Inc.

White, L. 1967. "The Historical Roots of Our Ecological Crisis." *Science* 155 (3767): 1203-1207.

Whitty, J. 2007. "Animal Extinction—the Greatest Threat to Mankind." *The Independent*, Monday, April 30, 2007. http://www.independent.co.uk/environment /animal-extinction--the-greatest-threat-to-mankind-397939.html, accessed 4/25/2011.

Wilkinson, R. 1996. *Unhealthy Societies—The Afflictions of Inequality*. London, UK: Routledge.

Willett, W.C. 2002. "Balancing Life-Style and Genomics Research for Disease Prevention." *Science* 296:695-698.

Williams, E. 2004. "Energy Intensity of Computer Manufacturing—Hybrid Assessment Combining Process and Economic Input-Output Methods." *Environmental Science and Technology* 38 (22):6166-6174.

Williams, E., R.U. Ayres and M. Heller. 2002. "The 1.7 Kilogram Microchip—Energy and Material Use in the Production of Semiconductor Devices." *Environmental Science and Technology* 36 (24):5504-5510.

Williams, R.H. 1987. "Exploring the Global Potential for More Efficient Use of

Energy" in *Resources and World Development*, edited by D.J. McLaren and B.J. Skinner. New York, NY: John Wiley.

Wilson, E.O. 2000. "Biodiversity at the Crossroads." *Environmental Science and Technology* 34 (5):123A-128A.

——. 2002. "The Bottleneck." *Scientific American* (February):82-91.

Winn, M. 1987. *Unplugging the Plug-In Drug*. New York, NY: Viking Penguin Inc.

Winner, L. 1977. *Autonomous Technology—Technics-out-of-Control as a Theme in Political Thought*. Cambridge, MA: MIT Press.

Winston, M.L. 1997. *Nature Wars—People Versus Pests*. Cambridge, MA: Harvard University Press.

Wolf, S. 1959. "The Pharmacology of Placebos." *Pharmacological Reviews* 11:689-704.

World Commission on Environment and Development. 1987. *Our Common Future*. Oxford, UK: Oxford University Press.

World Health Organization (WHO). 2000. *The World Health Report 2000—Health Systems: Improving Performance*.

Wright, D.H. 1990. "Human Impacts on Energy Flow through Natural Ecosystems, and Implications for Species Endangerment." *Ambio* 19 (4):189-194.

Wright, R. 2004. *A Short History of Progress*. New York, NY: Carroll and Graf Publishers.

Wuestenhagen, R., M. Wolsink and M.J. Buerer. 2007. "Social Acceptance of Renewable Energy Innovations—An Introduction to the Concept." *Energy Policy* 35:2683-2691.

Yankelovich, D. 1982. "Changing Public Attitudes to Science and the Quality of Life—Edited Excerpts from a Seminar." *Science, Technology, and Human Values* 7 (39):23-29.

Youngquist, W. 1997. *GeoDestinies—The Inevitable Control of Earth Resources over Nations and Individuals*. Portland, OR: National Book Company.

Zerzan, J. 2002. *Running on Emptiness—The Pathology of Civilization*. Los Angeles, CA: Feral House.

Zerzan, J., and A. Carnes, eds. 1991. *Questioning Technology—Tool, Toy, or Tyrant?* Gabriola Island, BC: New Society Publishers.

Ziman, J. 1971. "Social Responsibility (I)—The Impact of Social Responsibility on Science." *Impact of Science on Society* 21 (2):113-122.

Zinn, H. 1999. *A People's History of the United States—1492-Present*. New York, NY: HarperCollins Publishers Inc.

——. 2002. *You Can't Be Neutral on a Moving Train—A Personal History of Our Times*. Boston, MA: Beacon Press.

End Notes

1. Commoner (1971), pp. 33-39.
2. Schwartz (1971), p. 55.
3. Cited in Graedel and Allenby (1998), p. 37.
4. Cited in Peet (1992), p. 210.
5. Cited in Welford (1997), p. 1.
6. Schwartz (1971), p. 107.
7. Commoner (1971), p. 15.
8. Mayr (2002).
9. Commoner (1971), p. 41-45.
10. Commoner (1971), p. 41.
11. Commoner (1971), p. 43.
12. Ehrenfeld (1981), p. 126.
13. Ravetz (1990), p. 13.
14. Commoner (1971), p. 46.
15. Ellul (1976), p. 111.
16. Sinsheimer, Robert, "The Presumptions of Science," in Daly (1980), pp. 148-149.
17. Peet (1992), p. 78.
18. Schwartz (1971), p. 94; Mayr (2002); Ehrlich and Ehrlich (2009).
19. Joseph et al. (1949).
20. Starr, Chauncey, "Benefit Versus Risk," in Rhodes (1999), pp. 286-287.
21. Bacon (1620), pp. 343-371.
22. Commoner (1971), p. 189.
23. Commoner (1971), pp. 21, 23.
24. Stanford (2006).
25. Graedel and Allenby (1998), p. 39.
26. Ravetz (1996), p. 53.
27. Commoner (1971), pp. 183, 187.
28. Ravetz (1990), p. 270.
29. Winner (1977), p. 183.
30. Atkins (1984); Ayres (1998); Balzhiser et al. (1972); Prigogine (1961); Prigogine (1989). The reason the world does not disintegrate into chaos and disorder is related to the fact that Planet Earth, while being essentially a closed system with respect to the flow of matter, is an open system with respect to the flow of energy from the sun. As will be discussed in more detail in Chapter 6, the continuous flux of solar energy maintains the many complex structures observed on Earth, including life itself.
31. Daly (1996), p. 58.
32. Rifkin (1980), p. 123.
33. Ayres and Martinas (1995); Cleveland and Ruth (1997); Connelly and Koshland (1997); Faber et al. (1995); Glasby (1988); Huesemann (2001); Kuemmel (1989); O'Connor (1994); and Ruth (1993, 1995).

34. Huesemann (2001), more than one unit of "disorder" (entropy) is created under non-equilibrium conditions.
35. Georgescu-Roegen (1971); Georgescu-Roegen (1977).
36. Georgescu-Roegen, N., "The Entropy Law and the Economic Problem," in Daly (1980), p. 55.
37. Gruebler (1994).
38. Meadows et al. (2004), p. 113; Vitousek et al. (1997).
39. Carson (2005), p. 24.
40. Colborn et al. (1996).
41. Ehrlich and Ehrlich (1996), pp. 170, 171.
42. LeVay (2010); Colborn et al. (1996); Colborn and Clement (1992), pp. 403; Carpenter (1998).
43. Glennon (2009), p. 168.
44. co2now.org, accessed 7/4/2010.
45. Dukes (2003).
46. Intergovernmental Panel on Climate Change (IPCC), Working Group III Contribution, Climate Change 2007: Mitigation of Climate Change, ipcc .ch, accessed 7/5/2007.
47. Intergovernmental Panel on Climate Change (IPCC), Climate Change 2001 Synthesis Report, Summary for Policymakers, ipcc.ch, accessed 7/5/2007.
48. Jacobson (1990).
49. Huesemann (2006).
50. Huesemann (2006).
51. Myers and Kent (2001), p. 34.
52. Daily et al. (1997).
53. Schwartz (1971), p. 229; "Total Number of Species Estimated in the World," Current Results: Research News and Science Facts, currentresults.com/Environ ment-Facts/Plants-Animals/estimate-of-worlds-total-number-of-species .php, accessed 1/9/2011.
54. Wilson (2000).
55. May (1988), p. 1448.
56. Whitty (2007); Stork (2010).
57. Vitousek et al. (1997).
58. Quoted in Princen (2005), p. 36.
59. Hayes (1977), p. 97.
60. Jackson (1987), p. 103; Perelman (1977), p. v.
61. Berry (1977).
62. Robbins (1987), p. 357.
63. Ehrlich (1989), p. 13.
64. Youngquist (1997), p. 295.
65. Pimentel et al. (1995).
66. Postel (1999), p. 43.
67. Postel (1999), p. 262.
68. Pearce (2006).
69. Postel (1999); Reisner (1986).
70. See: noblis.org/missionareas/nsi/backgroundonchemicalwarfare/history ofchemicalwarfare/Pages/HistoryNerveGas.aspx; accessed 10/15/10.
71. Krebs (1992), p. 81; Wilson (1997), p. 12.
72. Colborn et al. (1996), p. 138.
73. Wilson (1997), p. 12.
74. Colborn et al. (1996), p. 138.

75. Carson, R., *Silent Spring*, in Rhodes (1999), p. 231.
76. Pimentel, D.H., et al., "Assessment of Environmental and Economic Impacts of Pesticide Use" in Pimentel and Lehman (1993), pp. 47–84.
77. Wilson (1997), p. 13.
78. Colborn et al. (1996).
79. Rifkin (1998), p. 115.
80. Wilson (1997), p. 136.
81. Shiva (1993), p. 108.
82. Rifkin (1998), pp. 71, 73.
83. Sinsheimer, R., "The Presumptions of Science," in Daly (1980), p. 150.
84. Project Censored, "Human Genome Project Opens the Door to Ethnically Specific Bioweapons," projectcensored.org/publications/2001/16.html, accessed 5/25/07.
85. Rifkin (1998), p. xiii.
86. Myers and Kent (2001), pp. 95, 96.
87. US Census Bureau (2000), Tables 1035 and 1038.
88. Myers and Kent (2001), p. 102.
89. Alvord (2000), p. 115.
90. Alvord (2000), p. 115.
91. Graedel and Allenby (1998), p. 129.
92. Myers and Kent (2001), p. 98.
93. Durning (1992), p. 83.
94. HighCountryNews.org, "Roadkill statistics," Vol. 37, No. 2, February 7, 2005, hcn.org/servlets/hcn.PrintableArticle?article_id=15268, accessed 3/20/09.
95. Myers and Kent (2001), p. 99; US Environmental Protection Agency (EPA), "U.S. Greenhouse Gas Inventory Reports, Inventory of U.S. Greenhouse Gas Emissions and Sinks: 1990–2004," April 2006, USEPA #430-R-06-002, epa.gov/climatechange/emissions/usinventoryreport.html, accessed 3/27/07.
96. Graedel and Allenby (1998), p. 136.
97. Graedel and Allenby (1998), p. 132; Myers and Kent (2001), pp. 95, 101.
98. Alvord (2000), p. 93.
99. Maddison et al. (1996), p. 118.
100. Korten (1995), p. 283.
101. Maddison et al. (1996), p. 84.
102. Graedel and Allenby (1998), pp. 4, 34.
103. Stockholm International Peace Research Institute (SIPRI), "Recent Trends in Military Expenditures," sipri.org/contents/milap/milex/mex_trends.html, accessed 3/27/07.
104. American Association for the Advancement of Science (AAAS), *AAAS Report XXXI, Research and Development FY 2007*, Chapter 2, "Historical Trends," Table I-11, aaas.org/spp/rd/rd07main.htm, accessed 3/27/07.
105. Korten (1995), p. 20; according to Zinn (1999, pp. 641, 642), ten times more civilians than combatants are killed in modern warfare.
106. White, M., "World War I Casualties," *Twentieth Century Atlas—Death Tolls*, users.erols.com/mwhite28/warstat1.htm; accessed 10/25/10; en.wikipedia.org/wiki/World_War_I_casualties, accessed 3/27/07.
107. White, M., "World War II Casualties," *Twentieth Century Atlas—Death Tolls*, users.erols.com/mwhite28/warstat1.htm; accessed 10/25/10; World War II Casualties, en.wikipedia.org/wiki/World_War_II_casualties, accessed 3/27/07.
108. Zinn (1999), pp. 422, 423.

109. Holmes (1989), pp. 227–230.
110. Schumacher (1997), p. 108.
111. Zinn (1999), p. 428.
112. Blum (2000), p. 106.
113. Dixon (1978), p. 4.
114. Commoner (1971), p. 189.
115. Kimbrell (1993), pp. 278, 293.
116. Illich (1975), p. 112.
117. Fries and Crapo (1981), p. 2.
118. Powles (1973), p. 13.
119. McKeown (1979), p. 5.
120. Dixon (1978), p. 55.
121. Temple, Norman, "Medical Research: A Complex Problem," in Temple and Burkitt (1994), p. 423.
122. Taylor (1979), p. 42.
123. Kohn et al. (2001); Starfield (2000).
124. Rifkin (1998), p. 233.
125. Konner (1993), p. 49.
126. Levy (1992), pp. 14, 30, 37.
127. Levy (1992), p. 7.
128. Konner (1993), p. 54.
129. Konner (1993), p. 53.
130. Levy (1992), p. 103.
131. Levy (1992), p. 225.
132. Union of Concerned Scientists, "Hogging It!: Estimates of Antimicrobial Abuse in Livestock" and "European Union Bans Antibiotics for Growth Promotion," ucsusa.org, accessed 4/13/2007.
133. Levy (1992), pp. 137–140.
134. National Institute of Health, "The Problem of Antimicrobial Resistance," April 2006, niaid.nih.gov/factsheets/antimicro.htm, accessed 4/13/2007.
135. Stokstad (2000).
136. Levy (1992), p. 183.
137. Callahan (1990), p. 169; Tenner (1996), p. 54.
138. Podrid (1989).
139. Jennett (1986), p. 25; Moss (1980), p. 57.
140. Taylor (1979), p. 126.
141. Colen (1986), p. 246.
142. Callahan (1990), p. 237.
143. Taylor (1979), p. 119.
144. Callahan (1990), pp. 242, 227.
145. National Coalition on Health Care, "Health Insurance Cost," nchc.org/facts/cost/shtml, accessed 4/13/07.
146. Deevey (1960); Jobling et al. (2003).
147. Deevey (1960); Jobling et al. (2003).
148. Deevey (1960).
149. Ehrlich and Ehrlich (1991).
150. Ehrlich and Ehrlich (1991, 2005); Grant (1992, 1996).
151. Pimentel and Pimentel (1996).
152. Campbell and Laherrere (1998); Romm and Curtis (1996).
153. Gaffin (1998).
154. Sieferle (2004).

155. McKeown (1988), p. 74.
156. McKeown (1988), p. 74.
157. Crow (1998), p. 12.
158. Crow (1973), p. 74.
159. Crow (1999).
160. Crow (1998), p. 11.
161. Crow (1997), pp. 83–85.
162. *Webster's New Collegiate Dictionary*, G&C Merriam Company, Springfield, MA, USA, 1979, p. 400.
163. Roszak et al. (1995).
164. Fox (1997), p. 16.
165. Fox (1997), p. 13.
166. Cited in Holmes (1989), p. 25.
167. Tudge (1999).
168. Callicott (1982); Mander (1991).
169. Chapman et al. (2000), p. 28.
170. Drengson (1995), p. 147.
171. Ellul (1976), p. 29.
172. Schwartz (1971), p. 30.
173. Sarewitz (1996), pp. 100, 103.
174. Rockstrom et al. (2009); Ehrlich and Ehrlich (2009).
175. Vitousek et al. (1997).
176. Vitousek et al. (1997).
177. Rojstaczer et al. (2001); Vitousek et al. (1986).
178. Wilson (2002).
179. Wackernagel and Rees (1996); Wilson (2002).
180. Huesemann (2006); Intergovernmental Panel on Climate Change (IPCC), *Special Report on Emission Scenarios*, ipcc.ch, 2001.
181. Youngquist (1997), p. 22.
182. Intergovernmental Panel on Climate Change (IPCC), *Third Assessment Report—Climate Change 2001: The Scientific Basis, Technical Summary of Working Group I*, ipcc.ch, 2001.
183. Approximately 40 percent of the world's food supply now comes from the 17 percent of cropland that is irrigated (Postel, 1999, p. 43).
184. Postel et al. (1996); Vitousek et al. (1997).
185. Postel (1999), p. 81.
186. Vitousek et al. (1997).
187. Postel (1999), p. 60.
188. Postel (1999), p. 80.
189. Glennon (2009).
190. Matson et al. (1997), Figure 4.
191. Vitousek et al. (1997).
192. Vitousek et al. (1997).
193. Klee and Graedel (2004).
194. Vitousek et al. (1997).
195. Youngquist (1997), p. 24.
196. Vitousek et al. (1997).
197. Colborn et al. (1996).
198. Rifkin (1998), pp. vii, 15.
199. Carson, Rachel, earthobservatory.nasa.gov/Features/Carson/Carson2.php, accessed 4/14/2011.

200. Dickson (1975), p. 82.
201. Dickson (1975), p. 73.
202. Princen (2005), p. 59.
203. Zinn (1999), p. 324.
204. Rothschild (1973), p. 34.
205. Maslow (1954).
206. Dickson (1975), p. 43.
207. Mander (1978), p. 67.
208. Cavanagh and Mander (2004), p. 236.
209. Mander (1978).
210. Gitlin (1972).
211. TNS Media Intelligence, "TNS Media Intelligence Forecasts 2.6 Percent Increase in U.S. Advertising Spending for 2007, January 8, 2007, tns-mi.com /news/01082007.htm, accessed 4/26/2007.
212. The Budget of the United States Government, Department of Education (PDF file), gpoaccess.gov/usbudget/fy07/browse.html, accessed 4/26/2007.
213. Kivel (2004), p. 136.
214. Mander (1991), p. 76; Kahneman et al. (2006), Table 4.
215. Kivel (2004), p. 136.
216. Cavanagh and Mander (2004).
217. Cavanagh and Mander (2004), p. 237.
218. Cavanagh and Mander (2004), p. 241.
219. Barsamian and Chomsky (2001), p. 152.
220. Mander (1978), p. 349.
221. Klare (2001); Westing (1986).
222. Parenti (1988).
223. Heinberg (2003), pp. 68–69; Westing (1986); Youngquist (1997); Zinn (1999), p. 359.
224. Heinberg (2003), p. 68–69; Youngquist (1997).
225. Parenti (1988), p. 38.
226. Parenti (1988), p. 12.
227. Commoner (1971), pp. 33–39.
228. Drengson (1995), pp. 157, 158.
229. Rosner (2004).
230. Ravetz (1996), p. 427.
231. Sarewitz (1996), p. 103.
232. Roszak, T., "White Bread and Technological Appendages," in Rhodes (1999), p. 309.
233. Mander (1991), p. 29.
234. Sarewitz (1996), p. 189.
235. Drengson (1995), p. 138.
236. Schwartz (1971), p. 274.
237. Zerzan (2002), p. 42.
238. Ravetz (1996), p. 427; McDermott, J., "Technology: The Opiate of the Intellectuals," in Bereano (1976), p. 81.
239. Rosner (2004).
240. Lappe et al. (1998), p. 59.
241. Schwartz (1971), pp. 76, 77.
242. Schwartz (1971), Chapter 4.
243. Huesemann (2003).
244. Durning (1992), p. 58.

245. Stunkel and Sarsar (1994), pp. 81, 82.
246. Skolnikoff, E., "Four Generalizations," in Rhodes (1999), p. 351.
247. Bereano (1976), p. 10.
248. Commoner (1971), pp. 39–41.
249. Tenner (1996), p. 72.
250. Huesemann (2001; 2003).
251. Ayres and Martinas (1995); Cleveland and Ruth (1997); Connelly and Koshland (1997); Faber et al. (1995); Glasby (1988); Huesemann (2001); Kuemmel (1989); O'Connor (1994); Ruth (1993).
252. Bedient et al. (1994).
253. Huesemann (2003).
254. Huesemann (2001).
255. Huesemann (2001).
256. Huesemann (2001).
257. Huesemann (2006).
258. Huesemann (2006).
259. Huesemann (2006).
260. Goodell (2010); Kintisch (2010a).
261. Huesemann (2006).
262. Behar, M, "How Earth-Scale Engineering Can Save the Planet," *Popular Science Magazine*, June 2005. popsci.com/environment/article/2005-06/how-earth-scale-engineering-can-save-planet, accessed 4/14/2011.
263. Teller et al. (1997).
264. Norman (1995).
265. Barsamian and Chomsky (2001), p. 34.
266. Melman (1984), p. 162.
267. Norman (1995), p. 243.
268. Holmes (1989), p. 257.
269. Clark (1972), p. 7.
270. Johnson (1984), p. 111.
271. Braun (1995), p. 91.
272. Schwartz (1971), pp. 202–215.
273. Tenner (1996), pp. 32–37.
274. McKeown (1988), Chapters 1 and 3.
275. Taylor (1979), p. 224.
276. Dubos (1959), pp. 179, 180.
277. Gura (2003), p. 849.
278. Silverman and Lee (1974), p. 273.
279. Taylor (1979), pp. 224, 225.
280. Bereano (1976), p. 11.
281. Stunkel and Sarsar (1994), p. 82.
282. Cited in Ehrenfeld (1981), p. 108.
283. Schumacher (1979), p. 98.
284. Princen (2005), p. 84.
285. Princen (2005), p. 54.
286. Princen (2005), pp. 53, 73.
287. Princen (2005), p. 54.
288. Gruebler (1994).
289. Maddison (1991), Table A.2.
290. Maddison (1991), Table 1.1.
291. Nelson (1996).

292. Braun (1995); Schumpeter (1934).

293. Jevons (1865).

294. Greening et al. (2000).

295. Huesemann and Huesemann (2008).

296. Saunders (1992; 2000).

297. Bentzen (2004); Binswanger (2001); Hertwich (2005).

298. Greene (1992); Greening et al. (2000).

299. Haas and Biermayr (2000).

300. Hannon (1975).

301. Ayres and van den Bergh (2005); Ayres and Warr (2005); Beaudreau (2005); Samuelson and Nordhaus (1989); Solow (1957).

302. Samuelson and Nordhaus (1989).

303. Huesemann and Huesemann (2008), Table 1.

304. Organization for Economic Cooperation and Development (OECD) and International Energy Agency (IEA) Report, *30 Years of Energy Use in IEA Countries*, International Energy Agency, Paris, France, 2004, iea.org/textbase/publications/free_new_Desc.asp?PUBS_ID=1260OECD/IEA (2004).

305. Huesemann and Huesemann (2008).

306. Allenby (1999), p. 54; National Resource Council (1992).

307. Myers and Kent (2001), pp. 95, 96; Graedel and Allenby (1998), p. 115.

308. Allenby (1999), pp. 34, 56, 57.

309. Allenby (1999), pp. 34, 56, 57.

310. Alvord (2000), p. 192; Lovins and Hunter-Lovins (1995).

311. Organization for Economic Cooperation and Development (OECD) and International Energy Agency (IEA) Report, *30 Years of Energy Use in IEA Countries*, International Energy Agency, Paris, France, 2004.

312. Huesemann and Huesemann (2008).

313. Herring (1999).

314. Huesemann and Huesemann (2008).

315. Adriaanse et al. (1997).

316. Huesemann and Huesemann (2008).

317. Hoffren (2006).

318. Marland, G., T.A. Boden and R.J. Andres, "Global, Regional, and National Fossil Fuel CO_2 Emissions (up to 2002)," Carbon Dioxide Information Analysis Center, Oak Ridge National Laboratory, mercury.ornl.gov/cdiac, accessed 4/14/2011.

319. Huesemann (2006).

320. CO_2 Now, co2now.org, accessed June 11, 2010.

321. Huesemann (2006); Intergovernmental Panel on Climate Change (2001), "Special Report on Emission Scenarios," grida.no/publications/other/ipcc _sr/?src=/climate/ipcc/emission/, accessed 4/14/2011.

322. Huesemann and Huesemann (2008).

323. Holtz-Kay (1997), p. 149; Alvord (2000), p. 17.

324. Alvord (2000), p. 17.

325. Alvord (2000), p. 17.

326. Holtz-Kay (1997), p. 149.

327. Alvord (2000), p. 93.

328. Maddison et al. (1996), p. 99.

329. Cowan, R.S., "Less Work for Mother?" in Teich (1993), pp. 333, 335.

330. Vanek (1974).

331. Ropke (1999).

332. Ropke (1999); Vanek (1974).

333. Cowan, R.S., "Less Work for Mother?" in Teich (1993), pp. 332, 333.

334. Ausubel and Gruebler (1995), p. 127.

335. Lappe (1984), p. 154; Perelman (1977), p. 71.

336. Vanek (1974).

337. Cowan, R.S., "Less Work for Mother?" in Teich (1993), pp. 330, 331; Vanek (1974).

338. Vanek (1974).

339. Cited in Durning (1992), p. 47.

340. Ropke (1999).

341. Ausubel and Gruebler (1995); Gruebler (1994), p. 62; Schor (1992).

342. Maddison (1991), Appendix C.

343. Schor (2005).

344. Schor (2005).

345. Schor (1992); Schor (2005), Table 2.

346. Mumford (1934), p. 396.

347. National Coalition on Health Care, "Health Insurance Cost," nchc.org /facts/cost/shtml, accessed 4/13/07.

348. Ginzberg (1990); Kosterlitz (1993), p. 1969.

349. Kosterlitz (1993), p. 1969.

350. Callahan (1990), pp. 54, 57.

351. Callahan(1990), pp. 63–65.

352. Callahan (1990), pp. 100, 221.

353. Balzhiser et al. (1972).

354. Jochem (2000).

355. Jochem (1991).

356. Balzhiser et al. (1972).

357. Nakicenovic and Gruebler (1993); Nakicenovic et al. (1998b).

358. Nakicenovic and Gruebler (1993).

359. Huesemann (2006).

360. Cleveland and Ruth (1999).

361. Herman, R., S.A. Ardekani, and J.H. Ausubel, "Dematerialization," in Ausubel and Sladovich (1989).

362. Dresner (2002), p. 104.

363. Cleveland and Ruth (1999), pp. 15, 40.

364. Cited in Reid (1995), p. 169.

365. Winner (1977), pp. 229, 242.

366. Stunkel and Sarsar (1994), p. 87.

367. Kimbrell (1993), p. 345.

368. McDonough and Braungart (2002), p. 65.

369. Ehrlich, P., and A. Ehrlich, "Humanity at the Crossroads," in H.E. Daly (1980), p. 43.

370. Daly, H., "The Steady-State Economy: Toward a Political Economy of Bio-physical Equilibrium and Moral Growth," in Daly (1980), p. 353.

371. McDonough and Braungart (2002), p. 65.

372. Allenby (1999), p. 29.

373. Huesemann and Huesemann (2008), Figure 2; Maddison (1991), Table 1.1.

374. Huesemann (2006); Intergovernmental Panel on Climate Change (IPCC), *Special Report on Emission Scenarios*, grida.no/publications/other/ipcc_sr /?src=/climate/ipcc/emission/, accessed 4/14/2011.

375. Korten (1995), p. 81.

376. World Commission on Environment and Development (1987), p. 9.
377. Reid (1995), p. 212.
378. Welford (1997), p. 69.
379. Welford (1997), p. 75.
380. Organization for Economic Cooperation and Development, OECD, (1998).
381. Welford (1997), p. 79.
382. Schmidheiny (1992), p. xii.
383. DeSimone and Popoff (1997).
384. Welford (1997), p. 29.
385. World Commission on Environment and Development (1987), p. 206.
386. Hoffren and Korhonen (2007).
387. Chertow (2001); Ehrlich and Holdren (1971); Graedel and Allenby (1995).
388. Graedel and Allenby (1995), pp. 7, 8.
389. Cleveland and Ruth (1999); Dresner (2002).
390. Organization for Economic Cooperation and Development (1998); Lovins et al. (1997).
391. President's Council on Sustainable Development (1996).
392. Huesemann (2006).
393. US Energy Information Administration, "Annual Energy Review 2009," eia.doe.gov/emeu/aer/pecss_diagram.html, accessed 4/14/2011.
394. Steinhart and Steinhart (1974), p. 313.
395. Pirog, R. "Food Miles: A Simple Metaphor to Contrast Local and Global Food Systems," Leopold Center for Sustainable Agriculture, Ames, Iowa, 2004, leopold.iastate.edu/pubs/staff/files/local_foods_HEN0604.pdf; Trainer (1995a), p. 21.
396. Steinhart and Steinhart (1974), p. 313.
397. Cited in Daly, H., "Steady-State Economics," in Merchant (1994), p. 101.
398. Campbell and Laherrere (1998); Heinberg (2003); Romm and Curtis (1996).
399. Schumacher (1979), p. 20.
400. Youngquist (1997).
401. There has been a long-standing debate among scholars about the definition of sustainability conditions. One view held by many economists is that "weak sustainability" is sufficient (Gutes, 1996; Pearce and Atkinson, 1993; Renning and Wiggering, 1997). According to the weak sustainability criterion, depletion of natural resources is acceptable as long as the aggregate stock of manufactured and natural assets is not decreasing; that is, human-made capital is used as a substitute for depleted natural capital (Wackernagel and Rees, 1995). This viewpoint is opposed by many biologists and environmentally minded economists who assert that natural capital stocks need to be held constant independent of human capital in order to guarantee "strong" sustainability (Costanza and Daly, 1992). Given that human capital cannot indefinitely substitute for an ever-declining stock of natural resources, it is clear that "strong" rather than "weak" sustainability is necessary to ensure that current economic activities can continue indefinitely without serious interruptions.
402. Ayres (1996a); Daly (1980); Daly (1996); Gutes (1996); Huesemann (2003); Hueting and Reijnders (1998); and O'Riodan (1993).
403. Huesemann (2003), Figure 1.
404. Allenby (1999); Ayres and Ayres (1996); Graedel and Allenby (1995).

405. It should be noted here that nuclear energy is neither renewable nor sustainable because uranium reserves are limited; breeder reactors are currently illegal in the United States because of concerns over nuclear weapons proliferation; and the disposal of radioactive waste is likely to cause environmental and health problems for future generations (Huesemann, 2006).

406. Huesemann (2006); Intergovernmental Panel on Climate Change (IPCC), *Special Report on Emission Scenarios*, 2001, grida.no/publications/other/ipcc_sr/?src=/climate/ipcc/emission/, accessed 4/14/2011; Kerr (2010).

407. Myers and Kent (2001).

408. Boyle (1996); Brower (1992); Hayes (1977); Lovins (1977).

409. Huesemann (2001); Huesemann (2003); Pimentel (2008); Trainer (1995b).

410. Huesemann (2006), Table 1.

411. Atkins (1984); Ayres (1998); Nicolis and Prigogine (1977).

412. Ayres and Martinas (1995).

413. Clarke (1994); Haefele (1981), Chapter 10; Holdren et al. (1980).

414. Clarke (1994).

415. Holdren et al. (1980), p. 248.

416. Intergovernmental Panel on Climate Change (2001), *Special Report on Emission Scenarios*, grida.no/publications/other/ipcc_sr/?src=/climate/ipcc/emission/, accessed 4/14/2011; Pimentel et al. (1994b).

417. Huesemann (2006), Table 1.

418. Huesemann (2006), Table 1.

419. Rojstaczer et al. (2001); Vitousek et al. (1986).

420. Ehrlich and Ehrlich (1981); Vitousek et al. (1986); Wright (1990).

421. Pimentel et al. (1994a).

422. US Department of Energy (2009), Energy Information Administration, eia.doe.gov/emeu/aer/pecss_diagram.html, accessed 4/14/2011.

423. Williams, R. "Roles for Biomass Energy in Sustainable Development," in Socolow et al. (1994).

424. Hoffert et al. (2002); Nakicenovic et al. (1998b).

425. Kheshgi et al. (2000).

426. Williams, R. "Roles for Biomass Energy in Sustainable Development," in Socolow et al. (1994).

427. Huesemann (2006), Table 1.

428. Pimentel et al. (1984a).

429. Brown, L.R., "Distillery Demand for Grains to Fuel Cars Vastly Understated: World May be Facing Highest Grain Prices in History", Earth Policy Institute, January 4, 2007. earth-policy.org/index.php?/plan_b_updates/2007/update63, accessed 4/14/2011.

430. Rosegrant et al. (2006).

431. Runge et al. (2003); Senauer and Sur (2001).

432. Nakicenovik et al. (1998b).

433. Giampietro et al. (1997).

434. Pimentel and Lehman (1993).

435. Pimentel et al. (1994b).

436. Pimentel et al. (1994b).

437. Pimentel et al. (1994a).

438. Pimentel et al. (1994b).

439. US Department of Energy (2009), Energy Information Administration, eia.doe.gov/emeu/aer/pecss_diagram.html, accessed 4/14/2011.

440. Pimentel et al. (1984b).

441. Pimentel et al. (1994b).

442. US Department of Energy (2009), Energy Information Administration, Renewable Energy, eia.doe.gov/emeu/aer/renew.html, accessed 4/14/2011.

443. Pimentel et al. (1984b); Pimentel et al. (1994b).

444. Turner (1999).

445. Holdren et al. (1980).

446. Elliott et al. (1991).

447. US Department of Energy (2009), Energy Information Administration, eia.doe.gov/emeu/aer/pecss_diagram.html, accessed 4/14/2011.

448. Elliott (1997), Chapter 11; Wuestenhagen et al. (2007); Kintisch (2010b).

449. Dohmen and Hornig (2004).

450. MacKay (2009).

451. Commoner (1971), p. 46.

452. It should be noted that prior to the Industrial Revolution, all agricultural societies were operating solely on solar energy, which allowed for only very limited economic activities and rather frugal lifestyles.

453. Trainer (1995a), pp. 117, 128.

454. Youngquist (1997).

455. Meadows et al. (2004), Table 3-2, p. 105.

456. Meadows et al. (2004), Table 3-2, p. 105; Skinner (1987), Table 4.

457. Sagoff (1995), p. 611; Simon (1996).

458. Cleveland and Ruth (1999); Herman, R., S.A. Ardekani, and J.H. Ausubel, "Dematerialization," in Ausubel and Sladovich (1989), p. 68.

459. Braun (1995), p. 165.

460. Daly (1996), p. 61.

461. Dresner (2002); Gutes (1996).

462. Ehrlich, P., and A. Ehrlich, "Humanity at the Crossroads," in Daly (1980), p. 41.

463. Ayres (1996a), p. 252.

464. Ayres, R.U., "Industrial Metabolism: Theory and Policy," in Allenby and Richards (1994), pp. 23–37.

465. Ayres, R.U., "Industrial Metabolism: Theory and Policy," in Allenby and Richards (1994), Table 4.

466. Duales System Deutschland AG, www.gruener-punkt.de, accessed 4/14/2011.

467. Ayres, R.U., "Industrial Metabolism: Theory and Policy," in Allenby and Richards (1994), pp. 23–37.

468. Ayres (1999); Bianciardi et al. (1993); Connelly and Koshland (1997).

469. O'Connor (1994).

470. Huesemann (2003), Figure 2.

471. Connelly and Koshland (1997).

472. Skinner (1987); Youngquist (1997).

473. Ayres (1994); Klee and Graedel (2004).

474. As Ernest Braun (1995, p. 162) notes, "Ever since the publication of *Limits to Growth* (Meadows, 1972) there has been no excuse for not realizing that exponential growth in consumption, population and pollution cannot be sustained in a world with finite resources and a finite capacity to digest the waste products of our civilization."

475. Daly (1996), p. 33.

476. Wilson (2002), p. 84.

477. Wackernagel and Rees (1996), pp. 55, 57.

478. Diamond (2005); Ehrlich and Ehrlich (2005); Tainter (1988); Wright (2004).

479. Meadows et al. (2004).

480. Meadows et al. (2004), pp. 164, 167.

481. Tainter (1988), p. 209.

482. See also Turner (2008), who showed that 30 years of historical data compare favorably with key features of the business-as-usual scenario called "standard run" in the Club of Rome's study "The Limits to Growth." Modeling of the "standard run" scenario predicted a global systems collapse midway through the 21st century.

483. Meadows et al. (2004), pp. 8, 234.

484. Diamond (2005); Wright (2004).

485. Tainter (1988).

486. Mumford (1934), p. 182.

487. Braun (1995), p. 22.

488. Cited in Leitenberg (1971), p. 129.

489. Cited in Dickson (1975), p. 12.

490. Strauss, L.L., "Too Cheap to Meter," in Rhodes (1999), p. 196.

491. Mander (1991), p. 22.

492. Cited in Dickson (1975), p. 35.

493. Beckerman (1996), p. 56.

494. Simon (1996), pp. 12, 73.

495. Harrison (2000), p. 104.

496. Diamond (2005); Tainter (1988).

497. Winner, L., "Mythinformation," in Zerzan and Carnes (1991), p. 164.

498. Stunkel and Sarsar (1994), p. 36.

499. Platt (1969).

500. Hopper (1991), pp. 33, 34.

501. Zerzan (2002), p. 76.

502. Ravetz (1990), p. 222; Stunkel and Sarsar (1994), p. 36.

503. Bury, J.B., "The Idea of Progress," ca. 1920, cited in Hopper (1991), p. 33.

504. Hopper (1991), p. 40.

505. Drengson (1995), p. 68.

506. Postman (1992), p. 60.

507. Susskind (1973), p. 12; Stunkel and Sarsar (1994), p. 14.

508. Hopper (1991), p. 53.

509. Marx, L., "Does Improved Technology Mean Progress," in Teich (1993), p. 5.

510. Olmsted, D., "On the Democratic Tendencies of Science," *Barnard's Journal of Education* (1855-1856), cited in Hughes (1975), p. 144.

511. Hopper (1991), pp. 33, 53, 66; Marx, L., "Does Improved Technology Mean Progress" in Teich (1993), pp. 7, 9.

512. Hamilton (2003), pp. 98, 101.

513. Taviss (1972).

514. National Science Foundation (NSF, USA), Science and Engineering Indicators—2004, Chapter 7, Science and Technology: Public Attitudes and Understanding, pp. 7-23, nsf.gov/statistics/seind04/c7/c7h.htm, accessed 4/14/2011.

515. Catton and Dunlap (1980).

516. Barbour (1980), p. 17; Deyo and Patrick (2005), p. 17; Norem and Chang, Chapter 16, in Chang (2001), pp. 360-362.

517. Postman (1992), p. 55.

518. Stewart and Bennett (1991), Chapter 6.

519. Fromm (1967), p. 129.

520. Konner (1993), pp. 22, 23.

521. Greene (1998), Power Law 27.

522. Postman (1992), p. 58.

523. Ehrman (1996), Ehrman (2005).

524. Yankelovich (1982). This is another similarity between technological optimism and religious faith. Religious beliefs also appear to be most strong
and fanatic when they relate to topics or items for which there is no factual
evidence whatever.

525. National Science Foundation (NSF, USA), Science and Engineering Indicators—2004, Chapter 7, "Science and Technology: Public Attitudes and Understanding," pp. 7-15, 7-16, 7-20, nsf.gov/statistics/seind04/c7/c7h.htm, accessed 4/14/2011.

526. National Science Foundation (NSF, USA), "Science and Engineering Indicators—2004, Chapter 7, Science and Technology: Public Attitudes and
Understanding," p. 7-15, nsf.gov/statistics/seind04/c7/c7h.htm, accessed
4/14/2011.

527. National Science Foundation (NSF, USA), "Science and Engineering Indicators—2004, Chapter 7, Science and Technology: Public Attitudes and
Understanding," p. 7-23, nsf.gov/statistics/seind04/c7/c7h.htm, accessed
4/14/2011.

528. According to Thomas Hobbes, life in primitive cultures was "nasty, brutish, and short." Thomas Hobbes, *Leviathan*, Chapter 8, oregonstate.edu/in
struct/phl302/texts/hobbes/leviathan-c.html#CHAPTERXIII, accessed
10/7/2010.

529. Mander (1991), Chapter 14.

530. Simon (1996), p. 37.

531. Simon (1996), p. 163.

532. Ehrlich, P., and A. Ehrlich, "The Cornucopian Fallacies," in Grant (1992),
p. 137.

533. See, for example, vanguard.com or fidelity.com.

534. Schumacher (1977), p .35; Thomas Hobbes, *Leviathan*, Chapter 8, oregon
state.edu/instruct/phl302/texts/hobbes/leviathan-c.html#CHAPTERXIII,
accessed 10/7/2010.

535. Schumacher (1977), p. 201.

536. For the sake of comparison, it is interesting to observe that most wild animals in nature appear to live reasonably content and happy lives without
the benefits of modern science and technology.

537. Mander (1991), pp. 144, 146, 246.

538. Gruebler (1994).

539. Schor (2005).

540. Schor (2005), Table 2.

541. Mander (1991), p. 255.

542. Mander (1991), pp. 248-252.

543. Runge et al. (2003), p. 17.

544. Deyo and Patrick (2005), p. 14.

545. Kranczer (1999); Sagan (1987), Table 1.1., p. 16.

546. McKeown (1979), p. 22; Oeppen and Vaupel (2002).

547. Sagan (1987), p. 79. Because life expectancies at birth increased primarily

as a result of reductions in childhood mortalities, increases in life expectancies at higher ages have been less pronounced. For example, a 15-year-old male in the United States in 1900 had an average life expectancy of about 61 years, whereas in 1996, he had an average life expectancy of about 75 years, a gain of only 14 years. At age 45, the male life-expectancy in the US was 69 years in 1900 and 77 years in 1996, a gain of only 8 years (Kranczer, 1999, Table 1).

548. Taylor (1979), pp. 14, 15.
549. Approximately 86 percent of the total reduction in the death rate from the beginning of the 18th century can be attributed to the decline in infectious diseases (McKeown, 1979, p. 33).
550. McKeown (1988), p. 89.
551. Taylor (1979), p. 11.
552. Fries and Capro (1981), p. 61.
553. Illich (1975), p. 17; McKinlay and McKinlay (1977), p. 421; Sagan (1987), p. 68.
554. Illich (1975), p. 17.
555. McKinlay and McKinlay (1977), p. 425.
556. Temple, N., "Organized Medicine: An Ounce of Prevention or a Pound of Cure," in Temple and Burkitt (1994), p. 385.
557. Sagan (1987), p. 81.
558. US National Center for Health Statistics, Vital Statistics of the United States, annual, National Vital Statistics Report NVSR, infoplease.com/ipa/A0922202.html, accessed 3/7/2008.
559. US Department of Health and Human Services, National Health Expenditure Data, Historical, Spreadsheet of National Health Expenditure (NHE) Summary including Share of GDP, 1960–2006, cms.hhs.gov, accessed 2/11/2008.
560. Wolf (1959), p. 689.
561. Beecher (1955), p. 1603; Wolf (1959), p. 692.
562. Weil (1998), p. 227.
563. Wolf (1959), p. 689.
564. Weil (1998), p. 227.
565. Wolf (1959), p. 694.
566. Beecher (1961), p. 91.
567. Weil (1998), p. 195.
568. Weil (1998), p. 209.
569. Beecher (1961), p. 91.
570. Weil (1998), p. 226.
571. Beecher (1961), p. 90; Konner (1993), p. 107.
572. Deyo and Patrick (2005), pp. 207, 208.
573. Beecher (1955).
574. Beecher (1961).
575. Fries and Crapo (1981), p. xii.
576. Fries and Crapo (1981), pp. 3, 71. The life span is 85 years plus/minus four years standard deviation.
577. The effect of medical treatments on increasing life expectancies is very limited. According to one estimate, even if cancers of all kinds were either prevented or permanently cured, the longevity of 40-year-old white males in the US would be increased by only two years (Jennett, 1986, p. 44).

578. Fries and Crapo (1981), p. 34.
579. Anonymous (2000).
580. Mumford, L., "The Technological Imperative," in Rhodes (1999).
581. Mander (1991), pp. 34, 35.
582. Mander (1991), p. 2.
583. Herman and Chomsky (1988), p. 298.
584. Herman (1992), p. 14.
585. Herman and Chomsky (1988), p. 5.
586. Herman and Chomsky (1988), p. 306.
587. Postman (1992), pp. 9, 12.
588. Ravetz (1990), p. 216.
589. Mander (1991), p. 29.
590. Dickson (1975), p. 42.
591. Herman (1992), p. 12.
592. Hughes (1975), p. 213.
593. Stunkel and Sarsar (1994), p. 97.
594. Mander (1991), p. 190.
595. Ravetz (1990), p. 153.
596. Ehrlich and Ehrlich (1996); Ravetz (1990), p. 285.
597. Juenger (1956), p. 22.
598. Ravetz (1990), p. 13.
599. Ellul (1976), p. 106.
600. Postman (1992), p. 5.
601. Postman (1992), p. xii.
602. Commoner (1971), p. 45-46.
603. Quote by Harvey Brooks of Harvard University, cited in Ravetz (1990), p. 14.
604. Rifkin (1998), p. 36.
605. Rifkin (1998), pp 36, 231.
606. Daly, H.E., "The Steady-State Economy: Toward a Political Economy of Bio-physical Equilibrium and Moral Growth," in Daly (1980), p. 353.
607. Boardman et al. (1996), pp. 7-11.
608. See, for example, Stone (1996).
609. Lowrance (1985), p. 136.
610. Dovers and Handmer (1995).
611. Barbour (1980), p. 171; Lowrance, W.W. "Choosing our Pleasures and Our Poisons," in Teich (1993), p. 184.
612. Mander (1991), p. 33.
613. Commoner (1971), p. 256; Mesthene (1970), p. 39.
614. Barbour (1980), p. 87.
615. Dresner (2002), p. 116.
616. Routley and Routley (1978), p. 137.
617. Barbour (1980), p. 84; Routley and Routley (1978), p. 150.
618. Barbour (1980), p. 85.
619. Des Jardins (1993), p. 92.
620. Routley and Routley (1978), pp. 160, 161.
621. Here we can also learn from Native Americans, specifically from members of the Iroquois Confederacy, who, following the Great Law of the Iroquois, carefully considered all potential consequences of their planned actions on the next seven generations.
622. Peet (1992), p. 119.

623. Alvord (2000), p. 111; Kimbrell et al. (1998); Dower (1997), p. 34.
624. Ferris and Wallace (2009).
625. Dickson (1975), p. 91.
626. Barsamian and Chomsky (2001), p. 185.
627. Korten (2009), p. 127.
628. Korten (1995), p. 77.
629. Perrow (1999).
630. Boardman et al. (1996), p. 15.
631. Ravetz (1990), p. 268.
632. Sclove (1995), pp. 5, 7.
633. Boardman et al. (1996), p. 15.
634. Colborn et al. (1996).
635. Green, H., "Adversary Processes in the Social Control of Science and Technology," in Bereano (1976), p. 542.
636. Boardman et al. (1996), p. 16.
637. Barbour (1980), p. 128; Huesemann (2002); Ravetz (1990), p. 296.
638. Boardman et al. (1996), pp. 20, 22.
639. Rayner, S., "Risk and Relativism in Science for Policy," in Johnson and Covello (1987).
640. Dake, K., "Myths of Nature: Culture and the Social Construction of Risk," in Johnson and Covello (1987).
641. Dietz et al. (1989).
642. Proctor (1995), p. 85.
643. Schnaiberg and Gould (1994), p. 228.
644. Barbour (1980), p. 164.
645. Douglas and Wildavsky (1982), p. 70.
646. Kelman, S., "Cost-Benefit Analysis: An Ethical Critique," in Glickman and Gough (1990), p. 132; Proctor (1995), p. 87.
647. Stunkel and Sarsar (1994), p. 96.
648. Peet (1992), pp. 137, 145.
649. Starr, C., "Social Benefit Versus Technological Risk." in Glickman and Gough (1990).
650. Commoner (1971), p. 206.
651. Barbour (1980), p. 176.
652. Boardman et al. (1996), p. 2; Stunkel and Sarsar (1994), p. 96.
653. Bush, C.G., "Women and the Assessment of Technology," in Teich (1993), p. 206.
654. Barbour (1980), p. 167.
655. This is a typical example of environmental racism.
656. Barbour (1980), p. 163.
657. Quoted by Lowrance, W.W., "Choosing Our Pleasures and Poisons," in Teich (1993), pp. 184, 185.
658. Barbour (1980), p. 202.
659. Environmental Literacy Council, "Automobiles," enviroliteracy.org/article.php/1127.html, accessed 4/4/2008.
660. Alvord (2000), p. 110.
661. Holtz-Kay (1997), p. 120; Alvord (2000), p. 110.
662. Delucchi (1996); Ketcham and Komanoff (1992); Litman (1992); MacKenzie et al. (1992); Moffet and Miller (1993); US Congress, Office of Technology Assessment (1994); Kimbrell et al. (1998).
663. Department of Energy, Energy Information Agency, *Monthly Energy Review,*

February 2008, Table 9.4, Motor Gasoline Retail Prices, US City Average, eia.doe.gov/emeu/mer/pdf/pages/sec9.6.pdf, accessed 5/10/2008.

664. According to Kimbrell et al. (1998, p. 106), gasoline in the 1990s was cheaper in real dollars than it has been for the past 60 years, and it was still cheaper than milk or Perrier water.

665. It has been estimated that the total amount of car driving in the United States would decrease by about 50 percent if drivers had to pay the correct market-price for transportation fuels (Alvord , 2000, p.111).

666. Holtz-Kay, (1997), p. 121.

667. Maddison et al. (1996), p. 150.

668. Kimbrell et al. (1998), p. 7.

669. Holtz-Kay (1997), p. 122.

670. US Census Bureau (2000), Statistical Abstracts of the United States, Chapter 7, Transportation—Land, Table 1048.

671. US Census Bureau (2000), Statistical Abstracts of the United States, Chapter 7, Transportation—Land, Table 1033.

672. US Census Bureau (2000), Statistical Abstracts of the United States, Chapter 7, Transportation—Land, Table 1048.

673. US Census Bureau (2000), Statistical Abstracts of the United States, Chapter 7, Transportation—Land, Table 1049.

674. DeNavas-Walt et al. (2003), Figure 3.

675. Tenner (1996), p. 264.

676. Pimentel (2008). According to a recent article in Science by Scharlemann and Laurance (2008),although many biofuels are associated with lower greenhouse gas emissions, their aggregate environmental costs are greater than gasoline.

677. Pimentel (2003).

678. Rosegrant et al. (2006).

679. Runge et al. (2003); Senauer and Sur (2001).

680. Farrell et al. (2006); Hill et al. (2006); Pimentel (2003); Pimentel and Patzek (2005); Wang (2005).

681. Wang (2005).

682. Pimentel (2003); Pimentel and Patzek (2005), Wang (2005).

683. Searchinger et al. (2008).

684. Searchinger et al. (2008).

685. Searchinger et al. (2008).

686. Searchinger et al. (2008).

687. Dixon (1978), p. 226.

688. Jennett (1986), p. 174; Lesser, H., "Technology and Medicine: Means and Ends," in Braine and Lesser (1988), Chapter 4.

689. Jennett (1986), p. 103.

690. Deyo and Patrick (2005), p. 39.

691. Taylor (1979), p. 255.

692. Konner (1993), p. 113.

693. McKinlay (1981).

694. It should be noted here that patients, after seeing new drugs and high-technology treatments advertised in the mass media, frequently ask their doctors to try these new therapies. Thus, doctors often have little choice but to adopt new medicines or procedures, even though they have not been rigorously tested for efficacy.

695. McKinlay (1981).
696. McKinlay (1981), p. 395.
697. Rogers (2003).
698. Konner (1993), p. 106.
699. Deyo and Patrick (2005), p. 11.
700. Deyo and Patrick (2005), p. 128.
701. Deyo and Patrick (2005), p. 138; McKinlay (1981), p. 382.
702. Silverman and Lee (1974), p. 40.
703. Deyo and Patrick (2005), p. 186.
704. Deyo and Patrick (2005), p. 39.
705. Sagan (1987), p. 77.
706. Simon Kuznets, the originator of the system of uniform national accounts in the United States, warned Congress in 1934, "The welfare of a nation can scarcely be inferred from a measurement of national income" (Hamilton, 2003, p. 13).
707. Cavanagh and Mander (2004), p. 198–199.
708. Cobb and Halstead (1996), as cited in Cavanagh and Mander (2004), p. 200.
709. Examples of the many unexpected negative consequences of new technologies and industrial activities are given in Chapter 2. Counter-technologies to compensate for these various negative effects are described in Chapter 4.
710. Korten (1995), p. 40.
711. Barnard et al. (1995).
712. Cited in Schumacher (1979), p. 194.
713. Cavanagh and Mander (2004), p. 202.
714. Brown et al. (1990), p. 8.
715. Cavanagh and Mander (2004), p. 199; Hamilton (2003), p. 55; Korten (1995), p. 40; Naess and Rothenberg (1989), p. 113; Reid (1995).
716. Cavanagh and Mander (2004), p. 199.
717. Brown et al. (1990), p. 189.
718. Daly (1996), pp. 101–102.
719. Daly and Cobb (1989).
720. Hamilton (2003), p. 55; Jackson and Marks (1999).
721. For additional information on GPI, see gpipacific.org/; gpiatlantic.org/; accessed 4/15/2011.
722. Cavanagh and Mander (2004), p. 205.
723. Cavanagh and Mander (2004), p. 205; Hamilton (2003), p. 60; Jackson and Marks (1999), p. 423.
724. Maslow (1954).
725. Max-Neef (1991).
726. Jackson (2005); Reid (1995), pp. 83, 84.
727. Des Jardins (1993), p. 228.
728. Jackson (2005).
729. Jackson and Marks (1999), p. 430.
730. Winner (1977), p. 73.
731. Mumford (1934), p. 104.
732. Dickson (1975), p. 88; Wachtel (1989), p. 287.
733. Braun (1995), p. 190.
734. Korten (1995), p. 151.
735. Cavannagh and Mander (2004); Hamilton (2003), p. 119.
736. Graedel and Allenby (1998), p. 27.

737. Kivel (2004), p. 136.
738. Cavanagh and Mander (2004), p. 237.
739. Korten (1995), p. 152; Korten (2006), p. 338; Mander (1991), p. 79.
740. Lane (2001), p. 179.
741. Mander (1991), p. 126.
742. Polak and McCollough (2006), p. 346.
743. Frey and Stutzer (2002).
744. Hamilton (2003), p. 33.
745. Argyle (1987), p. 95; Oswald (1997); Lutter (2007).
746. Frey and Stutzer (2002).
747. Diener and Seligman (2004); Hamilton (2003), p. 33; Lane (2001).
748. Diener and Seligman (2004); Korten (2006), p. 299.
749. Jackson (2005).
750. Frey and Stutzer (2002).
751. Frey and Stutzer (2002); Kasser (2002), p. 44; Myers and Diener (1996).
752. Frank (1999), p. 73; Frey and Stutzer (2002).
753. Diener and Oishi (2000).
754. Lane (2001), p. 3; Nesse and Williams (1994), p. 220.
755. Diener and Seligman (2004).
756. Jackson (2005).
757. Jackson (2005).
758. Diener and Seligman (2004).
759. Diener and Seligman (2004).
760. Diener and Seligman (2004).
761. Nesse and Williams (1994), p. 220.
762. Diener and Seligman (2004).
763. Jackson (2005).
764. Hamilton (2003), p. 11.
765. Durning (1992), p. 38; Lane (2001).
766. Jackson (2005).
767. Lane (2001), p. 147.
768. Wachtel (1989), p. 17.
769. Samuelson and Nordhaus (1989).
770. Frey and Stutzer (2002).
771. Argyle (1987), p. 149.
772. Lane (2001), p. 76.
773. Seligman (2002), p. 49.
774. Dresner (2002), p. 74.
775. Schor (2000).
776. Jackson (2005).
777. Jackson and Marks (1999).
778. Frey and Stutzer (2002); Schor (1992).
779. Frank (1999), p. 105.
780. Jackson (2005), p. 27.
781. Lane (2001); Schor (1992).
782. Durning (1992); Meadows et al. (2004), p. 262.
783. Drengson (1995), pp. 73, 125.
784. Daly (1980), p. 123.
785. Glendinning, C., "Recovery from Western Civilization," in Sessions (1995), p. 39.

786. Glendenning, C., "Technology, Trauma, and the Wild," in Roszak et al. (1995), p. 53.
787. Frank (1999), p. 183.
788. Mander (1978).
789. Kubey and Csikszentmihalyi (2002).
790. Glendenning, C., "Technology, Trauma, and the Wild," in Roszak et al. (1995), p. 46.
791. Burroughs and Rindfleisch (2002), p. 348.
792. Kasser (2002), pp. xii, 42.
793. Kasser (2002), p. ix.
794. Kasser (2002), p. 18.
795. Nickerson et al. (2007).
796. Polak and McCollough (2006).
797. Burroughs and Rindfleisch (2002), p. 365.
798. Nickerson et al. (2007).
799. Diener and Seligman (2004); Kasser (2002).
800. Kasser (2002), p. 18.
801. Burroughs and Rindfleisch (2002); Eckersley (2000); Kasser (2002); Nickerson et al. (2007).
802. Diener and Seligman (2004); Kasser (2002).
803. Lane (2001), p. 137.
804. Kasser (2002), p. 65.
805. Kasser (2002).
806. Kasser (2002), p. 93; Nickerson et al. (2007).
807. Nickerson et al. (2007).
808. Burroughs and Rindfleisch (2002), p. 365.
809. Kasser (2002), pp. 74, 78; Tatzel (2003).
810. Seligman (2002).
811. Seligman (2002); Sheldon and Lyubomisrky (2006).
812. Myers and Diener (1996), p. 71.
813. Seligman (2002).
814. Carlson (2006); Martin (2005); Sheldon and Lyubomirsky (2006).
815. Burns (1999).
816. Peale (2007).
817. Lane (2001), p. 269; Myers and Diener (1996), p. 71.
818. Frey and Stutzer (2002).
819. Frey and Stutzer (2002); Diener and Seligman (2004).
820. Frey and Stutzer (2002).
821. Braun (1995), p. 197; Hamilton (2003), p. 35; Lane (2001), p. 10; Seligman (2002).
822. Hamilton (2003), p. 35; Lane (2001), p. 23; Seligman (2002), p. 61.
823. Lane (2001), p. 6.
824. Argyle (1987).
825. Arygle (1987), p. 31.
826. Argyle (1987), p. 204; Diener and Seligman (2004).
827. Lane (2001), p. 169.
828. Argyle (1987); Hamilton (2003), p. 35; Seligman (2002); Peterson et al. (2005).
829. Csikszentmihalyi (1991).
830. Seligman (2002).

831. Barbour (1980), p. 43; Stunkel and Sarsar (1994), p. 60.

832. Mesthene (1970), pp. 28, 29, 49, 50; Rhodes (1999), p. 19.

833. Cotgrove (1982), p. 69.

834. Durning (1992); Wachtel (1989).

835. Trainer (1995a), p. 147.

836. Allport, F.H., "Leisure Worth Having?" in Rhodes (1999), p. 103.

837. Anderson and Bushman (2002); Cavanagh and Mander (2004).

838. Kivel (2004), p. 136.

839. Cavanagh and Mander (2004), p. 237; Kubey and Csikszentmihalyi (2002), p. 79; Lane (2001), p. 189.

840. Ellul (1976), pp. 378, 380.

841. Consider, for example, that in spite of today's numerous and readily available high-tech communication technologies, most serious business is still carried out in face-to-face meetings to establish trust and rapport.

842. Bell, D., "Five Dimensions of Post-Industrial Society," in Bereano (1976), p. 152.

843. Dickson (1975), p. 32; Lane (2001), p. 162; Schumacher (1979), p. 28.

844. Schumacher (1997), pp. 67, 69.

845. Horkheimer and Adorno, "The Concept of Enlightenment," in Merchant (1994), p. 47.

846. Mander (1978).

847. Mumford (1934), p. 273.

848. Marks (2001), p. 227.

849. Schumacher (1979), p. 40.

850. Dickson (1975), 183; Rifkin (1998), p. 230.

851. Mander (1991), p. 3.

852. Mander (1991), p. 35.

853. Cotgrove (1982), p. 68.

854. Sclove (1995), p. 62; Stunkel and Sarsar (1994), pp. 87, 88, 93.

855. Proctor (1991), p. 3.

856. Mander (1991), pp. 35, 36; Schumacher (1997), p. 108.

857. Callahan (1990).

858. Rifkin (1998).

859. Mander (1978), p. 269.

860. Mander (1978), pp. 270, 271, 321.

861. Mander (1978), pp. 277–280.

862. Mander (1978), p. 287.

863. Drengson, (1995), p. 187.

864. Dickson (1975), p. 171.

865. Stevens (1991), p. 16.

866. Dickson (1975), pp. 60, 63, 82, 183.

867. Proctor (1991), p. 117.

868. As was discussed in Chapter 7, it is taboo to question belief in technological progress. The myth of value-neutrality helps maintain this taboo by shielding technologies from valid criticism.

869. Dickson (1975), pp. 16, 61, 183.

870. Rifkin (1998), p. 230.

871. Dickson (1975), pp. 28, 29, 42, 184.

872. Dickson (1975), pp. 180, 181.

873. Drengson (1995), p. 82; Postman (1992), p. 142.

874. Ellul (1976), pp. 134, 140.
875. Winner (1977), pp. 28, 46, 58.
876. Dickson (1975), p. 46.
877. Winner (1977), p. 251.
878. Lowrance (1985), p. 36.
879. Sclove (1995), p. 19.
880. Graedel and Allenby (1998), pp. 4, 58, 62.
881. Sclove (1995), p. 103.
882. Schumacher (1979), p. 31.
883. Commoner (1971), p. 181.
884. Mander (1991), p. 29.
885. Ehrenfeld (1981), p. 97.
886. Mumford, L., "The Technological Imperative," in Rhodes (1999), p. 295.
887. Winner (1977), p. 73.
888. Lifton and Markusen (1990), p. 80.
889. Oppenheimer, J. Robert, "A Common Problem," in Rhodes (1999), p. 165.
890. Mumford, L., "The Technological Imperative," in Rhodes (1999), p. 294.
891. Fromm (1967), pp. 32, 33.
892. Mander (1991), pp. 29, 50.
893. Mesthene, E., "Tyrannies," in Rhodes (1999), p. 266.
894. Barbour (1980), p. 45; Dickson (1975), p. 89; Bell, D., "Five Dimensions of Post-Industrial Society," in Bereano (1976), p. 157.
895. Schumacher (1979), p. 50.
896. Barbour (1980), p. 45.
897. Roszak, T., "White Bread and Technological Appendages," in Rhodes (1999), p. 309.
898. Illich (1975), pp. 57, 92.
899. Sclove (1995), p. 240; Sarewitz (1996), p. 130.
900. Meadows et al. (2004), p. 113; Vitousek et al. (1997).
901. Mander (1978), pp. 351, 352.
902. Wachtel (1989), p. 269.
903. Korten (2006), p. 181.
904. Domhoff (1998), p. 1.
905. Korten (2009), p. 52.
906. Korten (2009), p. 55.
907. Wright (2004), p. 128.
908. Korten (1995), p. 83.
909. Schumacher (1979), pp. 21, 53.
910. Sclove (1995), p. 52.
911. Domhoff (1998), p. 163.
912. Dickson (1975), p. 46.
913. Dickson (1975), p. 46.
914. Dickson (1975), p. 46.
915. Naess and Rothenberg (1989), p. 94.
916. Dickson (1975), p. 11.
917. Melman (1984), p. 145; Sarewitz (1996), p. 123.
918. Braun (1995), p. 21.
919. Ferris and Wallace (2009).
920. Mander (1991), p. 133.
921. Blaisdell (2003), p. 138; Parenti (1988), p. 8; Trainer (1995a), p. 92.

922. Princen (2005), p. 54.
923. Korten (2009), p. 13.
924. Korten (2006), p. 216.
925. Maddison (1991); Huesemann and Huesemann (2008).
926. Brooks (2003).
927. Baye (2006), p. 7.
928. Welford (1997), p. 12.
929. Schumacher (1979), p. 26.
930. Mander (1991), p. 124.
931. Quoted in Allenby (1999), p. 251.
932. Korten (2006), p. 133.
933. Korten (1995), pp. 13, 212.
934. Reid (1995), p. 223.
935. Perelman (1977); Welford (1997), p. 140.
936. Runge (2003); Senauer and Sur (2001).
937. Pollan (2007).
938. Nestle (2002), pp. 1, 3, 4.
939. Nestle (2002), pp. 363, 364.
940. McKinlay (1977).
941. Deyo and Patrick (2005), pp. 29, 239; Taylor (1979), pp. 73, 94.
942. Relman (1980), p. 969.
943. Deyo and Patrick (2005), p. 138; Martensen (2008); Taylor (1979), pp. 73, 94.
944. Deyo and Patrick (2005), pp. 138, 140.
945. Weil (1998), p. 105.
946. Krimsky (2004); Weil (1998), p. 108.
947. Moss (1980), p. 429.
948. Silverman and Lee (1974), p. 40.
949. Allen (2007).
950. Proctor (1995), p. 55.
951. Moss (1980).
952. Proctor (1995), pp. 64, 266.
953. Proctor (1995), pp. 108.
954. Proctor (1995), pp. 245.
955. National Coalition on Health Care, Health Insurance Cost, nchc.org/facts
/cost/shtml, accessed 4/13/07; photius.com/rankings/who_world_health
_ranks.html, accessed 11/12/2008; photius.com/rankings/healthranks
.html, accessed 11/12/2008; photius.com/rankings/world_health_perfor
mance_ranks.htm, accessed 11/12/2008.
956. photius.com/rankings/who_world_health_ranks.html, accessed
11/12/2008; photius.com/rankings/healthranks.html, accessed 11/12/2008;
photius.com/rankings/world_health_performance_ranks.html; accessed
11/12/2008.
957. The World Factbook, Total health expenditures as percentage of GDP,
2002–2005, Country Rankings, photius.com/rankings/total_health_expen
diture_as_percent_of_gdp_2000_to_2005.html, accessed 4/15/2011.
958. photius.com/rankings/who_world_health_ranks.html, accessed
11/12/2008; photius.com/rankings/healthranks.html, accessed 11/12/2008;
photius.com/rankings/world_health_performance_ranks.html, accessed
11/12/2008.
959. Relman (1980), p. 969.

960. Melman (1970), p. 10.
961. Stockholm International Peace Research Institute (SIPRI), "Recent Trends in Military Expenditures," sipri.org/contents/milap/milex/mex_trends .html, accessed 3/27/07.
962. American Association for the Advancement of Science (AAAS) Report XXXI, Research and Development FY 2007, Chapter 2, Historical Trends, Table I-11, aaas.org/spp/rd/rd07main.htm, accessed 3/27/2007.
963. Johnson (2000), p. 218.
964. Parenti (1988), p. 171.
965. Herman (1992), p. 26.
966. Korten (2006), p. 315.
967. Fairhurst and Sarr (1996), p. 8.
968. en.wikipedia.org/wiki/Paradigm, accessed June 26, 2009.
969. Korten (2006), p. 76.
970. Meadows et al. (2004), p. 4.
971. Stunkel and Sarsar (1994), p. 48.
972. Ravetz (1996), p. 40; Domhoff (1998), p. 7.
973. Blaisdell (2003), p. 140.
974. Sclove (1995), p. 100.
975. Eagleton (1991), p. 52.
976. Korten (2006), pp. 237, 238.
977. Cotgrove (1982), p. 32.
978. Naess and Rothenberg (1989), pp. 66, 67.
979. Kuhn (1996), p. 150.
980. Holmes (1989), p. 286; Peet (1992), p. 162.
981. Kuhn (1996), p. 110.
982. Kuhn (1996).
983. Kuhn (1996), p. 95.
984. Sarewitz (2004), p. 390.
985. Elliott (1997), pp. 185, 186; Sarewitz (2004), p. 398.
986. Dietz et al. (1989), pp. 66, 67.
987. Cotgrove (1982), p. 88.
988. Korten (2006), p. 252.
989. Kuhn (1996), p. 152.
990. Kuhn (1996), p. 151.
991. Martin (2007); Barsam (2008).
992. Devall and Sessions (1985); Sessions (1995).
993. Des Jardins (1993), p. 192.
994. This means that, in the presence of uncertainty and lack of scientific evidence, caution must be exercised *before* a technology is deployed, not *after* it has caused widespread and possibly irreversible harm. According to an official Communication by the European Commission, "[The precautionary principle] covers those specific circumstances where scientific evidence is insufficient, inconclusive or uncertain and there are indications through preliminary objective scientific evaluation that there are reasonable grounds for concern that the potentially dangerous effects on the environment, human, animal or plant health may be inconsistent with the chosen level of protection." eur-lex.europa.eu/LexUriServ/LexUriServ.do ?uri=CELEX:52000DC0001:EN:NOT, accessed 4/17/2011.
995. MacKellar (1996), p. 146.

996. Speth (2009), see Chapters 4, 5, 6.
997. Daly, H.E., "The Steady-State Economy: Toward a Political Economy of Biophysical Equilibrium and Moral Growth," in Daly (1980), p. 329.
998. Mill (1848).
999. Daly (1996), p. 57.
1000. Meadows et al. (2004).
1001. Diamond (2005).
1002. Daly, H.E., "The Steady-State Economy: Toward a Political Economy of Biophysical Equilibrium and Moral Growth," in Daly (1980).
1003. Dresner (2002); Weizsacker and Jesinghous (1992).
1004. For completeness, it should be noted here that it is also possible for the total size of the human economy (GDP) to contract substantially if society decides that it is necessary to severely reduce matter-energy use and pollution to ensure environmental sustainability. If the reductions in matter-energy throughput and pollution are very large, it may not be possible for technological innovation to compensate fast enough to avoid a significant decline in material affluence (per capita GDP).
1005. Daly, H.E., "The Steady-State Economy: Toward a Political Economy of Biophysical Equilibrium and Moral Growth," in Daly (1980), p. 325.
1006. Meadows et al. (2004); Daly, H.E., "The Steady-State Economy: Toward a Political Economy of Biophysical Equilibrium and Moral Growth," in Daly (1980), p. 325.
1007. Daly, H.E., "The Steady-State Economy: Toward a Political Economy of Biophysical Equilibrium and Moral Growth," in Daly (1980), p. 325.
1008. Daly (1996), p. 165.
1009. Korten (2006), p. 343.
1010. Daly (1996), p. 167; Meadows et al. (2004), p. 256.
1011. Daly (1996), p. 210.
1012. Korten (2009), p. 135.
1013. Dixon (1978), p. 4; Illich (1975), p. 112; Kimbrell (1993), p. 293; McKeown (1979), pp. 5, 13.
1014. US Department of Health and Human Services, National Health Expenditure Data, cms.hhs.gov, accessed 2/11/2008.
1015. The World Factbook, Total Health Expenditures as percentage of GDP, 2002–2005, Country Rankings; photius.com/rankings/total_health_expenditure_as_pecent_of_gdp_2000_to_2005.html, accessed 4/15/2011; World Health Organization (WHO), 2000.
1016. Dixon (1978), pp. 4, 10; Weil (1998), pp. 48–55, 72, 113.
1017. McKeown (1988), pp. 39, 40, 94, 216; Nesse and Williams (1994); Temple and Burkitt (1994), p. 17.
1018. McKeown (1988), p. 89; Taylor (1979), pp. 14, 15.
1019. Dixon (1978), p. 143; Konner (1993), p. 71; Sagan (1987), pp. 149, 150, 166, 173, 177, 180; Waltner-Toews (2000), p. 663.
1020. Barnard et al. (1995); Epstein (1978), p. 23; Proctor (1995), Taylor (1979), p. 37.
1021. Willet (2002).
1022. Fries et al. (1993), p. 322.
1023. Dixon (1978), p. 186.
1024. Fries et al. (1993), p. 322.
1025. McKeown (1979), p. 103; Weil (1998), pp. 55, 72, 76, 77.
1026. Ford (1984), p. 72.

1027. Weil (1998), p. 55.
1028. Callahan (1990), pp. 67,152,176,177.
1029. Korten (2006), p. 44.
1030. Korten (2006), p. 45.
1031. Korten (2006), pp. 46-47.
1032. Korten (2006), pp. 46-47.
1033. Korten (2006), pp. 47-48.
1034. Note that the more we identify with others, the less likely we are to harm them.
1035. Elgin (1993), p. 127.
1036. Korten (2006), p. 41.
1037. Elliott (1997), pp. 170,175.
1038. Elliott (1997), Chapter 11; Wuestenhagen et al. (2007).
1039. Ayres (1996a).
1040. Vitousek et al. (1997).
1041. Weisman (2007), pp. 112-128.
1042. Schumacher (1997), p. 71.
1043. Brown (2008), Chapter 10; Sclove (1995), p. 138.
1044. Trainer (1995a).
1045. Sclove (1995), pp. 131-133.
1046. Cavanagh and Mander (2004), p. 149; Korten (2006), p. 343.
1047. Berry (1977), p. 52.
1048. Schumacher (1997), p. 8.
1049. Cavanagh and Mander (2004), p. 196.
1050. Schumacher (1979), pp. 21, 55.
1051. Schumacher, E. F., "The Age of Plenty—A Christian View," in Daly (1980), p. 134.
1052. Schumacher (1997), pp. 71, 73.
1053. Diener and Seligman (2004); Hamilton (2003), p. 155; Princen (2005), p. 141; Schumacher (1979), p. x; Sclove (1995), p. 43; Wachtel (1989), p. 253.
1054. Dickson (1975), p. 105.
1055. Santillo et al. (1998), p. 940.
1056. Dovers and Handmer (1995), p. 93.
1057. Rifkin (1998), p. 233.
1058. Sclove (1995), p. 219.
1059. Cavanagh and Mander (2004), p. 197.
1060. Costanza, R., "Balancing Humans in the Biosphere," in Grant (1992), p. 51.
1061. Feenberg (1999); Sclove (1995).
1062. Mander (1978), p. 353.
1063. Sclove (1995), p. 185.
1064. Sclove (1995), pp. 181,183.
1065. Sclove (1995), p. 194.
1066. Sclove (1995), p. 208.
1067. Sclove (1995), p. 194.
1068. Kleinman (2000).
1069. Sclove (1995), p. 211.
1070. Sclove (1995), p. 218.
1071. Sclove (1995), p. 218.
1072. Sclove (1995), p. 240.
1073. Sclove (1995), pp. 193,194, 202.

1074. Drengson (1995), pp. 143-145.
1075. It should be noted that peasants in India grow over 40 different crops on land that has been cultivated for more than 2,000 years without a decrease in yield, while remaining free of pests (Cavannah and Mander (2004), p. 210).
1076. Rose and Rose (1971), pp. 137-138.
1077. Zimmerman. et al., "Towards a Science for the People," in Bereano (1976), p. 525.
1078. Capra, F., "Systems Theory and the New Paradigm," in Merchant (1994), p. 340.
1079. Proctor (1991), p. 10.
1080. Welford (1997), p. 121.
1081. Lowrance (1985), pp. 3, 5, 6.
1082. Huesemann (2002); Ravetz (1990), p. 224.
1083. Rose and Rose (1971).
1084. Bradshaw and Bekoff (2001); Sarewitz (1996), p. 173; Shrader-Frechette & McCoy (1993), p. 99.
1085. Sarewitz (2004).
1086. Longino (1990), p. 84.
1087. Rose and Rose (1971), p. 144.
1088. Welford (1997), p. 121.
1089. Barnaby (2000).
1090. Proctor (1995).
1091. Proctor (1991), p. 225.
1092. Longino (1990); Proctor (1991).
1093. Longino (1990), pp. 94-100.
1094. Proctor (1991), p. 267.
1095. Douglas, M., and A. Wildavsky, "Assessment is Biased," in Johnson and Covello (1987), p. 69.
1096. Rikfin (1998), p. 197.
1097. Rifkin (1998), p. 200.
1098. Rifkin (1998), p. 201.
1099. Rifkin (1998), p. 202.
1100. Ho (1998), pp. 66, 101; Rifkin (1998), pp. 198-207; Rose and Rose (1971).
1101. Cummings (1991), p. 9; Korten (2006), pp. 271-280.
1102. Ravetz (1996), pp. 19, 160, 164.
1103. Proctor (1991), pp. 68, 70.
1104. Zimmerman, B., et al., "Towards a Science for the People," in Bereano (1976), p. 525.
1105. Rose and Rose (1971).
1106. Proctor (1991), p. 180.
1107. Proctor (1991), pp. 15, 231.
1108. Ravetz (1996), p. 424.
1109. Fuhrman (1979).
1110. Welford (1997), p. 243.
1111. Dixon (1973), p. 182.
1112. Ravetz (1996), p. 430.
1113. Guidotti (1994).
1114. Herman and Chomsky (1988), p. 305; Ravetz (1996), p. 430.
1115. Dickson (1975), p. 143; Ravetz (1996), p. 430.

1116. Von Hippel, F., and J. Primack, "Public Interest Science," in Bereano (1976), p. 507. In some cases, it has also been called "Science for the People" (see Beckwith (1992); Zimmerman, B., et al., "Towards a Science for the People," in Bereano (1976)).
1117. Dixon (1973), p. 182.
1118. Ravetz (1996), p. 436.
1119. Dixon (1973), p. 182; Von Hippel, F., and J. Primack, "Public Interest Science," in Bereano (1976), p. 513; Ravetz (1996), p. 425.
1120. Welford (1997), p. 230.
1121. Rose and Rose (1971), p. 148.
1122. See thebulletin.org, accessed 10/13/2010.
1123. Ravetz (1996), p. 157.
1124. Bradshaw and Bekoff (2001); Guidotti (1994); Proctor (1991), p. 240.
1125. See ucsusa.org, accessed 10/13/2010.
1126. Burhop (1971).
1127. Temple and Burkitt (1994), p. 433; Zimmerman, B., et al., "Towards a Science for the People," in Bereano (1976).
1128. See pcrm.org.
1129. See navs.org; aavs.org.
1130. Von Hippel, F., and J. Primack, "Public Interest Science" in Bereano (1976), p. 513.
1131. Forge (2008).
1132. Drengson (1995), p. 118.
1133. Ziman (1971).
1134. Collins (1972); Dixon (1973), p. 206; Leitenberg (1971).
1135. National Academy of Sciences, National Academy of Engineering, and Institute of Medicine, *On Being a Scientist: Responsible Conduct in Research*, The National Academies Press, 1995, nap.edu.
1136. Collins (1972).
1137. Collins (1972); Drengson (1995), pp. 118, 188.
1138. Lowrance (1985), p. 75.
1139. Ravetz (1996), p. 416.
1140. Ravetz (1990), p. 287; Ravetz (1996), p. 415.
1141. Forge (2000).
1142. Beckwith and Huang (2005); Susskind (1973) p. 118.
1143. Ravetz (1996), p. 44.
1144. Ravetz (1990), p. 210.
1145. Zimmerman, B., et al., "Towards a Science for the People," in Bereano (1976), p. 525.
1146. Schumacher (1979), p. 209.
1147. Domhoff (1998), p. 163; Kivel (2004), p. 98.
1148. Kivel (2004), p. 13; Zinn (1999), p. 649.
1149. Barsamian and Chomsky (2001), p. 80.
1150. Lifton and Markusen (1990), pp. 41, 98.
1151. Braun (1995), p. 32; Daly (1996), p. 198; Korten (1995), p. 212.
1152. Domhoff (1998), p. 194.
1153. Collins (1972).
1154. Alford (2002); Barsamian and Chomsky (2001), p. 170; Domhoff (1998), p. 194; Glazer (1991).
1155. Collins (1972).

1156. Drengson (1995), pp. 68, 189; Jonas (1973), p. 53.
1157. Russell (2000), Chapter 9, p. 147.
1158. Russell, "The Effect of Science on Social Institutions," in Hughes (1975), p. 238.
1159. Collins (1972).
1160. Capra, F., "Systems Theory and the New Paradigm," in Merchant (1994), p. 340.
1161. Edsall (1975), p. 691.
1162. Berlin, I., "Practical Virtue," in Rhodes (1999), p. 319; Lowrance (1985), p. 104.
1163. Von Neumann, J., "Global Effects," in Rhodes (1999), p. 205.
1164. Lowrance (1985), p. 104.
1165. Sinsheimer, R. L., "The Presumptions of Science," in Daly (1980), pp. 147, 158.
1166. Kempner et al. (2005), p. 854.
1167. Sinsheimer, R. L., "The Presumptions of Science," in Daly (1980), p. 157.
1168. Sinsheimer, R. L., "The Presumptions of Science," in Daly (1980), p. 157.
1169. Beckwith and Huang (2005).
1170. Drengson (1995), p. 117.
1171. Beckwith (1972); Beckwith and Huang (2005).
1172. Beckwith (1972).
1173. Commoner, B., "A New Duty," in Rhodes (1999), p. 251.
1174. Beckwith and Huang (2005).
1175. National Academy of Sciences, National Academy of Engineering, and Institute of Medicine, *On Being a Scientist: Responsible Conduct in Research*, The National Academies Press, 1995, pp. 20–21, nap.edu, accessed 4/16/2011.
1176. Lowrance (1985), p. 76.
1177. Galston (1972), p. 223.
1178. Collins (1972); Leitenberg (1971).
1179. Collins (1972); Edsall (1975), pp. 688, 692; Edsall (1981), p. 13.
1180. Calaprice (1996), p. 124.
1181. Zinn (1999), p. 649.
1182. Zimmerman, B., et al., "Towards a Science for the People," in Bereano (1976), p. 528.
1183. Herman (1992), p. 67.
1184. Lifton and Markusen (1990), p. 150.
1185. Clark (1972), p. 183.
1186. Ashkinazy (1972), p. 47.
1187. Writings by Advocates of Peace (2002), p. ix.
1188. For more information, see Huesemann (2006).
1189. Daly, H., "The Steady-State Economy: Toward a Political Economy of Biophysical Equilibrium and Moral Growth," in Daly (1980), p. 353.
1190. Murtaugh and Schlax (2009). The summed CO_2 emissions of 9,441 metric tons for a newborn child and its descendents assumes a constant-emission scenario (i.e., business as usual). But even in the most optimistic scenario tested by Murtaugh and Schlax (2009), which assumes an 85 percent reduction in global CO_2 emissions from 2000 to 2100, the summed CO_2 emissions of a newborn and its descendents are still 562 metric tons, i.e., larger than the CO_2 savings that can be gained by that above mentioned environmentally conscious consumer behavior.
1191. United States Census Bureau, US and World Population Clock, census.gov /main/popclock.html, accessed 10/19/2009.

1192. United Nations, Press Release, March 13, 2007, POP/952, "World Population Will Increase by 2.5 Billion by 2050," un.org/News/Press/docs/2007/pop952.doc.htm, accessed 10/19/2009.

1193. The World Fact Book—Country Comparison: Population, cia.gov/library/publications/the-world-factbook/rankorder/2119rank.html, accessed 4/24/2011; "Biggest US Cities by Population", biggestuscities.com, accessed 10/29/2009.

1194. The World Fact Book - Country Comparison: Population, cia.gov/library/publications/the-world-factbook/rankorder/2119rank.html, accessed 4/24/2011.

1195. United States Census Bureau, US and World Population Clock, census.gov/main/popclock.html, accessed 10/19/2009.

1196. Haub, Carl, "Our National Demographic Future: U.S. Population Could Reach 438 Million by 2050, and Immigration is Key," *Population Press*, Fall 2009, pp. 10-12, populationpress.org; source of data presented in *Population Press* article: Passel, J.S., and D.V. Cohn, *U.S. Population Projections 2005–2050*, Pew Research Center, Washington, D.C., pewhispanic.org/files/reports/85.pdf.

1197. "Biggest US Cities by Population", biggestuscities.com, accessed 10/29/2009.

1198. See Huesemann and Huesemann (2008) for a discussion of policy options.

1199. Wilson (2002), p. 86.

1200. Fanelli, Daniele, "Meat is Murder on the Environment." *New Scientist*, July 18, 2007, newscientist.com/article/mg19526134.500, accessed 10/12/2009; Organic Consumer Association, "Meat Production Spews More Greenhouse Gases than a Three-Hour Joyride," organicconsumers.org/articles/article_6165.cfm, accessed 10/24/2007.

1201. Eshel and Martin (2006), Table 3; Kanter, James. "A Vegetarian Diet Reduces the Diner's Carbon Footprint," *The New York Times*, Wednesday, June 6, 2007, nytimes.com/2007/06/06/business/worldbusiness/06iht-greencol07.4.6029437.html, accessed 4/25/11.

1202. Wilson (2002), p. 88.

1203. Chapman et al. (2000), pp. 172-173.

1204. Daily, Ehrlich, and Ehrlich (1994); Pimentel, D., "How many Americans can the Earth Support?" Population Press, populationpress.org/essays/essay-pimentel.html, accessed 4/20/2010.

1205. Pimentel, D., and M. Pimentel, "Land, Energy and Water," in Grant (1992), p. 30; Werbos, P.J., "Energy and Population," in Grant (1992), p. 49; Costanza, R., "Balancing Humans in the Biosphere," in Grant (1992), p. 58.

1206. Huesemann and Huesemann (2008), p. 808.

1207. Elgin (1993), pp. 48-49.

1208. Simon (1996), p. 37.

1209. Barbour (1980), p. 176.

1210. Deyo and Patrick (2005), pp. 268-271.

1211. Cavanagh and Mander (2004), p. 230; Chapman et al. (2000), p. 89; Korten (1995), p. 22; Korten (2009), Chapter 15; Trainer (1995a), pp. 141, 151.

1212. Kasser (2002), p. 103.

1213. Jackson and Marks (1999).

1214. Durning (1992), p. 137.

1215. Meadows et al. (2004), p. 262.

1216. Winn (1987).

1217. Sclove (1995), p. 205.

1218. Diener (2000); Diener and Seligman (2004); Hamilton (2003), pp. 213–216; Speth (2009), Chapter 6.
1219. Oswald (1997).
1220. Korten (2009), p. 121.
1221. Diener (2000); Diener and Seligman (2004).
1222. Diener and Seligman (2004).
1223. Mander (1991), p. 133.
1224. Jacobson (2006); Nestle (2002), p. 21.
1225. Zinn (2002), p. ix.
1226. Advocates of Peace (2002), p. 123; see also: famous-speeches-and-speech
    -topics.info/martin-luther-king-speeches/martin-luther-king-speech
    -beyond-vietnam.htm, accessed 3/15/2010.
1227. Hamilton (2003), p. 219.
1228. Chapman et al. (2000), p. 116.
1229. Hamilton (2003), p. 80; Welford (1997), p. 213.
1230. Hamilton (2003), p. 237; Trainer (1995a), p. 220.
1231. See revivevictorygarden.org.
1232. Brown (2008), p. 206.
1233. Hess (1979).
1234. Brown (2008), p. 207.
1235. Trainer (1995a), p. 37.
1236. Cited in Daly (1996), p. 198.
1237. World Commission on Environment and Development (1987), Chapter 11, p. 157.
1238. Melman (1970), p. 40; Ziman (1971).
1239. Prasad, R., "Nonviolent Resistance", in: Advocates of Peace (2002), p. 91.

Index

A

Acceptance of new technologies, 235-51
Acid waste, 80
Activism, 67-8, 239, 286, 325, 334, 335
Adaptive capacity, 8
Addictive behavior, 220-1
Advertising: budgets, 169; function of, 213; mass, 210, 212; and stimulation of wants, 210, 212, 213, 221
Affluence: drivers of, 98, 156-7; and efficiency improvements, 113; and happiness, 214-6, 217; and the Industrial Revolution, 117; and labor-saving technology, 105; limited, 283; material, 94-8; per capita, 94, 95, 118, 257; and renewable energy, 132; rising, 94-8, 94-8, 257-8; and steady-state economy, 282-4; and technological optimism, 151
Age of Reason, 50
Agent Orange, 33
Agribusiness, 261-2, 301. *see also* Industrialized agriculture
Agricultural productivity, 23
Agricultural revolution, 44, 117
Agricultural runoff, 298
Agricultural societies, 45, 97, 117, 148
Agriculture: industrial, 123, 261; industrialization of, 23, 44; Industrialized, 23-5, 123; irrigation. *see* Irrigation; organic, 261, 309-11; pre-industrial, 44, 123; prehistoric, 117; and profit maximization, 261-3; sustainable, 23; world, 128
AIDS, 40
Air pollution, 29, 51, 79, 79, 81. *see also* specific pollutants
Air-stripping, 78
Alienation, 31, 76, 87, 220, 231, 232
Alteration of biogeochemical cycle, 57, 59, 82
Alternatives, 175, 176, 242
American capitalists, 151
American exceptionalism, 151

Amish, 216
Analysis: cost-benefit analyses. *see* Cost-benefit analyses; root cause analysis, 90, 329
Animals: abuse of, 40, 52, 54, 291; deaths, 30; genetic modification of, 26, 27; patentable, 27
Anthropocentric optimum, 280
Anthropogenic carbon emissions, 101
Anti-vivisection societies, 325
Antibiotic resistance, 38-40, 86
Antibiotics, 38-40, 86, 161, 195
Anxiety, 76, 215
Appropriateness, 280-1, 286, 295-311, 300-4
Arable land, loss of, 24
Artificial environments, 46, 64, 232, 247
Assembly line, 61-2, 114, 230-1, 337
Assimilation capacity, 124, 296, 298
Atomic bomb, 32, 243, 324
Atrocities, 291
Authoritarianism, 63, 238, 248, 330
Authority, belief in, 153-4
Autocracy, 307
Automobiles: and autonomous technology, 243; and community life, 228-9; and cost-benefit analyses, 189-91; deaths and injuries caused by, 29; and human needs, 209, 213-4; hybrid cars, 99; and time saving, 104; total fuel use, 99; uncritical acceptance of, 189-91; unintended consequences of, 28-31; US automobile fleet, 29, 99
Autonomous technology, 241-3
Autonomy: personal, 63, 186-7, 195, 223, 226, 303; of science, 318; of technology. *see* Autonomous technology
Awareness, need for increase in, 289-94

B

Backfire, 96
Bacterial evolution, 39
Barrier-free design, 309

BASF, 81
Belief in progress, 148–52, 152–4
Benefits, versus cost. *see* Cost-benefit analyses
Biases: institutional, 183–5; of scientific theories, 317; of solar energy generation, 238; structural, 105; subjective, 314; of technology assessments and cost-benefit analyses, 173–205; toward the quantifiable and material, 114
Bicycle, 191, 302
Bio-accumulation, 20, 25, 59, 182, 299
Biocentric optimum, 280–1
Biochemical cycles, 82
Biodegradation, 6, 124, 298–9
Biofuels: and agricultural feedstocks, 192; biodiesel, 128, 192; ethanol from corn, 128, 193–4; and government subsidies, 192–3; hidden costs, 192–4
Biogeochemical cycles, 57–8, 82–3, 136
Biological diversity, 22, 278
Biological egalitarianism, 278
Biological evolution: and adaptation to environment, 9, 46; belief in, 153, 155; and bias, 317–8; discovery of, 5; exploitation of, 317–8; and natural selection, 6; pace of, 7, 10
Biological fitness, 5, 46–7
Biological inertness, 299
Biological weapons, 25, 28, 33
Biologists, molecular, 26
Biomass: conversion, 194; energy, 127–9, 133; photosynthetically fixed carbon energy food sources, 127; productivities, 129; and sustainability, 297
Biophysical limits, 280, 286
Bioremediation, 80–1
Biosphere, 28, 127
Biotechnologies, 59, 136. *see also* genetic engineering
Birds, 22, 25, 130
Bivalve mollusks
Boundaries. *see also* System boundaries: and cost-benefit analyses, 176–80; of energy analysis, 193; for life-cycle analyses, 297; planetary, 56; spatial and temporal of an analysis, 176–80
Breakdown products, 298
Breeding techniques, traditional, 25–6
Brundtland Report, 119
Bureaucracies, distant, 301

Business, 62, 179, 184, 251
Business potential, 308

C

Cancer: carcinogens, 265, 288; genes, 265; research, 265; war on, 35, 265
Cap and trade, 281
Capital: natural, and GDP, 202; substituting for labor, 95, 97
Capitalism, 150, 158, 179
Capping schemes, 281
Car culture, 29, 76, 101, 189, 242
Carbon: efficiency of use, 101; emissions, 101; fossil carbon release, 21
Carbon cycle, 57, 75
Carbon dioxide: atmospheric, 57, 75, 101, 298–9; atmospheric concentrations, 21, 57, 101; and entropy, 18; and ocean fertilization, 82–3
Carbon taxes, 297
Carcinogens, 265, 288
Carnot cycle, 110
Carrying capacity, xviii, 44
Cars. *see* Automobiles
Categorical Imperative, 278
Cause and effect, 3, 173, 275, 288
Causes: root, 87, 89, 90, 256, 321, 329; root cause analysis, 90, 329; systemic, 88; underlying, 77, 87, 88, 156–7
Centralization, xx, 238
Charity, 232, 293, 323
Chemical and biochemical reactions, 3–4, 33, 75
Chemical production, large-scale, 305
Chemical weapons, 24, 33
Chemotherapy, 38
Chlorinated organic compounds, 6, 298
Chlorinated pesticides, 80
Chlorinated solvents, 79, 80
Choice, 227, 315. *see also* freedom
Cholera, 161
Circle of concern, 278, 292
Circular biological transformation pathways, 136
Citizen representatives, 307, 308
Civilization: collapse of, 8, 11, 47, 244, 331; life-support services, 134; and technology, 146
Climate change: abrupt, 21–2; and automobiles, 30; in Europe, 22; global.

see Global climate change; irreversible, 21-2; and irreversible consequences, 9

Closing the materials loop, 124

Clothing, 209, 212, 218, 299

CO₂. *see* Carbon dioxide

Coal, 18, 57, 81

Cold War, 32, 53

Collapse: environmental, 116, 118, 120, 139; as response to climate change, 10; societal, 116, 118, 120, 139

Commodifying plants and animals, 238

Communication: and community life, 230; mass, 210, 212

Community, 227, 228, 229-30, 233

Community life, 167, 228

Community supported agriculture (CSA), 311

Compatibility, 238, 300, 307, 317

Compensatory response, 221

Compensatory satisfactions, 219, 303

Competition: justification of, 318; for resources, 44, 68-9; vs cooperation, 228, 318

Complete recycling, 134-7, 298

Complex, integrated systems, 13

Complex interdependencies of species, 4

Complexity, 75, 302

Composting facilities, 298

Compulsive behavior, 220

Computers, 62, 64

Concealment, of involvement, 245

Conditions for long-term sustainability, 122-5, 296

Conflicts: among stakeholders, 184-5; between food and energy, 128; between materialistic and spiritual values, 223, 293; over land use; between profit and social responsibility, 259-60; between worldviews and paradigms, 273-7, 315

Consciousness, orders of, 289-92

Consensus, 281, 301, 307, 314

Consequences: biological, 75, 77; delayed, 11, 182; difficult to detect, 177; difficult to predict, 11; ecological, 27; environmental, 17-23; irreversible, 8-12; negative unintended. *see* Negative consequences; short-term, 39, 89, 167, 177, 188; social, 23, 34, 189, 336; unavoidable negative. *see* Negative consequences; unintended. *see* Unintended consequences

Conservation, 96, 100

Conservation of energy principal, 17-8

Conservation of mass principal, 17-8, 78

Conspicuous consumption, 218

Constraints, thermodynamic and practical, 121

Consumer culture: and mass media, 68; and mass production, 210; and profit maximization, 211

Consumer goods and services: and efficiency, 304; negative impacts of, 233, 239, 299; and obsolescence, 212; and sustainability, 282, 300

Consumer preferences, 96

Consumerism, xxiii, 208-14, 232, 234, 239

Consumers, 210, 229, 249, 310

Consumption. *see also* Material consumption: acceleration of, xxvi; conspicuous, 218, 220; mass, 65, 68, 210, 221, 317; new avenues of, 95, 207; rising per capita, xviii-xix, 207; stimulation of, 96, 106, 112; of transportation fuel, 99

Contaminants. *see also* Pollutants: carbon dioxide. *see also* specific contaminants; dispersed, 79-80, 136; highly dispersed air pollutants, 79-80; intermediates, 80; metal, 80; metals, 29; particulates, 79

Contamination, toxic of land, air, water, 29, 80, 180

Contraceptive technology, xviii, 86, 233-4

Control. *see also* Exploitation; Manipulation: disguised, 63; legitimized, 63; of mass media, 169; of mind, 68; of nature, 26, 54, 56, 60; by technology, 61-3; of technology, 243, 248-51, 301, 306-9; by television, 64-8; undemocratic, 248-51

Cooperation, 337

Copper, 133, 136

Corals, 82

Coronary heart disease. *see* Heart disease

Corporate industrialized agriculture,
  311
Corporations: and control of technology, 248-51, 255; and the mass media,
  168-9, 219, 229; and rapid obsolescence, 211
Cost-benefit analyses: and automobiles, 189-91; and biofuels, 192-4; and
  boundaries, 176-80; ethics of, 187-9;
  institutional biases, 183-5, 188; limitations of, 174-6; and medical testing, 194-200; monetization of non-
  market values, 185-7; and objectivity/
  value-neutrality, 176-7, 183, 185; outcomes dependant on power and politics, 183-5, 188, 197; overview of, 174-
  6; perception of, 183-5; positive biases
  of, 173-205; prediction of impacts,
  180-2; selection of indicators, 180-2;
  steps in, 174-6
Cost-benefit balance, skewed, 49, 70
Cost-effectiveness, 36, 43, 106, 236, 288
Cost externalization, 179-80
Costs: environmental, 23, 177, 179, 190,
  192; externalized. see Externalization
  of costs; incentive to ignore, 188; internalized, 179, 188, 189-90, 281-2;
  private, 190; social. see Social costs;
  transference of, 51-2, 178, 202
Costs and benefits: different views of,
  183-5; relative importance of, 177, 183
Counter-technologies: defined, 72, 73;
  environmental, 77-83; medical, 86-8;
  military, 83-6; next generation of, 81;
  residue problems, 73-4; social fixes,
  72-3, 75-7, 83; unintended consequences of, 88-90
Craft technologies, 302, 304
Creativity, 114, 230
Critical analyses, 153, 324-5
Critical natural capital, 134
Critical science: nd engineering, 334-8;
  need for, 286; and social responsibility, 313-38
Critical thinking, 245
Criticism: of new technologies, 169-70;
  suppression of, 169-70, 319
Critique, 216, 314, 322
Crops: genetically engineered, 25, 27,
  261; productivity of, 45; rotation of,
  310

Cuba, 266
Cultural consciousness, 292
Cultural exchanges, 293
Cultural resistance, 309
Cultural values, 239-40, 313, 316
Culture: Car culture, 29, 76, 101, 189,
  242; Desert, 10; influences on, 64, 67-
  8; Western, xvii, 209
Cycles: biogeochemical, 57-8, 82-3, 136;
  of dissatisfaction, 216-7, 216-7; hydro-
  geological, 21, 126; of nature, 137, 139,
  148

D
Dams, 130-1
DDT, 6, 80
Dead zones, 298
Death, 165; acceptance of, 107, 108, 289;
  crusade against, 43, 107, 108, 165; de-
  fined, 166; and dying, 41-3; fear of,
  152; leading causes of, 76, 215, 288
Death rates, declining, 160-2
Debasement, 186
Decentralization, 300-1
Decision-makers' delayed responses,
  139
Decisions, legitimizing, 188
Decline: in fitness of future genera-
  tions, 46-7; of techno-optimism,
  169-72
Decoupling of economic production
  from material input, 111
Degradative pathways, 20
Dematerialization, 282, 300
Democracy: and control, 67, 249, 307;
  farce of, 249; and happiness, 224-5
Democratic control, 286, 306-7
Democratic control of technology,
  306-9
Democratic technological decision-
  making, 308
Denmark, 308
Dependency, 245-8
Depletion: of mineral resources, 133.
  see also specific minerals; of non-
  renewable energy sources, 45, 109,
  123; of renewables, 129, 134; of unre-
  coverable resource, 180, 329
Depletion quota, 281
Depression: and affluence, 215; as un-
  intended consequence, 76

Desert culture, 10
Desertification, 139
Design: for environmental sustainability, 295-300; participatory, 307-9; popular involvement in, 307; for social appropriateness, 300-4
Destruction: of the environment, 29, 70, 114; of traditional sources of happiness, 226-34
Deterministic views, 13
Developers, 274
Development, sustainable, 119-25
Devices: labor-saving, 60, 104-5, 254; medical, 198-200, 264
Diagnostic tests, 37, 41, 106
Diffusion, of a new technology, 10
Dilution, 78
Dioxins, 33
Diphtheria, 161
Disability-adjusted life years, 215
Disagreements. see Conflicts
Discharge limits, 281
Discounting the future, 176
Disease: degenerative, 162, 248, 262, 287; heart disease. see Heart disease; iatrogenic, 37-8, 86, 194-5; infectious, 161-2; lifestyle-caused, 76, 87, 287-8; and natural healing, 162; prevention, 36, 287-8
Disease causation, 36, 286-8
Disease vectors, 134
Dispersal, 29, 78, 82, 137, 299
Dispersed materials, 74, 79-80, 136, 297-9
Disrupting sensitive ecosystem functions, 137
Dissatisfaction: creating, 212; cycle of, 216-7
Dissipative loss, 135
Dissipative structures, 9, 126
Dissipative use, 124, 136, 298, 299
Distraction, constant, 233, 293
DNA, 23, 25, 27, 39
Dobe Bushmen, 159
Doctors, 42
Dominant values, 115
Domination, 56-60, 240
Drawdown, 58, 261
Drosophila, 47
Droughts, 21
Drug-resistant bacterial strains, 40

Drugs: marginally effective, 263; negative side effects, 38, 164; new, 197, 199, 200; testing for effectiveness, 197
Dynamic instability, 9
Dysentery, 161
Dysfunction, social and moral, 172, 202

E
Eco-efficiency: limits to, 121; and sustainable development, 119-22
Ecoagriculture, 261, 309-11, 310-1
Ecological disruptions by genetically engineered organisms, xxv, 27
Ecological footprint: vs. total land area, 138; US vs developing nations, 57
Ecological resilience, 8-9, 138
Ecological services, 134
Ecological tax, 282
Ecology, 27, 81, 121
Economic activity, 118
Economic development, 56, 216, 259
Economic expansion, 91, 101, 132
Economic globalization, 203, 212, 276
Economic growth: continual, 284; sustainable, 120; unlimited, 118, 137, 279
Economic output, 94-5, 97
Economic promise, 308
Economic resources, diversion of, 267, 309
Economic theory, 19, 134, 217
Economics: conventional, 201; neoclassical, 96, 97, 118; the problem of, 93; steady-state, 118, 135, 279-80; subsistence, 158
Economy: circular-flow economies, 124, 132; different view of, 279-85; free-market. see Free-market economies; future, 283; growth-oriented industrial economies, 122, 279; human, 118, 137, 279, 282; linear-flow economies, 124; steady-state, 118, 135, 279, 282-4
Ecosystems: adaptability of, 138; assimilation capacity of, 124, 277, 296, 298; collapse of, 22; the economy as a subsystem of, 137; finite and closed, 137; fragile desert ecosystems, 130, 131, 238; host, 137; regenerative capacity of, 296
Education, 66, 97, 293, 334
Effects: long-term, 181, 200; negative-positive nature of, 173
Efficiency: age of, 92; and automobile

fuel use, 99; of carbon use, 101; conversion efficiency, 109–10; eco-efficiency, 119–22; end-use, 110; fuel efficiency, 73, 95, 99–100; of health care, 93, 105–9; increasing, 93, 95, 112; of labor, 91, 111; labor-saving technology, 103–5; lighting, 100; of manufacturing, 104, 111, 114; of materials use, 100; maximal, 112; maximum, 109–10, 112, 286; measures of, 93; and medical progress, 105–9; optimization of technical, 113; of road networks, 101–2; supply-side, 110; and total energy use, 98; value-laden, 113

Efficiency improvements: and affluence, 113; counterproductive, 115; and limited resources, 93–4, 95, 98–109; limits to, 109–16, 122, 157, 282–3; and quality of life, 113–5; response to, 96, 112; and sustainability, 299–300; as techno-fixes, 73; thermodynamic and practical limits, 109–11; unintended consequences of, 112–6

Efficiency ratios, 92

Efficiency solutions, 112–6

Electronic communication technologies, 62, 209, 232

Electronic media, 64–5, 67, 239

Elements, cycling of, 3, 58, 82–3, 136

Embedded values, 236–7

Empathy, 114, 290

Endocrine disruptors, 20, 21, 182

Energy: embodied, direct and indirect, 97; fossil. see Fossil energy; fusion, xix–xx; net renewable, 192, 193; non-renewable, 45, 68, 95, 122–3, 125; nuclear. see Nuclear power; renewable. see Renewable energy; solar. see Solar energy; substituting for labor, 95, 97; total energy use, 93–4, 96, 98, 100; total primary energy use (TPEU), 98, 99

Energy and materials use, 111, 122

Energy balance, 193

Energy efficiency, and total energy use, 98

Energy installations, large-scale, 132

Engineering: geo-engineering, 82–3

Engines: internal combustion, 44, 109, 243; steam, 44, 56, 117

Enlightenment: and exploitation, 55; methods of knowledge acquisition, 12, 13; and religion, 232; and views on progress, 149–51

Entropy: disorder, 17–9, 79; and eco-efficiency, 121; high, 17–8, 137; high entropy wastes, 18, 137; low, 17–8, 137; neg-entropy, 19, 79; order, 18, 79

Environment: absorption capacity of, 93, 139; artificial, 46, 64, 232, 247; collapse of, 116, 118, 120, 139; consequences to, 17–23; deep sea, 82

Environmental counter-technologies, 77–83

Environmental disasters, 56

Environmental movement, 171, 309, 324

Environmental pollution, 6, 78, 212, 272, 277, 329

Environmental refugees, 21, 51

Environmental regulation, 11, 135, 177, 281, 299

Environmental sustainability: and critical science, 325; design criteria for, 295–300; economic threat to, 118, 257; and local organic agriculture, 309–11; and steady-state economy, 258

Environmentally appropriate technology, 286, 295–311

Environmentally friendly renewable energy, 125, 238, 286

Equilibrium, 9, 289

Equitable distribution, 187–8

Equity, 140, 187–8

Ethanol: from corn, 128, 193–4; from switchgrass, 194

Ethical sensitivities: limited, 51; suppression of, 53, 70, 233

Ethical vacuum, 260

Ethics: codes of, 326, 331–2; of cost-benefit analyses, 187–9; enforcement of, 332; professional ethics and social responsibilities, 330–8; and profit maximization, 259

Ethnic weapons, 28

Europe, 44, 52, 69, 99, 151, 215

European Organization for Economic Cooperation and Development OCED, 120

Evolution: biological. see Biological evolution; cultural, 10

Expectations, 217, 258, 272

Expertise, areas of, 51
Experts: and ignorance, 152, 251; and techno-optimism, 154-5
Exploitation: of animals, 52; beyond sustainable yield, 51, 133; counter-productive, 70; defined, 49; of future generations, 178; and military technology, 68-70; of natural resources, 50, 68; of nature, 8, 50, 53-6, 63, 232, 327; political, 63; of the poor, 291; of renewable resources, 51, 133; and safe distance, 50-3; technological, 9, 17, 53-6; of workers, 52, 60-4, 114, 230-1, 318
Exporting waste and pollution, 283
Externalization of costs: and automobiles, 189-90; and cost-benefit analyses, 49, 179-80, 189-90; and increased profits, 179, 188; intergenerational, 52, 179
Extinction: causes and effects, 22, 24, 51, 56, 127; causes, effects, and rates of, 22-3; measurement indicator, 182; rates of, 22
Extrapolation from historical trends, 156-7

F
Factor 4 Club, 121
Factor 10 Club, 110, 111
Factories: chemical, 300; and exploitation, 52, 60; purposes for, 60-1; and safe distance, 227
Factory farms, 52
Fairness, 47, 70
Faith: loss of, 170, 172; in medicine, 159, 162-3; in power of reason, 5; in progress, 145, 149, 152-4; religious faith, 13, 149; in technology, 71-2, 164, 170
False needs, 208
Family and community life, 167, 238
Farmers markets, 310
Farming. *see* Agribusiness; Agriculture
Farmland, 23, 129, 131, 156
Fertilizers: ammonia, 58; fossil fuel based, 44, 129; nitrogen, 58, 247, 298; phosphate, 123; synthetic, 58, 261
Financial support of research, 315
Finland, 100
First Nations of North America, 54
Fish, 82, 131

Fitness, decline in, 46-7
Floods, 21, 183, 202
Flue gasses, 80, 81
Flux, 126, 137
Food: dairy, 262; grains, legumes, vegetables, fruit, 262; land for food production, 128, 194; meat, 52, 262, 288, 343, 351; prices, increases in, 129; processed, 103, 213; refined, 262; vegetarian, 329, 343, 349
Food and Drug Administration, 197
Food industry, 262, 349
Food insecurity, 123, 129
Food production: efficient, 23; fuel inputs, 45, 123; and population growth, 44, 45, 117
Fordism, 62
Fossil carbon release, 21
Fossil energy: generation of, 237-8; massive use of, 130; substituting for labor, 44
Fossil fuels: burning of, 21, 57, 75; long-distance transport, 191, 203, 310; and transportation technologies, 50
Fossil water, Ogallala aquifer, 57-8
Free market, 105, 249, 257
Free-market economies: and competition for scarce resources, 69; exploitative nature of, 257; and happiness, 224; production of goods and services, 60; and profits, 255-6; and sustainability, 261
Freedom: economic, 224; loss of, 245-8; personal, 224; political, 224; of public speech, 170; scientific, 318, 333; and technological dependency, 245-8
Fresh water: and biomass, 129; depletion of, 123; human use of, 57-8
Friedman, Milton, 260
Fuels. *see also* biofuels; Fossil fuels: automobile fuel efficiency, 99; from biomass, 128; ethanol from corn, 128; limitations, effects of, 112
Funding structure for research and development, 315-6, 319, 327-8
Fusion power, xix-xx
Future generations: decline in fitness, 46-7; discounting in cost-benefit analysis, 178-9; and externalization of costs, 52; risks or damage, 179, 187

G

Gardening, 302, 310-1, 351
Gasoline, 190-1
GDP. *see* Gross domestic product (GDP)
Gene-line modification, 28
Gene pool, 178, 265
Gene splicing, 25
Genetic constitution, 224, 265
Genetic engineering: crops, 25, 27;
    military applications, 28; organisms,
    25, 27; in pharmaceuticals, 27; unin-
    tended side effects of, 25-8
Genetic pollution, 28
Genetic screening, 27
Genetic traits, transfer of, 25
Genetically modified organisms, 27
Genuine Progress Indicator GPI, 204
Geo-engineering, 82-3
Georgescu-Roegen, N., 19
Germany, 69, 125, 131, 135, 158
Giampeitro, M.S., 129
Gitlin, Todd, 66
Glennon, Robert, 21
Global climate change, 21-2, 79
Global warming, 194, 238, 343. *see also*
    Climate change
Globalization, economic, 203, 212, 276
GMO, 25, 27
Goals: illusory, 111; inherently un-
    achievable, 42; of society, iv-v, 85, 112-
    3, 115-6, 140
Goods and services. *see* Consumer
    goods and services
Government: and control of technol-
    ogy, 11, 254
Grain, 58, 262
Grant, Lindsey, iv
Greed, 221, 259, 318
Greek philosophy and worldview, 54,
    148, 216
Green, Harold, 182
Green revolution, xviii, 44
Greene, Robert, 154
Greenhouse gas emissions, 193-4
Greening, L. A., 96
Gross domestic product (GDP): and ef-
    ficiency of carbon use, 101; increase
    in, 94, 118, 121-2; as indicator of eco-
    nomic progress, 200-5; and labor pro-
    ductivity, 104; and steady-state condi-
    tions, 282-3

Gross world product (GWP), 101, 118,
    122
Groundwater, 58, 79, 81, 129
In group, 50, 291
Growth: economic. *see* Economic
    growth; sustainable economic, 120,
    121, 122; technology-induced, 98; un-
    limited, 118, 137, 279
Growth accounting, 97
Growth-oriented economies, 122, 279
Guidotti, T. L., 322

H

Haber-Bosch process, 58
Habitat: and automobiles, 29-30; loss
    of, 22
Half knowledge, 13, 304
Halstead, Ted, 201
Hamilton, Clive, 150-1, 216
Handmer, John, 304-5
Hannon, Bruce, 97
Happiness: and affluence, 214-6; cog-
    nitive factors, 224; genetic factors,
    224; and material goods, xviii, 208-
    23; predictors of, 225; sources of, 210,
    224-6; traditional sources of, 226-34;
    and work, 303
Harmony, 34, 54-5, 255, 287
Health: defined, 34; mechanistic ap-
    proach, 35; of the nation, 266, 287,
    289; preventative measures, 36, 287-
    8; promotion of, 287-8; public, 43, 160
Health care. *see also* Medicine: effi-
    ciency of, 105-9; escalating costs, 43,
    105-9, 195, 266, 286; and mortality
    rates, 161-2; and profit maximization,
    263-6; system performance, 266, 287
Health insurance, 93, 198
Heart disease: as cause of death, 37, 265;
    as lifestyle disease, 76, 87, 248, 262,
    288; and quality-adjusted life years,
    215; transplants, 107
Heating, direct solar energy, 126, 130
Heavy metals, 135, 298, 299
Hedonic adaptation, 217-9
Hedonic treadmill, 217-9
Heinberg, Richard, v
Heisenberg, Werner, 243
Hemoglobins, 82
Herbicides, 24, 25, 299
Herman, Edward, 168, 267

Hierarchical organization, 63, 153, 317, 326

High-fat, animal-based diet, 288

High-tech medicine: antibiotics, 38-40; effectiveness of, 36; and iatrogenic disease, 37-8; life-prolonging interventions, 41-3; unintended consequences of, 33-43

High-tech military technologies, unintended consequences of, 31-3

High-tech weaponry, consequences of, 31-3

High-technology warfare, 31-43, 53

Hobbes, Thomas, 157

Holdren, John, 126-7

Holistic and preventative approaches, 288-9

Holmes, Robert, 84

Homeostasis, 166

Hope: of affluence, 258; economic, 146; medical, 43, 121; in progress and science, 145, 150; unrealistic, 258, 284

Hormone-mimicking chemicals, 20

Hospital infections, 37, 38, 40

Housework, 103-4

Huesemann's Law of Techno-Optimism, 154

Hughes, Thomas Parke, 171

Human behavior, and sustainability, xix

Human domination of nature, 56-60

Human economy, size of, 137-8, 279, 282

Human overpopulation, consequences of, 44-5

Human population: and energy, 45; growth of, xxii, 11, 43-4, 117; irreversible consequences of, 11; and land use, 56-7; and steady-state economy, 283

Human-scale and simple technologies, 301-2

Humane and sustainable world, 115-6, 245, 273

Humane technology, 303

Humanism, 278

Hunger: extent of, current, 159; and food prices, 129; increase of, xvii; risk of starvation, 247, 262

Hunter-gatherers, 44, 45, 53, 86, 148, 159

Hupke's Constant, 104

Hybrid cars, 99

Hydrocarbon, 79

Hydroelectricity: generation of, 130-1; hydroelectric dams, 130-1

Hygiene, and mortality rates, 161

Hypercars, 99

Hypothesis testing, 12, 55, 315

I

Iatrogenic disease, 37-8, 86, 194-5

Ideology, 63, 150, 238, 254, 273, 318-9

Ignorance: of anthropological data, 156; collective, 15, 181, 182; and cost-benefit analyses, 181, 188; of history, 155-6; of ignorance, 14; and moral accountability, 327; and problem solving, 271; scientific, 14-5, 155, 181; and technological optimism, 154-9

Illich, Ivan, 34-5, 190-1

Immigration, 283

Immortality, quest for, 165-6

Immunization, 46, 161

Impacts. *see also* Consequences; Side effects: and cost-benefit analyses, 173, 180-2; monetization of, 185-7; positive and negative, 173, 181, 185

Imperial consciousness, 290

Improving on nature: assumption of, 5-7; and genetic engineering, 26; impossibility of, 11

Incineration, 80

Income: equitable distribution, 284-5; per capita, 215; and psychological well-being, 214-6, 219; threshold, 214

Index of Sustainable Economic Welfare (ISEW), 204

Indicators, appropriate, 180-2

Indigenous peoples, 9, 54, 158

Individualism, excessive, 228, 293

Individuality, 30, 62, 114

Industrial ecology, 121

Industrial processes: eco-efficient, 91, 121; and materials, 124, 137, 296; and safe distance, 50

Industrial Revolution: and economic production factors, 97, 118; effects of, 44; and exploitation, 52; and happiness, 227-8; and population growth, 117

Industrial societies, unsustainability of, 122-3

Industrialization, 158; of agriculture, 23, 44; and life expectancy, 46
Industrialized agriculture, 23-5, 123
Industry-funded studies, 316
Inequalities in income and wealth, 140, 250, 256, 256-7, 260, 284
Inevitability, illusion of, 170, 236
Infant mortality, 161
Infrastructure, 31, 104, 214, 242-3, 250, 302
Inherent unavoidability, 1-15
Inhumane conditions, 114
Injustice, 188, 273, 292
Innovations, control of, 60, 250, 251, 306, 309, 326
Institute of Medicine, 335
Institutional biases, 183-5
Insurance: liability coverage, 305-6; medical, 106
Insurance companies, 196, 305-6
Interconnectedness: and cause-and-effect relationships, 173; described, 3-5; and exploitation, 49; and medical care, 37, 287; and sustainable development, 295-300; and worldview, 278-9
Interdependence, 4, 242
Intergovernmental Panel on Climate Change, 21, 118, 125
Internal combustion engine, 28, 44
Internalization of costs, 179, 188, 189-90, 281-2
International Chamber of Commerce, 119
International Energy Agency (IEA), 98
International Food Policy Research Institute, 128-9
International Panel on Climate Change (IPCC), 122
International trade, 118, 283
Interventions, medical, 41-3
Inverse Power Law of Truthfulness, 168
Investments, return on, 198, 264
IPAT equation, iii, 120-1
Irreversible changes, 9-10, 21
Irreversible consequences, 8-12, 15, 21-2, 22-3, 32, 174
Irrigation: and agriculture, 57-8, 129; and groundwater, 58, 192; large-scale, 24
Isolation, 64, 227, 229

J
Jackson, T., 209, 218
Japan, 99, 158, 215
Jevons paradox, 95, 112. see also Rebound effect
Jevons, Stanley, 93, 95
Jochem, Eberhardt, 110
Johnson, James Turner, 85
Juenger, Georg Friedrich, 172
Justice: distributional, 187-8; distributive, 233; intergenerational, 178

K
Kant, Immanuel, 278
Kasser, Tim, 222-3
Kay, Jane Holtz, 102
Kempner, Joanna, 333
Kennedy, John F., 146
Kimbrell, Andrew, 114, 190
Knowledge: control of, 152-3; fragmentation of, 36, 37, 326; limited, 13; methods of acquisition, 12, 13; and progress, 149
Konner, Melvin, 38, 38-9, 153-4, 196
Korten, David: on cars and social interaction, 30-1; on consumer culture, 211; on externalization of costs, 179-80; on income distribution, 285; on neoclassical economics, 118-9; on orders of consciousness, 289-92; on paradigms, 276; on profit maximization, 257; on social responsibility, 260
Kuhn, Thomas, 274, 276

L
Labor: cheap, 70; and land, 44, 97; productivity, 104-5, 111
Labor force, control of, 60-4
Labor-saving devices, 60, 104-5, 254
Labor-saving technology, 103-5
Land, arable, 24
Land ethic, 278
Land use, 56, 128, 194
Landfill integrity, 78
Lane, Robert, 212-3, 226
Law: environmental, 177, 260; labor, 260
Law of entropy, 17-8
Laws of ecology: first law of ecology, 3; fourth law of ecology, 8; second law of ecology, 77; third law of ecology, 5-6
Leach, William, 211

Lead, 80
Lee, Philip, 198-9
Leisure time, 93-4, 158-9
Leopold, Aldo, 278
Levy, Stuart, 40
Life-cycle analysis, 174, 297
Life expectancy, xviii, 160, 165-6, 287
Life, prolonging, 41-3, 106, 107
Life-sciences companies, 26
Life span, 165-6
Life-support services, 134
Lifestyle: changes in, 43; choices in, 76, 87; and disease causation, 76, 87, 287-8
Lighting, 98, 100, 110
Limitations: of reductionism, 12-5; on resource use, 282; of resources, 93-4, 108-9; of scientific knowledge, 13, 26-7, 60, 182, 333
Living conditions, 46, 87, 157, 204
Local consumption, 300-1, 310
Local democratic control of organic farming, 310
Local energy systems, 302
Local financial resources, 300
Local production, 301
Long-distance transport, 191, 203, 310
Lorenz, Konrad, 53
Love, 114, 222, 278
Lovins, Amory, 99
Lowrance, William, 314-5
Luddites, 63, 169

M
Maasai, 215
Machines, and the control and exploitation of workers, 60-4, 230-1
Macro-economics, 97, 111
Maddison, David, 190
Magical consciousness, 290
Magnitude of human activities, 8, 137
Malthus, Thomas Robert, 119
Mander, Jerry: on biases, 238-9; on control of technology, 248-9; on decline of techno-optimism, 171; on democratic control, 306-7; on GDP, 200-1; on megatechnologies, 301-2; on pre-industrial life, 158; on precautionary principle, 305; on profit, 255, 259; on rationalizations, 170; on *Techno-Fix*, ii; on technological innovation, 167-8;

on television, 64, 65, 66, 67; on value neutrality, 236
Manhattan project, 243, 324
Manipulation: of attitudes and behavior, 65; of the masses, 64-8, 212, 249; political, 63
Manufacturing processes: and labor productivity, 104-5, 111, 114; small, flexible, 302
Maori proverb, 4
Market value, 185, 188
Marketing, 169, 211
Markets: distortion of, 180; foreign, 70; saturated, 210, 212
Marks, N., 209
Marx, Leo, 150
Maslow, Abraham, 63, 207-8
Mass and energy, addition or removal, 74
Mass consumption, stimulation of, 65-6
Mass media. *see also* Television: and advertising, 210, 212-3; consumption statistics, 66, 212; social purpose of, 168; and technological optimism, 167-9
Mass production: and consumer society, 210; and media influence, 65
Material affluence. *see* Affluence
Material consumption: and happiness, 65, 210, 217, 219-20; and steady-state economy, 283-4; US level of, 57
Material standard of living. *see* Standard of living
Material sufficiency, 140
Materialism: consumer culture. *see* Consumer culture; consumer lifestyle, 66, 68, 76, 83; defined, 221; materialistic individuals, 222-3; and value orientation, 222; and well-being, 221, 222-3
Materials cycle, 124, 135
Materials loop, 124, 135
Materials use, 100, 110-1, 122
Matter-energy throughput, 282, 286
Max-Neef, M., 208
Maximization of profits. *see* Profit maximization
Maximum autonomy, 226, 303
Maximum extraction rate, 281
Maximum scale, 280-1
Maximum sustainable yield, 133-4

Mayan society, 138
McDonough, William, 115
McKeown, Thomas, 160-1
McKibben, Bill, i
McKinlay, John, 196
Meadows, Donella, 23, 139, 140, 272
Meaning and purpose, 226
Measles, 161
Measurement indicators, 175, 180-2, 180-2
Mechanistic reductionist science: defined, 34; and health, 34-5; limitations of, 12-5
Mechanistic view, 286, 316
Media: corporate control of, 219; mass. see Mass media; propaganda. see Propaganda
Medical counter-technologies, 86-8
Medical device industry, 199
Medical efficiency, 105-9
Medical-industrial complex: and conquering death, 107; and disease prevention, 265, 288; and profit maximization, 263, 266
Medical innovation: seven stages of, 196-7; testing for effectiveness, 198-9
Medical practice, as cause of illness and death. see Iatrogenic disease
Medical practice, third leading cause of death, 37
Medical procedures. see Medical treatments
Medical products, new, 106
Medical techno-optimism, 159-67
Medical technologies, 41, 86-8, 159, 238
Medical tests, 109
Medical therapies. see Medical treatments
Medical treatments: inappropriate, 195; infinite demand for, 42, 106; life-prolonging interventions, 41-3, 107; new, 106, 199-200; testing for effectiveness, 162, 194-200
Medication errors, 37
Medications. see drugs
Medicine. see also Health care: advances in, 159-60, 198; and critical science, 325; diagnostic tests, 37, 41, 106; different view of, 286-9; doctors, 42, 76, 161, 196, 198, 263-4; family practice, 264; high-tech. see High-tech

medicine; history of, 162, 165; industrialization of, 263; medical and pharmacological research. see Research; medical costs. see Health care, escalating costs; medical establishment, 41-2, 86-7, 196, 199; and profit maximization, 198-9; as social fix, 86-7; and technological optimism, 159-67
Mega-investments, 301
Mega-management, 301-2
Mega-technologies, 173, 247, 301-2
Mental maps, 271, 277
Mental models, 271, 276
Mesthene, Emmanuel, 246
Metabolic suppression, 82
Methane, 82
Metric, technical, 91
Microbial biodegradation, 298
Middle Ages, 153, 154, 159
Midwest Business Group on Health, 195
Migration, of fish, 131
The military: counter-technologies, 83-6; expenditures, 31-2, 266-7; and GDP, 202; power, 69-70; and profit maximization, 266-7; research, 324; technologies, 28, 31-3, 68-70, 237; weapons. see Weapons
Mill, John Stuart, 280
Millennium Assessment of Human Behavior (MAHB), xix
Minerals: depletion of, 133. see also specific minerals; extraction of, 58; recycling of, 135-6
Mobility, 28, 30, 101, 209, 228
Moderation, 54-5
Modernization, 92, 233, 234
Molecular biology, 26, 59
Monetization: of benefits and costs, 185-7; of non-market values, 185-7, 188; underlying assumption of, 186
Money. see Affluence; Income
Moral accountability, 326
Moral awareness, 289-93
Moral development, stages of, 290
Mortality rates, 160-2
Moss, Ralph, 265
Muir, John, 4
Mumford, Lewis, 105, 145, 209-10, 233, 244
Mutagenicity, 20
Mutation, 5, 39, 46-7, 255

Mutually assured destruction, 32, 53
Myers, Norman, iii
Myopic engineering, 14
Myths: of autonomous technology,
 241-3; of value neutrality, 235-41,
 313-20

N
Nader, Ralph, 202
Naess, Arne, 274
Narratives, 148, 149, 273
National Academy of Engineering, 335
National Academy of Science, 187, 326,
 335
National Science Foundation, 155
Natural cycles, disturbing, 137
Natural gas, 57
Natural healing, 162
Natural resources: critical natu-
 ral capital, 134; depletion of, 134;
 exploitation of, 50, 68, 70, 94, 133;
 intergenerational impact, 51
Natural selection, 5-7, 46, 157
Nature: alienation from, 231-2; concept
 of, 317; control of, 26, 54, 56, 60; domination of, 55, 56-60; exploitation of,
 53-6, 232; improvement upon, 5-7, 11;
 resilience of, 9
Needs: basic human, 207-8, 209; false,
 artificial, 208; material, 209; non-
 material, 209; primary, 220; and technological innovation, 208-14; vs.
 wants, 208
Negative consequences. *see also* Unintended consequences: addition or
 removal of mass and/or energy, 74; biological, 75; chemical, 75; of counter-
 technologies and social fixes, 73-4,
 82-3; inherently unavoidable, 7-8, 11,
 17, 73, 304; inherently unpredictable,
 11, 73; removal of immediate, 233-4; of
 renewable energy conversion, 125-7
Negative effects of previous technologies, 72, 73-4, 77
Negative feedback cycles, feedback, 11
Negative imperatives, xx
Negative side-effects. *see* Negative consequences
Neo-Malthusianism, 119
Neoclassical growth theory, 95, 96, 97
Nestle, Marion, 262

Netherlands, 158
Neutralization, 80
New science, 34, 55
Newton, Isaac, 13
Nitrogen, 298
Nitrogen cycle, 3, 56, 57, 58
Nitrogen fertilizers, 58, 247, 310
Nitrous oxide, 18, 29, 79, 82
Nitrous oxides, 79, 82
Nixon, Richard, 265
Nobel, Alfred, 84-5
Non-market values, 184, 185-7
Non-material needs, 209, 219
Non-renewables: depletion of, 45, 109,
 123, 178; dissipative use of, 296-8;
 limiting use of, 282; non-renewable
 resources, 45, 58, 68, 92, 122-3, 296;
 recycling of, 134-7, 298; and renew-
 able substitutes, 133-4
Norms, 35, 113, 272, 290, 333
Novum Organum, 12, 34, 55, 149
Nuclear hazards, 51, 237
Nuclear power: accidents, 181; genera-
 tion of, 175, 237; information, 324;
 power plants, xix, 131, 175-6, 237; and
 sustainability, 296
Nuclear radiation, 181
Nuclear waste, 51, 146, 178, 296
Nuclear weapons, 32-3, 267, 324
Nutrition: and food production, 261,
 261262; and mortality rates, 46, 160,
 161; and technology, 86

O
Obesity, 87
Objectivity, 314, 319. *see also* Value neu-
 trality
Observer and observed, 4, 50, 55
Obsolescence, 211-2
Ocean fertilization, 82-3
Ogallala aquifer, 57-8
Oil spills, 29, 202
Olmstead, Denison, 150
Open systems, 9
Oppenheimer, J. Robert, 243-4
Opposition from powerful institutions,
 309
Optimization, 13
Optimum scale, 280-1
Organic agriculture, 261, 309-11
Organic chemicals: production of, 59;

release of anthropogenic, 30, 56, 101, 137
Out group, 50, 233, 291
Output per unit input, 92
Over-exploitation, 94, 133-4
Over-harvesting, 133
Overpopulation, human, 44, 75, 83, 86, 166, 234
Overshoot, xviii, 138-9, 140
Oxygen: dissolved, 74; transport mechanisms, 82

P
Paleolithic period, Upper, 44
Paradigms: conflicts between, 273-7; dominant, 275; power of, 271-3; shared, 316; shift in, 267, 271, 273-7, 285, 289
Paradox: of desires, 216-23; of ignorance, 14
Parenti, Michael, 69, 267
Participatory design, 307-9
Participatory R&D (research and development), 307-9
Passive acceptance, 245
Passivity, 170, 239, 249
Pasteur, Louis, 38
Patients, terminal, 41
Patrick, Donald, 197-8, 264
Pavement, 30
Peer-review process, 308, 314
Peet, John, 9, 186
Penicillin, 38-9
Perelman, Michael, 261
Periodic table, 58
Permanent war economy, 267
Perrow, Charles, 180-1
Perverse subsidies, 311
Pesticides: metal-containing, 38-9; synthetic, 299; use of, 24-5
pH, disturbance of, 82
Pharmaceutical industry: drugs vs. lifestyle, 82; and genetic engineering, 27; and profit maximization, 198-9, 264; research, 35
Phenotype, 5
Photosynthesis: and CO_2 from fossil fuels, 298; photosynthetic conversion into food biomass, 23, 45, 56, 123, 126, 130
Photosynthetically fixed carbon, 56, 127
Photovoltaic electricity generation, 131

Physical barriers, 52
Physicians, 37, 87, 264. see also Doctors
Physicians Committee for Responsible Medicine, 325
Piaget, 290
Pimentel, David, v
Placebos: and new drugs, 197; placebo effect, 160, 162-3; placebo response, 163-5; studies of, 164-5
Planck, Max, 276-7
Planetary boundaries, 56
Plasmids, 39
Plasticizers, 20
Plastics: fossil-fuel derived, 299; recycling of, 135, 297; and sustainable development, 125
Platt, John, 147-8
Pleistocene fauna, 53
Pleonexia, 216
Polar ice, 21
Pollutants. see also Contaminants: strategies for dealing with, 78
Pollutants, highly dispersed, 79-80
Pollution: environmental. see Environmental pollution; prevention of, 80, 286; sustainable, 298-9
Pope Paul VI, 234
Population: animals, 11; human. see Human population
Positional treadmill, 218, 219
Postman, Neil: on new technologies, 169; on progress, 149-50; on religious faith, 154
Poverty: and disease, 287; and happiness, 215; of nations, 283; of people, 129
Power. see also hydroelectricity; solar power; wind power: disproportionate, 260; distribution of, 254; fusion power, xix-xx; of worldviews and paradigms, 271-3
Power laws, 168
Precautionary principle, 279, 304-5
Prediction: inherently impossible, 11, 181; of potential impacts and selection of appropriate indicators, 180-2
President's Council on Sustainable Development, 121
Prevention: of diseases, 36, 43, 265; of unintended consequences, 304-6
Primary productivity, 21, 56, 127

Primitive culture, 54, 156, 157–8, 159
Princen, Thomas: on efficiency, 91; on profit, 256
Principle of reciprocity, 49, 70, 278, 290
Principle of subsidiarity, 301
Problems: human-caused, 71, 82; residue problems, 74; social problems as simple technical challenges, 75–6; technology-caused, xxii, 73
Proctor, Robert: on disease prevention, 265; on risk assessment, 185; on value-freedom, 318; on value neutrality, 237, 240, 319–20
Production: and consumption, 95, 118, 203, 207, 256; mass, 65, 210; organization of, 62, 63, 96; rising per capita, 95
Productivity, increasing labor, 95, 111
Professional ethics, 330–8
Professionalism, problem of, 328–30
Profit: meaning of, 255–8; as recent historical phenomenon, 256
Profit maximization: and agriculture and food, 261–3; and cost-benefit analysis, 188; and the development of new technologies, 258–61; and medical care, 198–9, 263–6; and military technologies and foreign policy, 266–7; and technological innovation, 211; and value neutrality, 241
Profit motive. *see also* Profit maximization: as driver of technological development, 253–67; of drug and medical industries, 198–200
Profits: reinvestment of, 256
Progress: belief in, 5, 72, 145–72; continual, 145; a dogma, 145; history of, 148–52; illusion of, 145, 172, 174; material, 150–2; redefined, 150, 172; technological, 146
Progress, universal human, 150
Propaganda: and economic power, 275; political, 67–8, 212; system-supportive, 168; and television, 67
Psychological disorders, 87–8, 215, 222
Psychological safe distance, 50–3. *see also* Safe distance
Psychological well-being: and income, 214–6; and materialism, 221, 222–3; and personal freedom, 224; in steady-state economy, 284
Public health, 43, 160

Public interest science, 322
Public speech, 170

Q
Qualitative improvement, 283
Quality-adjusted life years, 215
Quality of life, 41, 45, 113–5, 283, 283–4
Quality of work life, 286
Quasi-equilibrium, 9
Questioning and confronting, 169, 329–30

R
Radiation: and damage to gene pool, 178; and diseases, 175, 176, 178; from nuclear accidents, 181; from nuclear wastes, 51; and premature death, 175; solar, xix, 126, 130
Radioactive waste, 175, 237, 296
Rae, John, 102
Random mutation, 5
Rationalizations, 338
Ravetz, Jeremy: on critical science, 320–1, 322; on decline of techno-optimism, 172; on interconnectedness, 173; on moral accountability, 326–7; on professionalism, 328; on reductionism, 14; on research funding, 328; on scientific ignorance, 181
Raw materials: costs of, 156; low-entropy, 18, 137; sustainable, 122, 124, 283, 296
Reality, a different view of, 277–9
Rearranging effect, 78
Reason, science, and progress, 55
Rebound effect: categories of, 96; and consumption, 96, 112; defined, 95; direct, 96; indirect, 97; in medicine, 106; secondary, 96–7
Recharge, 58
Reciprocity, 49, 70, 278, 290
Recycling: classes of, 135–7; complete, 136; of non-renewable materials and wastes, 134–7; technical and economical potential for, 135
Redesigning for sustainability, 300
Reductionism: defined, 12, 34; limitations of, 12–5; scientific, 12–5, 26, 75, 304
Rees, William, ii, 138
Regulation, 11, 135, 177, 256, 281, 299

Reid, David, 119
Religion. see also Faith; Spirituality:
    Christianity, 148, 149, 152; compared
    to belief in progress, 152-4; and hap-
    piness, 225, 233; Judaism, 148; and
    materialism, 223; and money, 256;
    and suppression of criticism, 152-3,
    171; views of time, 148
Relman, Arnold., 263
Removal of toxics from wastewater, 81
Renewable energy: cost of, 125; envi-
    ronmental impacts, 125-32; environ-
    mentally friendly, 286; generation of,
    193, 296-7; and land limitations, 132;
    public opposition to, 132; and sustain-
    able development, 124, 296-7; tech-
    nologies, 125; types of, 130-2, 296
Renewable resources: depletion of, 123;
    distribution of, 68; exploitation of, 51;
    and sustainability, 124, 133, 281, 296,
    296-7
Renewable substitutes, 133-4
Renewable technologies, 125-32, 297-8
Renewables, 129, 134
Replacement of non-renewable materi-
    als with renewables, 133
Research: basic, 254, 265, 317, 319; bio-
    logical warfare, 334; efficiency, 99;
    funding of, 315-6, 327-8; interpreta-
    tion of results, 316; medical, 36, 151,
    166, 265, 287, 325, 329; military, 32,
    324, 335, 336; morally dubious, 327,
    328, 333, 336; pharmaceutical, 35;
    scientific, 36, 151, 286, 318, 326, 332;
    topics, choice of, 325, 333
Research and Development (R&D),
    307-9
Resilience: of nature, 9
Resistance: to antibiotics, 38-40
Resources: control of, 68, 113; deple-
    tion of, 45, 109, 123, 129, 134, 329; im-
    ported, 301; limited, 93-4, 98-109;
    non-renewable. see Non-renewables;
    power used to extract, 56; reducing
    the use of, 94, 98, 121, 281-2; renew-
    able. see Renewable energy; Renew-
    able resources; scarce, 44, 301; uneven
    distribution of, 68; unit limited, 92
Responsibility: collective, 325, 326; and
    critical science, 313-38; diffusion of,
    326; escape from, 286, 314, 326, 327;

professional, 321, 330-4; of scientists,
    313-4, 319, 325-8; social. see Social re-
    sponsibility
Reverence, 54, 278
Reverse adaptation, 112-3, 115
Revolution: agricultural, 44, 117; scien-
    tific and technological, xxvii, 43-6,
    173, 317
Rifkin, Jeremy: on biases, 317; on ge-
    netic engineering, 26, 28, 59; on lack
    of knowledge, 27; on order/disorder,
    18; on value neutrality, 240
Risk: difficulty of defining, 81; invol-
    untary, 186; lesser-known, 79; market
    value of, 186; quantifying, 81; reduc-
    tion of, 81; transfer of, 78, 179; volun-
    tary, 186; well-known, 79
Risk assessment, 174, 176, 184, 185. see
    also Cost-benefit analyses
Rivers, 57-78, 126, 182, 298, 299
Roads, 101-2, 189
Roman society, 138, 148, 254
Rose, Hillary and Steven, 315-6, 324

S
Safe distance, 50-3, 70, 233, 291
Scale: of exploitation, 8; of human eco-
    nomic activities, 280-1; of socially ap-
    propriate technologies, 301
Scarlet fever, 161
Schmidheiny, Stephan, 120
Science: different view of, 285-6; di-
    rection of, 285, 286, 316, 320; history
    of, 275, 335; a new critical, 286, 320-5,
    334-8; Novum Organum, 12, 34, 55, 149;
    reductionist, 12-5, 26, 75, 304; and
    social responsibility, 313-38; value-
    laden laden character of, 237, 313-20
Science and engineering professionals,
    330-4
Science and technology: current direc-
    tion, 140; democratic control of, 306-
    9; a different view of, 285-6; hidden
    value of, 320, 325; redirecting, 140, 321,
    322, 324, 325, 338; value-laden charac-
    ter of, 237, 313-20
Scientific data, 286
Scientific ignorance, 14-5, 155, 181
Scientific inquiry, 286
Scientific knowledge, fragmentation,
    36, 326

Scientific literacy, 155, 250

Scientific merits, 308

Scientific method, 12, 50, 60, 322

Scientific neutrality and objectivity, cloak of, 319

Scientific process, 155

Scientific reductionism, 12-5, 26, 75, 304

Scientific research, 36, 151, 286, 318, 326, 332

Scientific revolution, 43-6, 173, 317

Sea level rise, 21

Second law of thermodynamics, 109-10: defined, 17; and dispersed materials, 79, 80, 136; and environmental impact, 18, 121, 126; and knowledge/ignorance, 14

Self-sufficiency, 248, 301, 302

Selfishness, 193, 221, 259

Sensitivity analysis, 176

Separateness. *see also* Safe distance: illusion of, 4, 70, 277, 295; worldview of, 277-9, 289, 320

Separation: between humans and nature, 50; spatial, 51, 70; temporal, 51, 53, 70, 177

Sequestration of contaminants: carbon dioxide, 81-2; geologic reservoirs, 81-2; oceans, 82

Sexual orientation, 20-1

Side effects, 296. *see also* Consequences: drug, 38; environmental, 124; negative, 80, 304; physical, chemical, biological, 38, 74; unintended. *see* Unintended consequences

Simplicity, 301

Sloanism, 211-2

Social alienation, 31, 76, 87, 220, 231

Social control, 254

Social costs: of biofuels, 192; and cost-benefit ratio, 177; and efficiency, 23; externalization of, 179, 188-90; future, 178

Social fixes, 72-3, 75-7, 83, 86, 86-7, 88-90

Social interaction: and automobiles, 30-1

Social responsibility: of engineering professionals, 330-8; and profit maximization, 260; of scientists, 313-38, 330-8

Social status: expressed through con-

sumption, 213, 214, 218; of strangers, 218

Socialized consciousness, 290-1

Socially appropriate technologies, 286

Socially appropriate technology, 286, 300-4

Society: classless, 257

Soil: depletion of, 24, 123

Soil mining, 24

Solar energy: control of, 238, 296; ideological bias, 238; negative impacts of, 125-7; photosynthetic conversion into food biomass. *see* Photosynthesis; and present energy demands, 125; thermal, 130

Solar power: solar panels, xx, 130

Solutions: counter-technologies. *see* Counter-technologies; efficiency, 91-116; material affluence, 94-8; techno-fixes. *see* Techno-fixes; technological progress, 94-8

Soybeans, 192

Speciation, 22

Species extinction, 9. *see* Extinction

Speed of technological change, 8, 227

Spiritual consciousness, 289-93

Spirituality. *see also* Religion: decline of, 232-3; and moral awareness, 289-93; spiritual disciplines, 293

Stakeholders, 183-5, 188

Standard of living, 68, 104-5, 138, 139, 147, 217, 258

Standards, 114, 234, 273, 325, 328, 331

Standing, 174

Starr, Chauncey, 10

Status quo, 116, 329-30

Status symbols, 218-9

Steady-state conditions, 117, 282-3

Steady-state economy, 118, 135, 279-80

Stress, 76, 87, 225, 233

Subjectivity, 114, 177, 274, 314, 315

Subsidiarity, principle of, 301

Subsidies, 125, 192-3, 281-2, 297, 311

Substitutes: capital for labor, 95; metal, 133-4; renewable, 133-4

Suicide, 215

Sulfur dioxide, 18, 79

Sunlight, deflection of, 83

Superfund sites, 201

Supermarkets, 103

Suppression: of citizen activism by

media, 67-8, 239; of ethical sensitivities; metabolic, 82; of social criticism, 169-70, 319

Surgery: as counter-technology, 86; invasive, 76; and placebo response, 164-5; unnecessary, 37

Sustainability: challenges to, 125-37; conditions for, 122-5, 296; design criteria for, 295-300; environmental. see Environmental sustainability; and human behavior, xix; long-term, 120, 122-5; and steady-state environmental economy, 118; strong, 124

Sustainable development: conditions for, 122-5; and eco-efficiency, 119-22

Sustainable economic growth. see Sustainable development

Sustainable energy future, 296

Sustainable energy generation condition, 124, 296

Sustainable material use condition, 124, 135, 296

Sustainable pollution, 298-9

Sustainable waste discharge condition, 124, 296

Sustainable yield, 51, 133, 281, 297, 300

Sustainable yield criterion, 124, 296

Suzuki, David, i

Sweden, 308

Switchgrass, 194

Symptoms: of disease, 39, 76, 164, 288; and social fixes, 77, 88; of social problems, 75

Synthetic herbicides and pesticides, 123, 299

Synthetic organic chemicals, 19-21, 25, 124

System boundaries: and energy balance, 193; selection of, 176-80

Systems, 79: closed, 17; of domination, 240; open, 9

T

Taboos, against challenging faith in science, 169

Take-back effects, 96

Taoists, 13

Taxation, 282

Taylor, Frederick W., 61

Taylorism, 61-2, 337

TB, 161

Technical experts, 251, 301, 307, 308

Technical problems, iv, 75, 120, 300

Techno-addiction, 221

Techno-fixes: categories of, 72-3; counter-technologies. see Counter-technologies; efficiency improvements, 73; faith in the effectiveness of, 71-2; ineffectiveness of, xxiii, 321; limited ability of, 72; social fixes. see Social fixes

Techno-optimism. see Technological optimism

Technocracy, 248-9

Technological benefits, 177, 182, 187, 188

Technological challenges, xxvi

Technological change: direction of, 73; rate of, 8, 277

Technological democratization, 309

Technological dependency, 245-8

Technological determinism, 36, 63, 241, 254

Technological development: as a political process, 115, 254; as a social process, 253-5

Technological efficiency, 92, 94-8

Technological exploitation of nature, 53-6

Technological imperative, 41, 210, 243-5

Technological innovation: appeal of, 167; consumerism and materialism, 94, 208-14; critical analysis of, 320; drivers of, 253-67; and needs and wants, 208-14; and profit maximization, 211

Technological lock-in, 242

Technological megasystems, 235, 242

Technological optimism: and affluence, 110-1; and belief in progress, 72, 145-72; and buying time, 88; decline in, 169-72; degree of, 151; and efficiency improvements, 109; and end-use efficiency, 110; extrapolations of, 156; and ignorance, 154-9; and the mass media, 167-9; medical, 159-67; pervasiveness of, xxiii; and recycling, 136; and technological innovation, 167-8

Technological progress: benefits of, 147; defined, 146; and rising material affluence, 94-8

Technological revolutions, unintended consequences of, 43-6

Technologies. *see also* specific technologies: communications technologies, 230; and control of labor force, 61-3; development of, 56, 63, 248, 257; electronic communication technologies, 62, 209, 232; for exploiting resources, 56; future, 298, 300, 304, 307, 320; hasty introduction of, 11, 199; manufacturing technologies, 242; of marginal or no benefit, xxvi, 174; medical, 41, 86-8, 159, 238; military technologies, 32, 68-70, 85-6, 237, 266-7; and nature, alienation from, 231-2; negative sides of, 168; new, 120, 167; office technologies, 63; polluting technologies, 83, 145, 177; treatment technologies. *see* Treatment technologies

Technology: advanced, xxiii, 51, 68, 70; assessment of, 189; autonomous, 241-3; democratic control of, 306-9; different view of, 285-6; environmentally appropriate, 286, 295-311; environmentally sustainable, 295-311; excessive, xxiv; and exploitation, 49-53; exploitation and fairness of, 49-70; and happiness, 226-34; incompatibility with democracy, 238, 306; innovative, 11, 42, 55, 56, 177, 185; limitations of, 4, 156; and markets, 140; political role of, 254; record of, xx; socially appropriate, 286, 295-311; and spirituality, 232-3; uncritical acceptance of, 235-51; undemocratic control of, 248-51; unintended consequences of. *see* Unintended consequences; and value neutrality, 235-41, 313

Technology assessment, 188-9. *see also* Cost-benefit analyses

Television: biases of, 238-9; modification of culture by, 228, 229-30; as tool for social control and manipulation, 64-8; viewing statistics, 66

Temperature, global mean, 21

Tetraethyl lead, 29

Therapy: cognitive, 224; and effectiveness, 196-7, 200; vs. prevention, 36

Thermodynamic limits, 94, 121

Thermodynamics: first law of, 17, 18; second law of. *see* Second law of thermodynamics; and sustainable development, 19

Thermohaline ocean circulation, 21, 21-2

Third world, 69

Three-stage decision process, 308

Time: buying time, 88, 265; cyclical view of, 148-9; linear view of, 148-9

Tobacco industry, 265, 319

Tohono O'odham (Papago) Indians, 10

Topsoil loss, 24, 123, 202

Total carbon emitted (CARBON), 101

Total factor productivity (TFP), 97

Total fuel energy (TFE), 99-100

Total hours worked, in free market, 105

Total material requirements (TMR), 100

Total passenger kilometers (TPK), 99

Toxic metals, 58, 80

Toxic wastes, 52, 139

Traffic congestion, 30, 76, 101-2

Transportation: and cost-benefit analyses, 189; fuel demand, 191; infrastructure, 31, 104, 190, 242-3; long-distance, 101, 301, 310

Treatment technologies, 79-80, 80-1

Trends, simplifying and projecting, 157

Tuberculosis, 161

TV. *see* Television

Typhoid, 161

U

Unavoidable negative effects of technology. *see* Negative consequences

Uncritical acceptance of technology, 235-51

Undemocratic control of technology, 248-51

Underlying causes, 77, 87, 88, 156

Unemployment, 23, 204

Unethical behavior, 51

Unintended consequences: of the automobile, 28-31; causes of, 73; climate change. *see* Climate change; of counter-technologies and social fixes, 88-90; of efficiency solutions, 112-6; environmental, 17-23; of genetic engineering, 25-8; of high-tech medicine, 33-43; of high-technology warfare, 31-3; and ignorance, 182; of industrialized agriculture, 23-5; prevention of, 304-6; of technological revolutions, 43-6, 226-34

United Kingdom, 158

United States, 99, 158, 215
Unpredictability, 3–15, 73
Unsustainabiity: of industrial agriculture, 123; irrigation, 24, 58; reasons for, 122–3
Upper Paleolithic period, 44
Urban sprawl, 31
US Atomic Energy Commission, 146
US Office of Technology Assessment, 27
Utilitarianism, 36, 152, 174, 187, 209, 216
Utility, collective, 93

V

Vacuum cleaner, 103
Value conflicts, 113
Value-free science, 237
Value judgments, 280, 315, 319
Value neutrality: myth of, 235–41, 313–20; and suppression of criticism, 170, 240
Value orientation: and funding, 316; materialistic, 222; of objective facts, 314
Values: awareness of, 321; changes in, 140; changing, 226–8; embedded, 236–7; materialistic, 113, 222; personal and cultural, 112, 113; societal, 112–3, 140, 240, 286; socio-cultural, 239–40, 313, 316
Veblen, Thorstein, 218
Vegetarian, 329, 343, 349
Vespasian, 254
Vested interests, 182
Video technologies, 64, 239
Violence, 53, 170, 238, 267
Virtual reality, 65, 67
Von Neumann, John, 331
Vote, 249, 257

W

Wachtel, Paul, 216
Wackernagel, Mathis, 138
Wants: becoming needs, 213–4; vs. needs, 208; paradox of, 216; stimulation of, 212–3
War, and counter-technologies, 84–5
Warfare: atomic, 32; biological, 25, 28; chemical, 24; high-technology, 31–43, 53
Washing machine, 103

Waste: absorption capacity, 57, 93, 138–9, 280; discharge of, 296, 298–9; and entropy, 18–9, 137; exporting, 283; treatment of, 300; waste heat, 74, 110
Wastewater treatment, 81
Water: fresh water. see Fresh water; groundwater, 58, 79, 80, 129; quality of, 81, 281
Water mining, 58
Wealth: inequalities in, 140, 250, 256, 256–7, 260, 284
Weapons: biological, 25, 28, 33; chemical, 24, 33; ethnic, 28; nuclear, 32–3, 84, 267; and safe distance, 53, 233; systems, 32, 83, 202; and technology, 32–3, 83–6
Weinberg, Alvin, 76
Well-being. see also Happiness: and income, 214–6; and materialism, 221, 222–3, 227; subjective, 214
Whistle-blowers, 330, 337
Whooping cough, 161
Wildlife habitat, 29, 51
Willingness to pay, 186
Wind power, 131–2, 238
Work: conditions of, 114; mindless, 303; satisfying and meaningful, 226, 231, 284, 303
Work-time reduction, 105, 158–9
World Commission on Environment and Development, 119
World Energy Council, 129
World gross product, 22
World Health Organization, 266
World War I, 32, 69
World War II, 32, 69
Worldviews: changes in, 273–7; and conduct of science, 316; conflicts between, 273–7; Greek, 54; mechanistic, 34–5; need for different, 271–94; power of, 271–3

Y

Years of life prolonged per unit health care expenses, 105

Z

Zahavi's Law, 104
Zinc oxide, 74, 136

About the Authors

MICHAEL HUESEMANN received his Ph.D. in chemical engineering from Rice University. He has conducted both experimental and theoretical research in environmental biotechnology for more than 25 years and has published more than 50 peer-reviewed journal articles, book chapters, and proceedings papers on a wide range of topics. His fields of interest include engineering, biotechnology, environmental science, policy analysis, sustainability, critical science, economics, and business administration.

JOYCE HUESEMANN received her Ph.D. in statistics from Rice University and holds an M.A. in anthropology. Her fields of interest include population genetics, evolutionary genetics, ethnography, statistics and public policy. She has taught at several universities, and participates actively in a number of environmental, wildlife protection, and companion animal organizations.

If you have enjoyed *Techno-Fix*,
you might also enjoy other

BOOKS TO BUILD A NEW SOCIETY

Our books provide positive solutions for people who
want to make a difference. We specialize in:

**Sustainable Living ◆ Green Building ◆ Peak Oil
Renewable Energy ◆ Environment & Economy
Natural Building & Appropriate Technology
Progressive Leadership ◆ Resistance and Community
Educational and Parenting Resources**

New Society Publishers
ENVIRONMENTAL BENEFITS STATEMENT

New Society Publishers has chosen to produce this book on recycled paper made
with 100% post consumer waste, processed chlorine free, and old growth free.

For every 5,000 books printed, New Society saves the following resources:[1]

40	Trees
3,587	Pounds of Solid Waste
3,947	Gallons of Water
5,148	Kilowatt Hours of Electricity
6,521	Pounds of Greenhouse Gases
28	Pounds of HAPs, VOCs, and AOX Combined
10	Cubic Yards of Landfill Space

[1]Environmental benefits are calculated based on research done by the Environmental Defense Fund and
other members of the Paper Task Force who study the environmental impacts of the paper industry.

For a full list of NSP's titles, please call 1-800-567-6772 or check out our web site at:

www.newsociety.com

NEW SOCIETY PUBLISHERS